SPECIAL PAPERS IN PALAEONTOLOGY NO. 89

DEVONIAN SPORE ASSEMBLAGES FROM NORTHWESTERN GONDWANA: TAXONOMY AND BIOSTRATIGRAPHY

BY

PIERRE BREUER *and* PHILIPPE STEEMANS

with 56 figures and 1 table

THE PALAEONTOLOGICAL ASSOCIATION
LONDON

April 2013

CONTENTS

[Special Paper in Palaeontology, 89, 2013, pp. 5–163]

Abstract: Devonian spores from 16 subsurface successions in Saudi Arabia and North Africa are systematically documented to characterize assemblages for the northern margin of western Gondwana. The taxonomic study provides the identification and description for 205 spore taxa and forms the basis for a refined biostratigraphy and palaeogeography. Most species are illustrated by multiple specimens to document morphological variation and the effects of taphonomy. Numerous species have considerable morphological variability and require examination of larger populations to become fully understood. These spores show intergrading morphological variation and were grouped into morphons. Although a majority of spore species were previously described, many others are new and endemic to north-western Gondwana. Forty-two species (*Acinosporites tristratus*, *Alatisporites? trisacculus*, *Biornatispora elegantula*, *B. microclavata*, *Brochotriletes crameri*, *B. tenellus*, *B. tripapillatus*, *Camarozonotriletes asperulus*, *Chelinospora carnosa*, *C. condensata*, *C. densa*, *C. laxa*, *C. vulgata*, *Coronaspora inornata*, *Cristatisporites streelii*, *Cymbosporites variabilis*, *C. variegatus*, *C. wellmanii*, *Cyrtospora tumida*, *Diaphanospora milleri*, *Dibolisporites tuberculatus*, *D. verecundus*, *Dictyotriletes hemeri*, *D. marshallii*, *Elenisporis gondwanensis*, *Emphanisporites laticostatus*, *E. plicatus*, *Geminospora convoluta*, *Grandispora maura*, *Granulatisporites concavus*, *Hystricosporites brevispinus*, *Raistrickia commutata*, *R. jaufensis*, *Retusotriletes atratus*, *R. celatus*, *Samarisporites tunisiensis*, *Verrucosisporites nafudensis*, *V. onustus*, *V. stictus*, *Zonotriletes brevivelatus*, *Z. rotundus* and *Z. venatus*), five varieties (*Cymbosporites stellospinosus* var. *minor*, *C. variabilis* vars *variabilis*, *densus* and *dispersus*, and *Dictyotriletes biornatus* var. *murinatus*) and seven combinations (*Ambitisporites asturicus*, *Camarozonotriletes retiformis*, *Cristatisporites reticulatus*, *Cymbosporites ocularis*, *Dibolisporites gaspiensis*, *Grandispora stolidota* and *Jhariatriletes emsiensis*) are newly proposed. As the reference spore zones usually used and defined in Euramerica are not all recognized in the Gondwanan coeval sections, a new biostratigraphical scheme based on the own characteristics of the spore assemblages described here is proposed. It consists of nine assemblage zones, nine interval zones and one acme zone spanning from the upper Pragian to lower Frasnian.

Key words: Trilete spores, cryptospores, taxonomy, biostratigraphy, Devonian, Gondwana.

Tʜᴇ Devonian palynology of the Arabian Plate remains incompletely known despite recent significant studies (Loboziak and Streel 1995*b*; Steemans 1995; Al-Hajri *et al.* 1999; Al-Hajri and Owens 2000; Loboziak 2000; Al-Ghazi 2007; Breuer *et al.* 2007*c*). To further document north-western Gondwanan assemblages, palynological samples from two additional boreholes from North Africa (Libya and Tunisia) were re-examined to compare with the Saudi Arabian Devonian succession (Fig. 1). The palynology of these sections was first examined by Loboziak and Streel (1989), Streel *et al.* (1990) and Loboziak *et al.* (1992*a*), but additional study, in the light of results from Saudi Arabian material, allows a more detailed understanding of Devonian palynology including endemic species, from the northern margin of western Gondwana.

The palynostratigraphic studies of Gondwanan spores rely mainly on biostratigraphical zonations established for Euramerica (Richardson and McGregor 1986; Streel *et al.* 1987). Endemic species, which are not useful for intercontinental correlations, were not incorporated in their zonations, but can be of a great utility for refined local

and regional biozonations. It is essential to develop a biostratigraphical zonation combining endemic spores for the local correlations and more cosmopolitan species for interregional biostratigraphy. A new biozonation based on the own characteristics of the spore assemblages described here is thus developed for more accurate local, regional and intercontinental correlations in north-western Gondwana.

GEOLOGICAL SETTING

Regional geological background

Through much of the Palaeozoic, the broad passive margin of northern Gondwana, which bordered the Rheic Ocean, included present day North Africa, Arabia, Turkey, central and north-west Iran, Afghanistan, India and some minor plates (Scotese 2000). Much of this region was intermittently covered by shallow epeiric seas that bordered lowlands, which represented the low-relief

PIERRE BREUER

Saudi Aramco, Geological Technical Services Division, Biostratigraphy Group, 31311 Dhahran, Saudi Arabia; e-mail: pierre.breuer@aramco.com

PHILIPPE STEEMANS

NFSR Senior Research Associate, Palaeobiogeology, Palaeobotany, Palaeopalynology, University of Liège, Belgium; e-mail: p.steemans@ulg.ac.be

doi: 10.1111/pala.12031

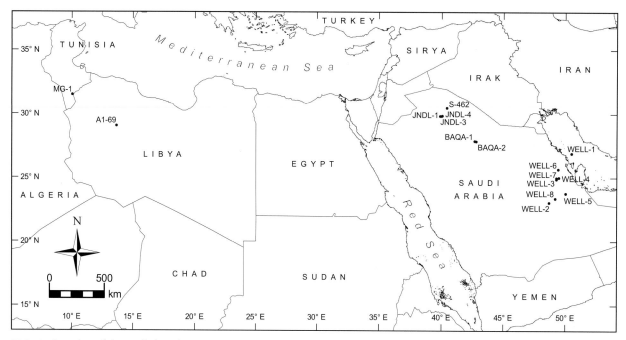

FIG. 1. Location of the studied sections.

erosion surface formed on Precambrian basement. Consequently, shallow-marine, littoral and fluvial sand, silts and muds were deposited along the margin of northern Gondwana from the end of the Neoproterozoic.

During the Devonian, the Arabian Plate lay in southern tropical latitudes (Beydoun 1991). Much of Arabia was subaerially exposed, with shallow seas extending over the remaining area. Substantial Devonian deposits accumulated, essentially as broad transgressive–regressive cycles: a Lower Devonian retrogradational cycle (fining-upward) and a progradational cycle (coarsening-upward) through much of the Middle–Upper Devonian. Thin Lower Devonian limestone intervals indicate that brief periods of limited clastic input occurred resulting in the shoreward advance of an outboard carbonate ramp in north-western Saudi Arabia (Sharland et al. 2001).

In North Africa, a transgressive-dominated sequence succeeded the Late Silurian hiatus with the development of a vast, regionally continuous, fluvial system notably in the Ghadames Basin, thinning to the north and locally onto intraplatform structural arches (Boote et al. 1998). The sediments were derived from a south-eastern source. Epeirogenic activity, reflecting the initial collision between Gondwana and Laurussia, increased in Middle to Late Devonian causing increased stratigraphical complexity. Eifelian–earliest Givetian uplift and erosion terminated the previous sequence. It was followed by a widespread marine transgression, grading up into a series of stacked depositional cycles, each strongly influenced by intraplatform highs. The Middle and Late Devonian cycles are made up of regressive, fluvial-dominated delta systems each with an

erosional upper surface, in places incised and capped by extensive transgressive marine shales, limestones and iron oolites (Boote et al. 1998).

Lithostratigraphy

Saudi Arabia. The Devonian strata of Saudi Arabia occur within a more or less conformable package of Upper Silurian – Lower Carboniferous deposits, which are subdivided into the Tawil, Jauf and Jubah formations (Fig. 2; Steineke et al. 1958; Powers et al. 1966; Powers 1968; Meissner et al. 1988). Below a regional disconformity, Silurian deposits of the underlying Qalibah Formation are present (Al-Hajri et al. 1999). Several regional unconformities separate the Upper Silurian – Lower Carboniferous package from higher strata: either the sub-Unayzah Unconformity (from the Permo–Carboniferous Unayzah Formation) or the sub-Khuff Unconformity (from the Middle Permian Khuff Formation), particularly on structural highs such as the Central Arabian Arch and the Ghawar structure. Current age calibrations of these deposits suggest that the sediments of the Tawil Formation, occurring directly above the disconformity, were deposited during the Ludlow (late Silurian) to early Pragian (Early Devonian; Stump et al. 1995; Al-Hajri and Paris 1998; Al-Hajri et al. 1999). Breuer et al. (2005b, 2007c) referred the Jauf Formation to upper Pragian – upper Emsian stages. The overlying Jubah Formation extends up into the lower Tournaisian (Al-Hajri et al. 1999; Clayton et al. 2000) where it is not removed by the

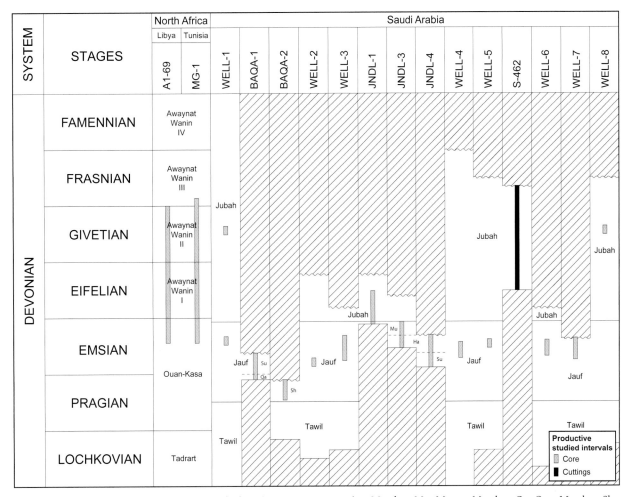

FIG. 2. Stratigraphical successions of the studied sections. Ha, Hammamiyat Member; Mu, Murayr Member; Qa, Qasr Member; Sh, Sha'iba Member; Su, Subbat Member.

Hercynian erosion. Additional biostratigraphical evidence is provided by different macrofossils groups collected at surface exposures (Boucot *et al.* 1989; Forey *et al.* 1992). Trilobites and conodonts indicate that the uppermost Sha'iba and Qasr members (lower Jauf Formation) are Pragian to lower Emsian, and brachiopods suggest that the Hammamiyat Member (upper Jauf Formation) is upper Emsian (Boucot *et al.* 1989). In Saudi Arabia, Devonian sedimentary rocks are only known to be exposed in the type outcrop area of north-western Saudi Arabia (Nafud Basin) and in the south-west (e.g. Wajid outcrop), adjacent to the exposed Precambrian Arabian Shield. Devonian deposits occur extensively in the subsurface, not only in the north-west but also in eastern Saudi Arabia.

These formations reflect, in order of succession, shifts in sedimentation from siliciclastics (Tawil Formation) to mixed siliciclastics and carbonates (Jauf Formation) and a return to siliciclastics (Jubah Formation). The Upper Silurian – lowest Carboniferous package is generally conti-

nental to near-shore shallow marine but there are regional-scale facies changes. The Jauf Formation changes from marine in north-western Saudi Arabia to marginal marine/continental in eastern, central and southern regions (Al-Hajri *et al.* 1999; Al-Hajri and Owens 2000). The alternating siliciclastics and carbonates of the Jauf Formation only in north-western Saudi Arabia are used to subdivide this formation into five members in ascending stratigraphical order: the Sha'iba, Qasr, Subbat, Hammamiyat and Murayr members. The five members constitute a conformable succession according to Wallace *et al.* (1996, 1997); their lithological descriptions and sedimentary interpretations are summarized in Al-Husseini and Matthews (2006).

North Africa. The Devonian strata present in the subsurface of Ghadames Basin occur within a more or less conformable package that is subdivided into the Tadrart, Ouan-Kasa, Awaynat Wanin I, Awaynat Wanin II and Awaynat Wanin III formations (Fig. 2). From outcrops, these lithological units were recently redescribed by Ben Rahuma *et al.* (2008) and interpreted in terms of sequence stratigraphy. Massa (1988) studied these formations and discussed their age assignment from numerous wells. The Tadrart and Ouan-Kasa formations are Lower Devonian according to macrofauna, although Streel *et al.* (1990) demonstrated diachronism for these units. The Awaynat Wanin I to IV formations are considered as being Eifelian, Givetian, Frasnian and Famennian, respectively, on the basis of faunal assemblages (Fig. 2). The rich spore assemblages from the Ghadames Basin allowed establishing a detailed palynostratigraphic zonation (Massa and Moreau-Benoit 1976; Moreau-Benoit 1988, 1989; Loboziak and Streel 1989; Loboziak *et al.* 1992*a*) and partially confirmed the age stated by Massa (1988; see below).

These lithological units are dominated by siliciclastic sediments. North-westward, along a structural dip section, correlation shows an overall thinning of the sandstone units and thickening and deepening of the shaly units. Marine limestones are mainly known along the northern flank of the Ghadames Basin. The environments of the Devonian deposits show a wide range of facies from continental, fluvial, lagoonal and intertidal to open marine facies. These deposits relate to the transgressive-dominated sequence that succeeded the uplifting and erosion of the southern flank of the basin by Caledonian tectonic activity during the Late Silurian (Boote *et al.* 1998).

REVIEW OF DEVONIAN SPORE PALYNOLOGICAL STUDIES

Saudi Arabia

Hemer and Nygreen (1967) were the first to report on Devonian spore assemblages from Saudi Arabia. These assemblages were isolated from cuttings samples of a 1341-ft-thick Devonian succession in the S-462 borehole, northern Saudi Arabia (Fig. 1), that were re-examined here. They interpreted this interval to represent a non-marine extension of the Jauf Formation present in outcrops approximately 70 km away (but which is now considered correlative with the Jubah Formation). Based on spore assemblages, they subdivided the strata into four zones, interpreted to range from Middle (probably Givetian) to Upper Devonian. Figure 3 compares and summarizes Devonian spore biozonations defined and/or applied in the different palynostratigraphic studies in Saudi Arabia.

Loboziak and Streel (1995*b*) examined Devonian age cuttings from TRBH-1 borehole also in northern Saudi Arabia, with additional data from the DMMM-45 and SDGM-211 wells from eastern Saudi Arabia. Based on taxa common to Euramerica, they applied the Devonian

spore zonation developed for Western Europe by Streel *et al.* (1987). In all three wells, they assigned the spore assemblages from the uppermost Tawil and lower Jauf formations to the Emsian. Higher in the sequence, in the Jauf Formation in DMMM-45 and the Subbat Member in TRBH-1, they recognized an upper Emsian spore assemblage. As few characteristic spore taxa occurred above this assemblage in the lower Jubah Formation, Loboziak and Streel (1995*b*) considered these strata to be late Emsian – early Givetian in age. Stratigraphically above this, in TRBH-1, they assigned assemblages to Givetian biozones already known in Western Europe.

Steemans (1995) published reports of Devonian spore assemblages from cuttings in the DMMM-45 and UDYN-1 wells. In DMMM-45, the spores were recovered from the lower part of the Tawil Formation and assigned to Lochkovian–Pragian (Lower Devonian). Those recovered from the Jauf Formation were dated as Emsian. In UDYN-1, Steemans (1995) reported Lochkovian spore assemblages from the Tawil Formation and Givetian–Frasnian spore assemblages from the upper Tawil to lower Jauf formations, which are now understood to be the result from downhole caving.

Al-Hajri *et al.* (1999) published an operational palynological zonation developed by Saudi Aramco for Devonian strata of Saudi Arabia. Because it was developed for oil industry application, and based largely on cuttings samples, this zonation is based primarily on first downhole occurrences of taxa (i.e. extinctions), although it also incorporates first common downhole occurrences/co-occurrences and acme zones. The biozonation was age-calibrated based on comparisons with the established spore zonations of Richardson and McGregor (1986) and Streel *et al.* (1987).

From the Devonian succession studied by Hemer and Nygreen (1967; see above), Loboziak (2000) described spore assemblages from Jubah Formation cuttings of S-462 over the interval 1465–2806 ft, also restudied here. The oldest assemblages were interpreted as lower lower Eifelian and the youngest upper lower Frasnian, based on comparisons with the spore biostratigraphy of Streel *et al.* (1987).

Clayton *et al.* (2000) recognized uppermost Devonian – lowermost Carboniferous spore assemblages from the uppermost Jubah Formation in wells of eastern Saudi Arabia. In HRML-51, 'Strunian' (latest Famennian) assemblages characterized by *Retispora lepidophyta* (Kedo) Playford, 1976 were recovered. Latest Famennian and earliest Tournaisian assemblages occurred in ABSF-29 typified by common representatives of *Verruciretusispora famennensis* (Kedo) Massa *et al.*, 1980 and *Indotriradites explanatus* (Luber) Playford, 1991, respectively. This work clearly demonstrated that in some locations the Jubah Formation extended into the Carboniferous.

STAGES	Hemer and Nygreen (1967)	Loboziak and Streel (1995b)	Steemans (1995)	Al-Hajri et al. (1999)	Loboziak (2000)	Loboziak (2000)	Clayton et al. (2000)	Breuer et al. (2007c)	Al-Ghazi (2007)
FAMENNIAN				D0			*R. lepiodphyta* Assemblage / *V. famennensis* Assemblage		
FRASNIAN	Zone I / ? / Zone II		Frasnian - Givetian	D1	TA-BJ/BM	*ovalis-bulliferus* / *optivus-triangulatus*			
GIVETIAN	Zone III / ? / Zone IV	TA / AD-Lem		D2	Lem / AD / Mac	*lemurata-magnificus* / *devonicus-naumovii*		AD-Lem (pars)	
EIFELIAN		AP		D3 / A	AP (pars)	*velata-langii* (pars)		AP-Pro (pars)	
EMSIAN		FD-Min / AB	AB	B / D3/D4				FD-Min / AB or FD	*annulatus - sextantii*
PRAGIAN			Pa / W / PoW / BZ-E	A / D4				PoW-Su	
LOCHKOVIAN			MN-Si to Z / MN-R	B					

FIG. 3. Chart comparing biozonation from the main studies on Devonian from Saudi Arabia.

More recently, Breuer *et al.* (2005*b*, 2007*c*) described spore assemblages and introduced new species from the Jauf and Jubah formations in five fully cored shallow core holes (BAQA-1, BAQA-2, JNDL-1, JNDL-3 and JNDL-4) from northern Saudi Arabia (Fig. 1). Based on cosmopolitan index species, the stratigraphical distribution of these new taxa was compared to the Devonian West European zonation of Streel *et al.* (1987). Although a late Pragian–Givetian age was suggested for this sequence by Breuer *et al.* (2005*b*, 2007*c*), it appears, in the light of these new results, that the top of the sequence is not younger than lower upper Eifelian. From an exploration borehole in northern Saudi Arabia, Al-Ghazi (2007) referred the partly cored Jauf Formation to the Emsian Stage by correlating with the spore zonations of Richardson and McGregor (1986) and Streel *et al.* (1987).

Finaly, Marshall *et al.* (2007) described for the first time two Givetian megaspore species from the Jubah Formation quite similar to the species found in Arctic Canada. This study demonstrated that plants producing megaspores were able to achieve dispersal over a wide range.

North Africa

The most significant Devonian palynological studies of North African Palaeozoic basins are summarized herein. Figure 4 compares Devonian spore biozonations defined and/or applied in the different studies. Jardiné and Yapaudjian (1968) and Magloire (1968) first reported on spore assemblages from the Devonian of North Africa; these were from core samples of petroleum wells drilled in the Illizi (Polignac) Basin (western Algeria) and in the Bechar Basin (eastern Algeria), respectively. These authors established local biozonations based on acritarch, chitinozoan and spore assemblages in 'Gedinnian' (Lochkovian) to Givetian strata (Jardiné and Yapaudjian 1968) and in Lochkovian and 'Siegenian' (Pragian) strata (Magloire 1968). The two papers provided stratigraphical ranges and illustrations of spore species, but no formal taxonomy was proposed.

Subsequently, Massa and Moreau-Benoit (1976) published a palynological synthesis of the Devonian of the western Libyan portion of the Ghadames Basin. Samples

AUTHORS / STAGES	Magloire (1968)	Jardiné and Yapaudjian (1968)	Massa and Moreau-Benoit (1976)	Streel et al. (1988)	Boumendjel et al. (1988)	Loboziak and Streel (1989)	Moreau-Benoit (1989)	Moreau-Benoit et al. (1993)	Rahmani-Antari and Lachkar (2001)
FAMENNIAN			Palynozone 11	D-VII	Assemblage 4		Palynozone 11	*pusillites-lepidophytus*	ST1
			Palynozone 10				?	*flexuosa-cornuta*	DS3
			Palynozone 9				Palynozones 9-10		DS2
FRASNIAN			Palynozone 8	D-VI		IV			
			Palynozone 7			BM	Palynozone 8	*ovalis-bulliferus*	DS1
GIVETIAN		d	Palynozone 6	D-V	Assemblage 3	TCo-BJ(?)	?	*optivus-triangulatus*	
						TA		*lemurata-magnificus* (pars)	DM2
		c	Palynozone 5			Lem / AD	Palynozone 7		
EIFELIAN		Zone IX		D-IV		Mac	Palynozone 6		
		b	Palynozone 4	D-III			Palynozone 5		DM1
		a		D-II		Vel / AP / Pro	Palynozone 4	*douglastowense-eurypterota*	
EMSIAN		c / Zone VIII b / a	Palynozone 3	D-I			Palynozone 3	*annulatus - sextantii*	DI2
PRAGIAN	Biozone J	Zone VII / Zone VI	Palynozone 2		Assemblage 2		Palynozone 2		DI1
LOCHKOVIAN	Biozone I / Biozone H	d / c / Zone V b / a	Palynozone 1				Palynozone 1 / Palynozone 0	?	

FIG. 4. Chart comparing biozonation from the main studies on Devonian from North Africa.

from numerous deep boreholes drilled during oil prospecting provided the study material. They assigned a Pragian–late Famennian age to the 11 spore assemblages that were recognized from the Tadrart Formation to the Tahara Formation. Moreau-Benoit (1979, 1980b) formally described the spore species from the 1976 synthesis. Further data about the systematics and stratigraphical distribution of Middle and Late Devonian spore species appeared later in Massa and Moreau-Benoit (1985) and Moreau-Benoit (1988, 1989). Moreau-Benoit (1989) reinterpreted the age of the biozonation previously proposed in Massa and Moreau-Benoit (1976). This synthesis, which is comprised of a series of publications by the same authors, constitutes the most important record of Devonian spores from North Africa.

Paris et al. (1985) presented preliminary results from a project, which illustrated all of the stratigraphically significant species encountered in a comprehensive study, on the Ordovician to Late Cretaceous palynology of north-eastern Libya. They recorded Emsian–Famennian spore assemblages from cuttings and core samples from exploration wells situated in Cyrenaica (north-eastern Libya).

This first publication gave only brief documentation of the different assemblages without presenting the stratigraphical ranges of the spores. In the final phase of this project, Streel et al. (1988) correlated these assemblages with the well-established Devonian Western European zonation of Streel et al. (1987). In addition, they proposed a recalibration of the western Libyan Palynozones 4–8 of Massa and Moreau-Benoit (1976) with emphasis on the significance of the evolutionary appearance of *Geminospora lemurata*. Palynozones 4–6 were restricted to the Eifelian by Streel et al. (1988). These zones were previously dated as Eifelian to upper Givetian. Palynozones 7–8, initially interpreted to be Frasnian by Massa and Moreau-Benoit (1976), were also reassigned by Streel et al. (1988) but to the lower Givetian.

Boumendjel et al. (1988) examined core samples of TRN-3 borehole drilled in the Illizi Basin. Only 'Siegenian' (Pragian) and early Givetian spore assemblages were recognized. Other data from the Illizi Basin were published on the Emsian to uppermost Famennian stratigraphical interval (Coquel and Latrèche 1989; Moreau-Benoit et al. 1993). Moreau-Benoit et al. (1993) applied the biozonation developed by Richardson and McGregor (1986) to the spore succession from the Illizi Basin.

Loboziak and Streel (1989) systematically studied the most important species encountered in four boreholes from the Ghadames Basin. Their results were compared with the zonation of Streel *et al.* (1987). In the A1-69 borehole (Fig. 1), the first appearances of characteristic spores showed the same stratigraphical succession as in the Ardenne-Rhenish biozonation. Chronostratigraphic correlations made using this biozonation gave an Emsian to latest Frasnian age for the studied Libyan samples. From the 55 recorded species, almost 90 per cent were also found in Western Europe according to Loboziak and Streel (1989).

Streel *et al.* (1990) reviewed the relationship between spores, faunas (i.e. brachipods and conodonts) and megafloras for the Lower and Middle Devonian in Libya. They demonstrated the diachronism of the Tadrart and Ouan-Kasa formations across the Ghadames Basin. At the northern margin of the basin, the lower part of the Tadrart Formation in MG-1 borehole (Fig. 1) is within the lowermost part of the Lochkovian, not at its base. At the southern margin, the Tadrart and Ouan-Kasa formations are probably no older than upper Emsian or possibly lower Eifelian. The Awaynat Wanin I Formation is often absent in the south or strongly reduced in thickness.

Devonian spores from the Tunisian MG-1 well on the northern margin of the Ghadames Basin were compared to available faunal data in Loboziak *et al.* (1992a). The age of the various formations drilled was revised. The authors concluded that the Emsian and the Eifelian are represented by an increased sedimentation rate, resulting in thicker sections, than present for the other stages.

Rahmani-Antari and Lachkar (2001) is the only palynological study on Devonian material from Morocco. The authors studied four wells and one outcrop section in order to document palynomorphs from the Devonian. Spores, acritarchs as well as chitinozoans allowed the subdivision of the Devonian into eight palynozones despite some stratigraphical gaps in the succession.

A most recent significant palynological study was carried out by Rubinstein and Steemans (2002). It concerned A1-61 borehole situated on the northern margin of the Ghadames Basin. The Silurian/Devonian boundary was recognized in the Tadrart Formation. Ludlow to early Lochkovian spore assemblages were described, compared and correlated with spore zonations established for the type sequences of the Welsh Borderland, and those previously described from Libya.

Finally, de Ville de Goyet *et al.* (2007) and Steemans *et al.* (2011b) studied a rich and diverse Middle Devonian megaspore assemblages from the A1-69 borehole (Fig. 1). Seventeen megaspore taxa were elaborately described, illustrated and compared to coeval Euramerican assemblages.

MATERIAL AND METHODS

Location of studied sections

The reference core holes for the Jauf–Jubah succession are located in north-western Saudi Arabia (Fig. 1). They are BAQA-1, BAQA-2, JNDL-1, JNDL-3 and JNDL-4 and S-462 borehole that in combination give a nearly complete section through the Jauf and Jubah formations (Fig. 2). The preliminary palynological results from the five core holes were published by Breuer *et al.* (2005b, 2007c). Also in the same general area is the S-462 borehole, which was previously studied by Hemer and Nygreen (1967) and then by Loboziak (2000; see above). The second study area is located in eastern Saudi Arabia and comprises cores from eight deep exploration wells (WELL-1, WELL-3, WELL-2, WELL-4, WELL-5, WELL-6, WELL-7 and WELL-8; Fig. 1). The studied intervals are also from the Jauf and Jubah formations but were not previously published (Fig. 2).

Palynology preparations from the two boreholes from the Ghadames Basin in North Africa were re-examined. A1-69 borehole was drilled on the southern flank of the basin (Libya) while MG-1 borehole is located on the northern flank in Tunisia (Fig. 1). Formations studied in ascending stratigraphical order are the Ouan-Kasa, Awaynat Wanin I, Awaynat Wanin II and Awaynat Wanin III formations (Fig. 2). Previous Devonian palynological results of these boreholes were published in Loboziak and Streel (1989) and Loboziak *et al.* (1992a; see above).

Material

Saudi Aramco provided 440 samples from various Saudi Arabian core holes and exploration wells. Among these, 295 were fossiliferous and were used for palynological analysis. The remaining 145 were either barren or contained very few biostratigraphically significant palynomorphs. The Laboratory of Palaeontology of the University of Lille (France) provided the North African samples. Of the 96 available samples from North Africa, 84 samples yielded palynomorphs and were usable for palynological study. In total, 350 core samples and 29 cutting samples (S-462) were productive and used in this study. All samples were collected from dark-coloured, fine-grained, shaly sandstones, siltstones and shales.

Palynological processing

The samples from BAQA-1, BAQA-2, JNDL-3 and JNDL-4 were processed by the Centre for Palynology of the University of Sheffield (UK). Samples from A1-69, JNDL-1, MG-1, S-462, WELL-1, WELL-2, WELL-4,

WELL-5, WELL-6, WELL-7 and WELL-8 and most of the samples from WELL-3 were processed in the Laboratory of 'Palaeobiogeologie, Palaeobotanique et Palaeopalynologie' at the University of Liège (Belgium). Saudi Aramco processed the remainder of the WELL-3 samples and some additional samples from BAQA-1, BAQA-2, JNDL-3 and JNDL-4. New palynological slides from the residues of MG-1, dating from the previous studies (see above), were prepared for this study in Liège.

All samples were processed according to standard palynological laboratory methods (Streel 1965). Each sample was crushed, and 10–25 g were demineralized in 10 per cent HCl and 40 per cent HF. The residue of the most thermally mature samples was oxidized in 65 per cent HNO3 and KClO3 (Schultze solution) and sieved through a 10-μm mesh. Subsequently, a hot bath in 25 per cent HCl eliminated the remaining fine neoformed fluorided particles. The residue of all samples was sieved through a 10-μm mesh. The final residue was mounted on palynological slides using Euparal or Eukit resin. One to four slides were made for each productive sample.

Taxonomy

The term 'spore' is used throughout the following text as all propagule produced by embryophytes. They include cryptospores, monolete spores, trilete spores and megaspores. Some spore species were previously described; others are new and endemic to north-western Gondwana and consequently defined herein.

For stratigraphical correlation, it is essential to consistently identify spores; but differences between taxa may be so slight that intermediates can often be found between varieties, species or even genera, usually regarded as distinct. Many cases of intergradations are common in Devonian spores (Playford 1983; Steemans and Gerrienne 1984; Richardson *et al.* 1993; Marshall 1996; Breuer *et al.* 2005*a*, *b*, 2007*a*, *b*). The variability in spore morphology often obscures the taxonomic limits that were originally defined. The concept of the morphon was introduced by Van der Zwan (1979) to address this taxonomic issue. A morphon is defined as '...a group of species united by continuous variation of morphological characters' (Van der Zwan 1979, p. 11). The main idea of the morphon concept can be refined as follows: the apparent morphological continua may represent spore variation in a particular known or hypothetical natural plant species or group of related species. Evolutionary convergence may, however, cause morphological similarities between spores that do not necessarily reflect links between their parent plants. During Palaeozoic times, different plant groups often produced spores of similar morphology (Gensel 1980; Fanning *et al.* 1992; Wellman 2009). In contrast,

and further complicating the problem, cases where a single sporangium can produce two different genera of trilete spores were recognized (Habgood *et al.* 2002).

The present taxonomic study of large populations reveals that many defined spore species have considerable overlap in their morphologies. The morphological variations of spores can be attributed to phylogenetic evolution, ontogeny (maturation of sporangia) or taphonomic factors (Breuer *et al.* 2007*a*). This occurs at a time of explosive evolution of early land plant floras (Steemans *et al.* 2012). Some species show continuous morphological intergradations with other genera/species and, therefore, are grouped into morphons (Table 1). Morphons centred on the following nominal species are introduced herein: *Apiculiretusispora brandtii, Archaeozonotriletes variabilis, Chelinospora vulgata, Cristatisporites reticulatus, Cymbosporites catillus, Diaphanospora milleri, Grandispora incognita, G. protea, Samarisporites eximius, Synorisporites papillensis* and *Verrucosisporites scurrus.* They are briefly discussed in the systematic descriptions, and Table 1 summarizes the morphological features used to characterize the morphons introduced herein. Morphons are regarded as informal groupings of taxa under the Botanical Code; their circumscribing characters and the lists of their constituent taxa are subject to emendation as work continues to further understand these and other morphologically intergrading forms.

The morphon concept is considered complementary to the typological approach of traditional binomial nomenclature in palynology. It emphasizes the similarity of the morphological characters more than the differences and integrates morphological trends, which are space and/or time dependent, but also sensitive to various environmental conditions. The morphon may facilitate the interpretation of species with similar morphological characters in terms of a more natural grouping (Breuer *et al.* 2007*a*). The morphological variability presented by spores may reflect intraspecific variability or biological evolution. Indeed, it may be influenced locally by other parameters such as state of preservation (e.g. *Diaphanospora milleri* Morphon), sedimentary sorting (Jäger 2004) and/or reworking of simpler and older morphotypes into assemblages containing more complex and younger ones (Breuer *et al.* 2005*a*).

Systematic descriptions

The spore populations of 205 rare-to-common species are described and measured in detail in this contribution. Their morphological variability is illustrated in Figures 5–52 by several specimens for each species when possible and synonymies are listed for each species. The terminology used to describe the morphology of spores is defined in Wellman and Richardson (1993), Richardson (1996) and Punt *et al.* (2007). The Erdtman's (1952) wall stratification

TABLE 1. Spore morphons and their characterizing features referred to in the text.

Morphon name	Constituent taxa	Morphological characters
Apiculiretusispora brandtii	*Apiculiretusispora brandtii* *Cymbosporites asymmetricus* *Rhabdosporites minutus*	Densely spaced, small coni, grana and spinae; exine partially or completely detached from nexine at the equator.
Archaeozonotriletes variabilis	*Archaeozonotriletes variabilis* *Cyrtospora tumida* *Lophozonotriletes media*	Patinate spores; laevigate to variably sculptured with scattered coarse, rounded protuberances and flat-topped verrucae.
Chelinospora vulgata	*Chelinospora condensata* *Chelinospora densa* *Chelinospora laxa* *Chelinospora vulgata*	Patinate spores; distal region sculptured with loosely distributed or brain-like convoluted muri; laesurae simple; subcircular to triangular amb.
Cristatisporites reticulatus	*Cristatisporites reticulatus* *Cristatisporites streelii*	Zonate spores; distal region sculpured with fold-like ridges, bearing spinae.
Cymbosporites catillus	*Cymbosporites catillus* *Cymbosporites cyathus*	Patinate spores; distal region sculptured with densely packed grana, coni or spinae.
Dictyotriletes biornatus	*Cymbosporites variabilis* var. *densus* *Cymbosporites variabilis* var. *dispersus* *Cymbosporites variabilis* var. *variabilis* *Dictyotriletes biornatus* var. *biornatus* *Dictyotriletes biornatus* var. *murinatus*	Patinate spores; distal region sculptured of coni, discrete or partly fused in elongate elements, evenly distributed to organized in a reticulum pattern; simple laesurae; subcircular to subtriangular amb.
Diaphanospora milleri	*Diaphanospora milleri* *Retusotriletes celatus*	Strongly folded sexine possible; dark subtriangular apical area; subcircular amb.
Grandispora incognita	*Grandispora incognita* *Grandispora naumovae*	Camerate spores; distal region sculptured with slender spinae with flared bases.
Grandispora protea	*Grandispora douglastownensis* *Grandispora protea*	Camerate spores; distal region sculptured with biform or parallel-sided spinae.
Samarisporites eximius	*Samarisporites angulatus* *Samarisporites eximius* *Samarisporites praetervisus*	Zonate spores; distal region sculpured with discrete spinae, sometimes arranged in rugulae; sub-circular to triangular amb.
Synorisporites papillensis	*?Knoxisporites riondae* *Synorisporites papillensis*	Proximal papillae; distal region, laevigate, irregularly verrucate or showing an annulus.
Verrucosisporites scurrus	*Dibolisporites farraginis* *Dibolisporites uncatus* *Verrucosisporites premnus* *Verrucosisporites scurrus*	Varied, spaced or partially fused, evenly or asymmetrically distributed, coni/spinae/bacula/verrucae; simple laesurae; subcircular amb.

based on purely morphological criteria is more appropriate for Palaeozoic palynology and consequently applied here to identify the layers of the exine. The comparison of each described species with possible similar forms is discussed. The stratigraphical and geographical distribution of species are summarized almost exclusively from the major part of Devonian literature. Although the age of the spore assemblages, where the species described here also occur, are cited according to the Epochs and Ages defined by the International Commission on Stratigraphy, they were are not reappraised. Genera and species are arranged in the artificial morphographic system of suprageneric categories (*anteturmae* and *turmae*) devised by Potonié (1956, 1970) and modified by others, but within these categories, genera, species and varieties are arranged in alphabetical order for ease of use.

As sample depths are expressed in metres for the previous studies on MG-1 (Streel *et al.* 1990; Loboziak *et al.* 1992*a*), this convention is kept herein to allow easy comparisons between the different studies. Samples from other studied sections (A1-69 and Saudi Aramco sections) were originally measured in feet.

Repository of material. All material from JNDL-1, MG-1, S-462, WELL-1, WELL-2, WELL-3, WELL-4, WELL-5, WELL-6, WELL-7 and WELL-8 is housed in the collections of Laboratory of 'Palaeobiogeologie, Palaeobotanique et Palaeopalynologie' of the University of Liège. All these palynological slides have a five-digit reference number, except the third palynological slide of samples from JNDL-1, which have a three-digit reference number preceded by 'PPM'. From BAQA-1, BAQA-2, JNDL-3 and JNDL-4, one set of palynological slides is stored in the collections of the Laboratory of 'Palaeobiogeologie, Palaeobotanique et Palaeopalynologie', and each has a five-digit reference number. A second set of slides for these four core holes is housed in the Centre for Palynology of the University of Sheffield, and the slides have a three-digit reference number preceded by '03CW'. The palynological slides from A1-69 are from the collections of the Laboratory of Palaeontology of the University of Lille and are referenced by a five-digit number. These were previously studied by Loboziak and Streel (1989).

SYSTEMATIC PALAEONTOLOGY

Anteturma CRYPTOSPORITES (Richardson *et al.*) Richardson, 1988 emend. Steemans, 2000

Genus ARTEMOPYRA Burgess and Richardson, 1991 emend. Richardson, 1996

Type species. *Artemopyra brevicosta* Burgess and Richardson, 1991.

Artemopyra inconspicua Breuer *et al.*, 2007c
Figure 5A–B

2007c *Artemopyra inconspicua* Breuer *et al.*, p. 42, pl. 1, figs 1–5.

Dimensions. 36(48)57 µm; 20 specimens measured.

Occurrence. BAQA-1, JNDL-1, JNDL-3 and JNDL-4; Jauf (Subbat to Murayr members) and Jubah formations; *ovalis-biornatus* to *svalbardiae-eximius* zones. A1-69; Awaynat Wanin II Formation; *lemurata* Zone. MG-1; Awaynat Wanin I Formation; *incognita* Zone.

Artemopyra recticosta Breuer *et al.*, 2007c
Figure 5C–D

2006 *Artemopyra?* spp. Wellman, pl. 20, fig. i.
2007c *Artemopyra recticosta* Breuer *et al.*, p. 43, pl. 1, figs 6–12.

Dimensions. 34(51)70 µm; 36 specimens measured.

Occurrence. BAQA-1, JNDL-1, JNDL-3, JNDL-4, S-462, WELL-2, WELL-3, WELL-4, WELL-7 and WELL-8; Jauf (Subbat to Murayr members) and Jubah formations; *ovalis-biornatus* to *triangulatus-catillus* zones. A1-69; Ouan-Kasa, Awaynat Wanin I and Awaynat Wanin II formations; *lindlarensis-sextantii* to *triangulatus-catillus* zones. MG-1; Ouan-Kasa, Awaynat Wanin I, Awaynat Wanin II and Awaynat Wanin III formations; *annulatus-protea* to *langii-concinna* zones.

Previous records. From upper Pragian – lower Emsian of Paraná Basin, Brazil (Mendlowicz Mauller *et al.* 2007); upper Eifelian – middle Givetian of Parnaíba Basin, Brazil (Breuer and Grahn 2011); Emsian from Saudi Arabia (Al-Ghazi, 2007); and upper Pragian – ?lowermost Emsian of Scotland (Wellman 2006).

Genus CYMBOHILATES Richardson emend. Breuer *et al.*, 2007c

Type species. *Cymbohilates horridus* Richardson, 1996.

Cymbohilates baqaensis Breuer *et al.*, 2007c
Figure 5E–F

2007c *Cymbohilates baqaensis* Breuer *et al.*, p. 43, pl. 1, figs 13–19.

Dimensions. 30(36)40 µm, 16 specimens measured.

Ocurrence. BAQA-1, BAQA-2, JNDL-1, JNDL-3, JNDL-4, WELL-3, WELL-4, WELL-6 and WELL-7; Jauf and Jubah formations; *papillensis-baqaensis* to *svalbardiae-eximius* zones. MG-1; Ouan-Kasa Formation; *svalbardiae-eximius* Zone but occurrences are probably reworked.

Cymbohilates comptulus Breuer *et al.*, 2007c
Figure 5G–I

? 1988 *Gneudnaspora* sp. (Chibrikova) Balme, p. 17, pl. 3, fig. 15.
? 1996 *Cymbohilates amplus* Wellman and Richardson, p. 55, pl. 10, figs 1–3.

FIG. 5. Each figured specimen is identified by borehole, sample, slide number and England Finder Co-ordinate location. All figured specimens are at magnification ×1000 except where mentioned otherwise. A–B, *Artemopyra inconspicua* Breuer *et al.*, 2007c. A, JNDL-4, 182.5 ft, 68636, H45; hilum exhibits a pseudotrilete mark. B, BAQA-1, 227.1 ft, 66784, D34. C–D, *Artemopyra recticosta* Breuer *et al.*, 2007c. C, WELL-7, 13689.7 ft, 62319, L51/4. D, JNDL-1, 172.7 ft, 60845, T-U36. E–F, *Cymbohilates baqaensis* Breuer *et al.*, 2007c. E, BAQA-2, 134.4 ft, 03CW137, P40/1. F, BAQA-2, 133.0 ft, 03CW136, T29. G–I, *Cymbohilates comptulus* Breuer *et al.*, 2007c. G, WELL-3, 14214.1 ft, 66839, O57/2. H, BAQA-1, 376.4 ft, 03CW119, U25/1. I, MG-1, 2258 m, 62948, S48/2. J–K, *Cymbohilates heteroverrucosus* Breuer *et al.*, 2007c. J, JNDL-1, 162.3 ft, 60841, G32/4. K, 167.8 ft, 60843, X29/1. L–N, *Cymbohilates* sp. 1. L, MG-1, 2631.2 m, 62551, K29/2. M, MG-1, 2631.2 m, 62552, N45. N, MG-1, 2631.2 m, 62552, T43/1. O. *Dyadaspora murusattenuata* Strother and Traverse, 1979. BAQA-2, 57.2 ft, 66817, H40.

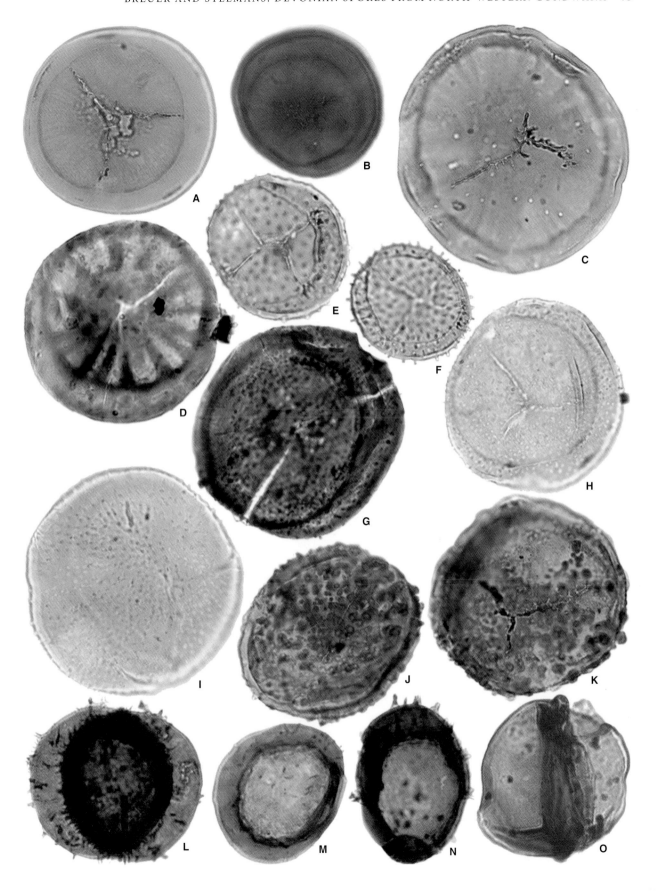

2007c *Cymbohilates comptulus* Breuer *et al.*, p. 43, pl. 2, figs 1–7.

Dimensions. 48(58)70 μm; 28 specimens measured.

Occurrence. BAQA-1, BAQA-2, JNDL-3, JNDL-4, WELL-3 and WELL-6; Jauf Formation (Sha'iba to Hammamiyat members); *papillensis-baqaensis* to *lindlarensis-sextantii* zones.

Cymbohilates cymosus Richardson, 1996

1996 *Cymbohilates cymosus* Richardson, p. 34, pl. 6, figs 3–6; pl. 7, figs 1–4.

Dimensions. 62 μm; one specimen measured.

Occurrence. BAQA-2; Jauf Formation (Sha'iba Member); *papillensis-baqaensis* Zone.

Previous record. From lower–middle Lochkovian of England (Richardson, 1996).

Cymbohilates heteroverrucosus Breuer *et al.*, 2007c
Figure 5J–K

2007c *Cymbohilates heteroverrucosus* Breuer *et al.*, p. 47, pl. 2, figs 9–12; pl. 3, figs 1–2.

Dimensions. 45(53)62 μm; 34 specimens measured.

Comparison. *Hispanaediscus? irregularis* Wellman and Richardson, 1996 has less variation in the size of the verrucae and is only sculptured outside the contact surface. It does not show tears on the proximal face.

Occurrence. JNDL-1, JNDL-3, JNDL-4 and WELL-7; Jauf (Hammamiyat and Murayr members) and Jubah formations; *lindlarensis-sextantii* to *svalbardiae-eximius* zones.

Cymbohilates sp. 1
Figure 5L–N

Description. Amb is sub-circular. A curvatura 0.5–1 μm wide delimits a circular to sub-circular, smooth hilum. The hilum radius either equals or is approximately seven-tenths of the amb radius. Exine is sculptured subequatorially and distally with widely spaced and irregularly distributed spines 1–3.5 μm high, 0.75–2.5 μm wide at base. In some specimens, spines are very few in number.

Dimensions. 37(39)42 μm; five specimens measured.

Occurrence. MG-1; Ouan-Kasa Formation; *svalbardiae-eximius* Zone but occurrences are probably reworked.

Genus DYADASPORA Strother and Traverse, 1979

Type species. *Dyadaspora murusattenuata* Strother and Traverse, 1979.

Dyadaspora murusattenuata Strother and Traverse, 1979
Figure 5O

1979 *Dyadaspora murusattenuata* Strother and Traverse, p. 15, pl. 3, figs 9–10.

Dimensions. 45–52 μm; two specimens measured.

Occurrence. BAQA-2 and WELL-3; Jauf Formation (Sha'iba Member); *papillensis-baqaensis* to *lindlarensis-sextantii* zones.

Previous records. Widely reported from Silurian through Lochkovian assemblages; e.g. Argentina (Rubinstein and Toro 2006); Brazil (Steemans *et al.* 2008), North Africa (Rubinstein and Steemans 2002; Spina and Vecoli 2009); Pennsylvania, USA (Strother and Traverse 1979; Beck and Strother 2008); Saudi Arabia (Steemans *et al.* 2000b; Wellman *et al.* 2000a); Sweden (Hagström 1997; Mehlqvist 2009); and UK (Wellman 1993; Wellman *et al.* 2000b).

Genus GNEUDNASPORA Balme, 1988 emend.
Breuer *et al.*, 2007c

Type species. *Gneudnaspora kernickii* Balme, 1988.

Gneudnaspora divellomedia (Chibrikova) Balme, 1988 var.
divellomedia Breuer *et al.*, 2007c
Figure 6A–B

1988 *Gneudnaspora divellomedium* (Chibrikova) Balme, p. 125, pl. 3, figs 1–7.
non 1991 *Laevolancis divellomedia* (Chibrikova) Burgess and Richardson, p. 607, pl. 2, figs 4, 6.
2007c *Gneudnaspora divellomedia* (Chibrikova) Balme, 1988 var. *divellomedia* Breuer *et al.*, p. 48, pl. 3, figs 3–9.

Dimensions. 32(54)82 μm; 17 specimens measured.

Occurrence. BAQA-1, BAQA-2, JNDL-1, JNDL-3, JNDL-4, WELL-2, WELL-3, WELL-4, WELL-5, S-462, WELL-6, WELL-7 and WELL-8; Jauf and Jubah formations; *papillensis-baqaensis* to *langii-concinna* zones. A1-69; Ouan-Kasa, Awaynat Wanin I and Awaynat Wanin II formations; *lindlarensis-sextantii* to *triangulatus-catillus* zones. MG-1; Ouan-Kasa, Awaynat Wanin I, Awaynat Wanin II and Awaynat Wanin III formations; *lindlarensis-sextantii* to *langii-concinna* zones.

Previous records. From Emsian – lower Frasnian of Australia (Balme 1988; Hashemi and Playford 2005); upper Eifelian – middle Givetian of Parnaíba Basin, Brazil (Breuer and Grahn 2011); upper Pragian – lower Emsian of Paraná Basin, Brazil (Mendlowicz Mauller *et al.* 2007); upper Emsian – lower Eifelian of Russian Platform (Avkhimovitch *et al.* 1993); and Emsian of Saudi Arabia (Al-Ghazi 2007).

Remarks. Contrary to the *G. divellomedia* (Chibrikova) Balme, 1988 var. *minor* Breuer *et al.*, 2007c, *G. divellomedia* var. *divellomedia* only occurs in Devonian strata.

Gneudnaspora divellomedia (Chibrikova) Balme, 1988 var. *minor* Breuer *et al.*, 2007c
Figure 6C–E

1973	*Archaeozonotriletes* cf. *divellomedium* Chibrikova; Richardson and Ioannides, p. 280, pl. 8, fig. 10–11.
1978	*Hispanaediscus* sp.; McGregor and Narbonne, p. 1296, pl. 1, figs 20–22.
1979	*Archaeozonotriletes* cf. *chulus nanus* Richardson and Lister; Holland and Smith, pl. 2, figs 7–9.
1979	'smooth-walled inaperturate spore'; Strother and Traverse, p. 14, pl. 3, fig. 5.
1993	*Archaeozonotriletes* cf. *divellomedium* Chibrikova; Moreau-Benoit *et al.*, pl. 1, fig. 2.
2007c	*Gneudnaspora divellomedia* (Chibrikova) Balme, 1988 var. *minor* Breuer *et al.*, p. 48 (*cum syn.*), pl. 3, figs 10–16.

Dimensions. 28(31)34 μm; 21 specimens measured.

Occurrence. BAQA-1, BAQA-2, JNDL-1, JNDL-3, JNDL-4, WELL-3, WELL-5, WELL-6 and WELL-7; Jauf and Jubah formations; *papillensis-baqaensis* to *svalbardiae-eximius* zones. A1-69; Ouan-Kasa Formation; *lindlarensis-sextantii* Zone. MG-1; Ouan-Kasa Formation; *svalbardiae-eximius* Zone but occurrences are probably reworked.

Previous records. Gneudnaspora divellomedia var. *minor* has been reported in Ordovician (Vecoli *et al.* 2011), widely in the Silurian through Middle Devonian palynofloras from many parts of the world.

Genus TETRAHEDRALETES Strother and Traverse emend. Wellman and Richardson, 1993

Type species. Tetrahedraletes medinensis Strother and Traverse emend. Wellman and Richardson, 1993.

Tetrahedraletes medinensis Strother and Traverse emend. Wellman and Richardson, 1993

1993	*Tetrahedraletes medinensis* Strother and Traverse emend. Wellman and Richardson, p. 165 (*cum syn.*), pl. 2, figs 8, 10–12.

Dimensions. 40(50)58 μm; 14 specimens measured.

Comparison. In *Cheilotetras caledonica* Wellman and Richardson, 1993, the equatorial crassitude of individual spores is drawn out into distinct extended flanges, and there are no discernible lines of attachment between adjacent spores. *Rimosotetras problematica* Burgess, 1991 encompasses tetrads composed of loosely attached, frequently partially separated, laevigate, alete or indistinctly trilete spores.

Occurrence. BAQA-2, JNDL-3, JNDL-4, WELL-2, WELL-3, WELL-4 and WELL-7; Jauf Formation (Sha'iba to Hammamiyat members); *papillensis-baqaensis* to *lindlarensis-sextantii* zones but most of occurrences are probably reworked.

Previous record. Tetrahedraletes medinensis has been reported worldwide in Ordovician through Lower Devonian.

Anteturma SPORITES Potonié, 1893
Turma MONOLETES Ibrahim, 1933

Genus ARCHAEOPERISACCUS Naumova emend. McGregor, 1969

Type species. Archaeoperisaccus menneri Naumova, 1953.

Archaeoperisaccus cf. *A. rhacodes* Hashemi and Playford, 2005
Figure 6F

cf. 2005	*Archaeoperisaccus rhacodes* Hashemi and Playford, p. 388, pl. 13, figs 10–11; pl. 14, figs 4, 7–8.

Description. Amb is generally oval to irregular. The length of the short axis is equal to three-quarters of the long axis. Laesura is marked by fold-like labra *c.* 2–4 μm thick. Central body radius equals about three-fifths to three-quarters of the amb radius. Ornamentation is closely and regularly distributed, consisting of coni and spines, up to 3 μm high and 2 μm in basal diameter.

Dimensions. 74–115 μm; two specimens measured.

Remarks. Archaeoperisaccus rhacodes Hashemi and Playford, 2005 is slightly different than the specimens figured here; it is larger (108–136 μm) and has conate elements, which can be loosely distributed. A larger population of

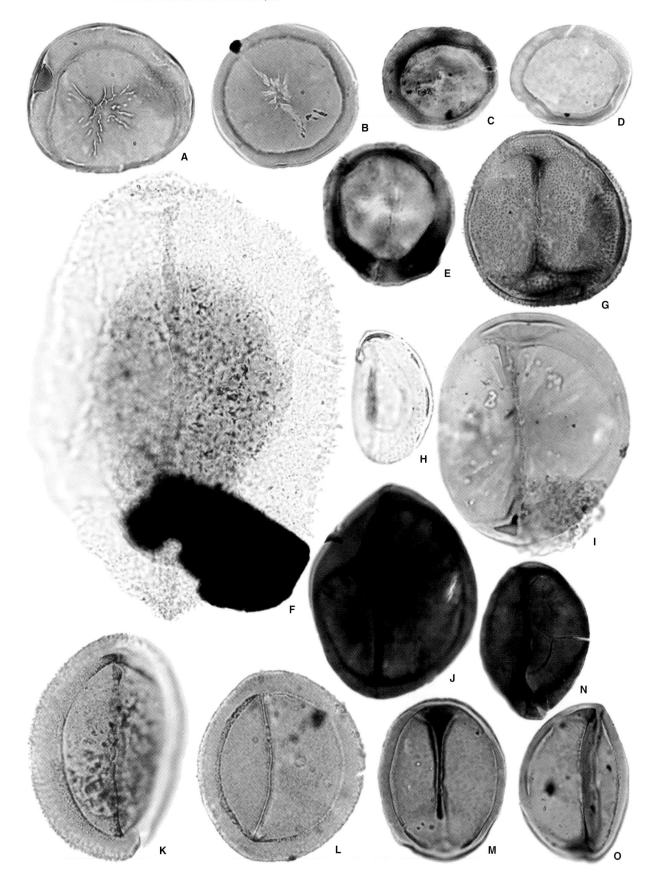

the form described here is needed to correctly place this taxonomically.

Comparison. Archaeoperisaccus verrucosus Pashkevich, 1964 is verrucate and therefore sculpturally distinct from *Archaeoperisaccus* cf. *A. rhacodes* Hashemi and Playford, 2005. *A. opiparus* Owens, 1971 has a greater overall length, a strongly elevated fold-like labra, thicker nexine and biform sculptural elements. *A. oblongus* Owens, 1971 has a more elongate amb and a coarser sculpture. *A. indistinctus* Lu Lichang, 1980 is scabrate.

Occurrence. MG-1; Awaynat Wanin II Formation; *undulatus* to *langii-concinna* zones.

Genus DEVONOMONOLETES Arkhangelskaya, 1985

Type species. Devonomonoletes microtuberculatus (Chibrikova) Arkhangelskaya, 1985.

Devonomonoletes sp. 1
Figure 6G

Description. Amb is sub-circular. Contact faces are laevigate. Laesura equals one-half to three-quarters of the amb radius. Laesura is distinct, straight to slightly curved, terminating in well-defined curvaturae. Sculptured outside of the contact area with evenly distributed spines 0.5–1.5 μm high and 0.5–1 μm apart.

Dimensions. 36(49)58 μm; five specimens measured.

Comparison. Devonomonoletes microtuberculatus (Chibrikova) Arkhangelskaya, 1985 is clearly oval-shaped and is sculptured with small tubercules. Furthermore, indistinctly outlined darkened muri are present where the distal and proximal sides meet.

Occurrence. BAQA-1, BAQA-2 and JNDL-4; Jauf Formation (Sha'iba to Hammamiyat member); *papillensis-baqaensis* to *lindlarensis-sextantii* zones.

Genus LATOSPORITES Potonié and Kremp, 1954

Type species. Latosporites latus Potonié and Kremp, 1954.

Latosporites ovalis Breuer et al., 2007c
Figure 6M–O

2006   ?monolete spore; Wellman, pl. 19, fig. h.
2007c *Latosporites ovalis* Breuer *et al.*, p. 52, pl. 12, figs 8–12.

Dimensions. 32(48)62 μm; 22 specimens measured.

Comparison. Laevigatosporites sp. A *in* Hashemi and Playford (2005) has laesura equal to two-fifths to one-half of the long axis. Wellman (2006) found a monolete spore similar to *L. ovalis* in the Rhynie Outlier, Scotland.

Occurrence. BAQA-1, BAQA-2, JNDL-1, JNDL-3, JNDL-4, WELL-1, WELL-2, WELL-3, WELL-4, WELL-5, WELL-6 and WELL-7; Jauf and Jubah formations; *ovalis-biornatus* to *svalbardiae-eximius* zones but specimens recovered from the Jubah Formation are probably reworked.

Previous records. From upper Pragian – lower Emsian of Paraná Basin, Brazil (Mendlowicz Mauller *et al.* 2007); Emsian from Saudi Arabia (Al-Ghazi 2007); and upper Pragian – ?lowermost Emsian of Scotland (Wellman 2006).

Latosporites sp. 1
Figure 7A–B

Description. Amb is oval. The length of the short axis equals eight-tenths to nine-tenths of the long axis. Laesura is distinct, simple and straight, and generally equals to more or less eight-tenths of the long axis and is terminated by well-defined curvaturae. Exine entirely laevigate, 0.5–1.5 μm thick.

Dimensions. 38(43)49 μm; three specimens measured.

Comparison. Latosporites ovalis Breuer *et al.*, 2007c is labrate, more elliptical and thicker, resulting in a sturdy appearance.

Occurrence. WELL-1; Jubah Formation; *lemurata-langii* Zone. A1-69; Awaynat Wanin I Formation; *svalbardiae-eximius* Zone. MG-1; Ouan-Kasa Formation, Awaynat Wanin I and Awaynat Wanin II formations; *svalbardiae-eximius* to *triangulatus-catillus* zones.

FIG. 6. Each figured specimen is identified by borehole, sample, slide number and England Finder Co-ordinate location. All figured specimens are at magnification ×1000 except where mentioned otherwise. A–B, *Gneudnaspora divellomedia* (Chibrikova) Balme, 1988 var. *divellomedia* Breuer *et al.* 2007c. A, BAQA-2, 50.8 ft, 03CW127, V35/2. B, BAQA-2, 50.2 ft, 03CW126, V24. C–E, *Gneudnaspora divellomedia* (Chibrikova) Balme, 1988 var. *minor* Breuer *et al.*, 2007c. C, WELL-7, 13614.6 ft, 62379, E28/1. D, JNDL-1, 172.7 ft, PPM007, M49/3. E, WELL-7, 13613.2 ft, 62370, J33/1. F, *Archaeoperisaccus* cf. *A. rhacodes* Hashemi and Playford, 2005. MG-1, 2264 m, 62950, G41/3. G, *Devonomonoletes* sp. 1. BAQA-2, 133.0 ft, 03CW136, E29. H, *Devonomonoletes* spp. MG-1, 2212.5 m, 62530, Q27. I–J, *Emphanisporites rotatus* McGregor emend. McGregor, 1973. I, MG-1, 2693 m, 62961, L37. J, WELL-3, 14188.5 ft, 60547, N35/1. K–L, *Geminospora lemurata* Balme emend. Playford, 1983. K, MG-1, 2180 m, 62972, U-V32. L, MG-1, 2160.6 m, 62746, V41. M–O, *Latosporites ovalis* Breuer *et al.*, 2007c. M, BAQA-2, 50.8 ft, 03CW127, T30/2. N, WELL-3, 14195.3 ft, 66837, R61. O, BAQA-2, 50.2 ft, 03CW126, H26/1.

Genus RETICULOIDOSPORITES Pflug *in* Thomson and Pflug, 1953

Type species. *Reticuloidosporites dentatus* Pflug *in* Thomson and Pflug, 1953.

Reticuloidosporites antarcticus Kemp, 1972
Figure 7C–E

1972 *Reticuloidosporites antarcticus* Kemp, p. 117, pl. 56, figs 1–13.

Dimensions. 37(47)60 μm; 13 specimens measured.

Occurrence. BAQA-1; Jauf Formation (Subbat Member); *milleri* Zone.

Previous records. From the Horlick Formation of Antarctica (Kemp 1972), the age of which is considered as Pragian by Troth *et al.* (2011) based on correlation of chitinozoans and spore assemblages from South America.

Turma TRILETES Reinsch, 1881 emend. Dettmann, 1963

Genus ACINOSPORITES Richardson, 1965

Type species. *Acinosporites acanthomammillatus* Richardson, 1965.

Acinosporites acanthomammillatus Richardson, 1965
Figure 7F–G

1925 Type I Lang, pl. 1, fig. 21.
1965 *Acinosporites acanthomammillatus* Richardson, p. 577, pl. 91, figs 1–2; text-fig. 6.

Dimensions. 63(71)80 μm; eight specimens measured.

Occurrence. JNDL-1; Jubah Formation; *svalbardiae-eximius* Zone. A1-69; Awaynat Wanin I and Awaynat Wanin II Formation; *svalbardiae-eximius* to *lemurata-langii* zones. MG-1; Awaynat Wanin I and Awaynat Wanin II formations; *rugulata-libyensis* to *lemurata-langii* zones.

Previous records. *Acinosporites acanthomammillatus* is eponymous for the upper Eifelian – lower Givetian AD Oppel Zone of Western Europe (Streel *et al.* 1987). *A. acanthomammillatus* occurs from Eifelian into Frasnian and has been widely reported, e.g. Argentina (Amenábar 2009), Australia (Hashemi and Playford 2005), Bolivia (Perez-Leyton 1990), Brazil (Loboziak *et al.* 1988; Loboziak *et al.* 1992*b*; Breuer and Grahn 2011), Canada (Owens 1971; McGregor and Uyeno 1972; McGregor and Camfield 1976, 1982), China (Gao Lianda 1981), Germany (Riegel 1968, 1973; Loboziak *et al.* 1990), Iran (Ghavidel-Syooki 2003), Libya (Paris *et al.* 1985; Streel *et al.* 1988; Moreau-Benoit 1989), Poland (Turnau 1996), Russian Platform (Avkhimovitch *et al.* 1993) and Scotland.

Acinosporites apiculatus (Streel) Streel, 1967
Figure 7H

1964 *Verrucosisporites apiculatus* Streel, pl. 1, fig. 13.
1967 *Acinosporites* cf. *apiculatus* (Streel); Streel, p. 36, pl. 3, figs 38–39.

Dimensions. 57(85)125 μm; seven specimens measured.

Occurrence. JNDL-1, JNDL-3, JNDL-4, WELL-1, WELL-3, and WELL-6; Jauf (Hammamiyat and Murayr members) and Jubah formations; *lindlarensis-sextantii* to *svalbardiae-eximius* zones. A1-69; Ouan-Kasa, Awaynat Wanin I and Awaynat Wanin II formations; *annulatus-protea* to *lemurata-langii* zones. MG-1; Ouan-Kasa and Awaynat Wanin I formations; *annulatus-protea* to *rugulata-libyensis* zones.

Previous records. *Acinosporites apiculatus* is eponymous for the upper Emsian – upper Eifelian AP Oppel Zone of Western Europe (Streel *et al.* 1987). *A. apiculatus* has been reported from upper Emsian through lower Givetian from Belgium (Streel 1964, 1967), Brazil (Loboziak *et al.* 1988; Melo and Loboziak 2003; Breuer and Grahn 2011), Germany (Riegel 1968, 1973; Tiwari and Schaarschmidt 1975; Loboziak *et al.* 1990), Luxembourg (Steemans *et al.* 2000*a*) and Morocco (Rahmani-Antari and Lachkar 2001).

Acinosporites eumammillatus Loboziak *et al.*, 1988
Figure 7I–K

1966 *Acinosporites* sp. McGregor and Owens, pl. 3, figs 9–10.

FIG. 7. Each figured specimen is identified by borehole, sample, slide number and England Finder Co-ordinate location. All figured specimens are at magnification ×1000 except where mentioned otherwise. A–B, *Latosporites* sp. 1. A, MG-1, 2241 m, 62964, V41/2. B, MG-1, 2639 m, 62779, X47. C–E, *Reticuloidosporites antarcticus* Kemp, 1972. C, BAQA-1, 308.3 ft, 62246, L29. D, BAQA-1, 308.3 ft, 62247, K46. E, BAQA-1, 308.3 ft, 62246, N44. F–G, *Acinosporites acanthomammillatus* Richardson, 1965. F, JNDL-1, 162.3 ft, 60841, M33. G, JNDL-1, 156.0 ft, 60839, F31. H, *Acinosporites apiculatus* (Streel) Streel, 1967. A1-69, 1850 ft, 26967, U32/2. I–K, *Acinosporites eumammillatus* Loboziak *et al.*, 1988. I, MG-1, 2520 m, 62594, H27/4. K, MG-1, 2520 m, 62593, H47. 11, MG-1, 2520 m, 62593, T49/1. L, *Acinosporites lindlarensis* Riegel, 1968. JNDL-4, 182.5 ft, 03CW220, M-N42.

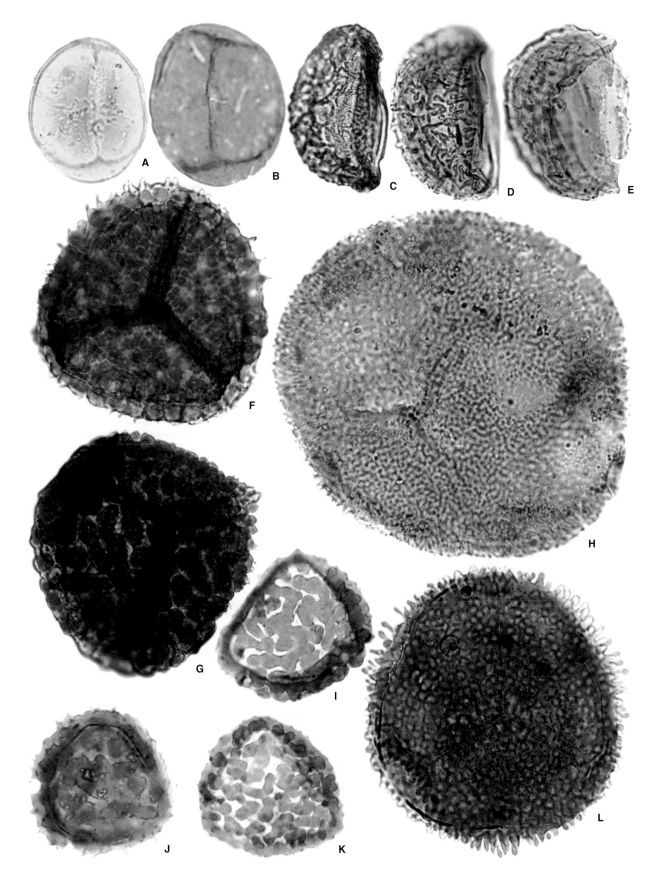

1968 *Acinosporites* sp. B Riegel, p. 89, pl. 19, figs 8–10.
1987 *Acinosporites* sp. Burjack *et al.*, pl. 1, fig. 6.
1988 *Acinosporites eumammillatus* Loboziak *et al.*, p. 354, pl. 1, figs 4–11.

Dimensions. 36(41)44 μm; seven specimens measured.

Comparison. *Acinosporites acanthomammillatus* Richardson, 1965 is clearly more massive. Its ornamentation is higher and wider (2–8 μm) and surmounted by spines, 1–3 μm high and 1 μm wide at base. They are very closely distributed to anastomosed. Other species of *Acinosporites* Richardson, 1965 have commonly biform ornamentation, but they do not show, in general, this typical feature of mammae.

Occurrence. MG-1; Awaynat Wanin I Formation; *rugulatus* Zone.

Previous records. From upper Givetian – lower Frasnian of Argentina (Ottone 1996); upper Frasnian of Paraná Basin, Brazil (Loboziak *et al.* 1988); Emsian of Canada (McGregor and Owens 1966); and Eifelian of Germany (Riegel 1968).

Acinosporites lindlarensis Riegel, 1968
Figure 7L

1966 Unidentified spore McGregor and Owens, pl. 5, fig. 17.
1968 *Acinosporites lindlarensis* Riegel, p. 89, pl. 19, figs 11–16.
? 1969 Indeterminate Cramer, pl. 2, fig. 27.
1972 *Aneurospora* cf. *heterodonta* (Naumova); Streel, p. 206, pl. 2, figs 1–7.
1973 *Geminospora treverica* Riegel, p. 97, pl. 16, figs 4–7.
? 1975 *Cymbosporites cyathus* Allen; Tiwari and Schaarschmidt, pl. 15, fig. 4.
1976 *Acinosporites lindlarensis* Riegel var. *lindlarensis*; McGregor and Camfield, p. 6, pl. 5, figs 2–3.
1976 *Acinosporites lindlarensis* Riegel var. *minor* McGregor and Camfield, p. 8, pl. 5, figs 4–5.
? 1991 *Dibolisporites* sp. cf. *echinaceus* (Eisenack) Richardson; Grey, pl. 1, fig. 9.

Dimensions. 63(75)91 μm; 20 specimens measured.

Remarks. A separated inner body (probably only attached proximally) is regarded by some authors (McGregor and Camfield 1976) of doubtful significance at the generic level. This feature is not mentioned as a criterion circumscribing *Acinosporites* by Richardson (1965). Indeed, some *A. acanthomammillatus* described by Richardson (1965) occasionally have a discernible central body. Therefore, the presence of this feature in *A. lindlarensis* is not considered sufficient to exclude it from the genus.

Acinosporites lindlarensis shows continuous morphological variation of the distribution, size and shape of the ornamentation. Morphotypes with widely distributed sculpture intergrade with others characterized by more crowded distal sculpture. Richardson *et al.* (1993) addressed this issue by defining the *A. lindlarensis* Morphon. It contains different types and subtypes that are difficult to recognize in the present material.

Comparison. McGregor and Camfield (1976) have shown that *Geminospora treverica* Riegel, 1973, characterized by a completely separated inner body, cannot be separated from *A. lindlarensis*. *A. acanthomammillatus* Richardson, 1965 has close similarities (Richardson *et al.* 1993), but it is distinguished from *A. lindlarensis* by a more massive appearance due to commonly coarser ridges.

Occurrence. BAQA-1, JNDL-1, JNDL-3, JNDL-4, S-462, WELL-1, WELL-3, WELL-4, WELL-6, WELL-7 and WELL-8; Jauf (Subbat to Murayr members) and Jubah formations; *lindlarensis-sextantii* to *triangulatus-catillus* zones. A1-69; Ouan-Kasa, Awaynat Wanin I and Awaynat Wanin II formations; *annulatus-protea* to *lemuratalangii* zones. MG-1; Ouan-Kasa, Awaynat Wanin I and Awaynat Wanin II formations; *annulatus-protea* to *langii-concinna* zones.

Previous records. *Acinosporites lindlarensis* has an almost worldwide distribution and has been widely reported from Emsian through Frasnian palynofloras; e.g. Algeria (Moreau-Benoit *et al.* 1993), Argentina (Le Hérissé *et al.* 1997; Amenábar 2009), Belgium (Lessuise *et al.* 1979; Gerrienne *et al.* 2004), Bolivia (Perez-Leyton 1990), Brazil (Loboziak *et al.* 1988; Melo and Loboziak 2003; Breuer and Grahn 2011), Canada (McGregor and Owens 1966; McGregor and Uyeno 1972; McGregor 1973; McGregor and Camfield 1976, 1982), Germany (Riegel 1968, 1973; Loboziak *et al.* 1990), Iran (Ghavidel-Syooki 2003), Libya (Streel *et al.* 1988; Moreau-Benoit 1989), Poland (Turnau 1986, 1996; Turnau *et al.* 2005), Scotland (Marshall and Allen 1982; Marshall and Fletcher 2002) and Georgia, USA (Ravn and Benson 1988).

Acinosporites tristratus sp. nov.
Figure 8A–C

Derivation of name. From *tristratus* (Latin), meaning three-layered; refers to the structure of the spore wall.

Holotype. EFC N43/2 (Fig. 8A), slide 60849.

Paratype. EFC J30/1 (Fig. 8C), slide 60845; JNDL-1 core hole, sample 177.0 ft.

Type locality and horizon. JNDL-1 core hole, sample 172.7 ft; Jubah Formation, Domat Al-Jandal, Saudi Arabia.

Diagnosis. An *Acinosporites* with three layers visible. Distal surface sculptured with coarse biform elements joined to form ridges of varied length.

Description. Amb is sub-circular to sub-triangular. Three wall layers are visible. Laesurae are straight, simple or bordered by labra up to 4 μm in total thickness and extending generally to the outer margin of the second layer. Endonexine is (*c.* 1 μm) not always discernible, ectonexine 1–2.5 μm thick, laevigate, sexine 1.5–3.5 μm thick, sculptured proximo-equatorially and distally with biform processes (bulbous base with apical spine). Sculptural elements on the distal hemisphere are round in plan view, 3 –7 μm wide at their base, 2–8 μm high, joined to form ridges of varied length. The longest elements occur subequatorially at the outer limit of the contact area. Proximal surface granular.

Dimensions. 59(71)85 μm; nine specimens measured.

Remarks. According to Richardson (1965), an inner body is not a feature circumscribing *Acinosporites* (see above). Therefore, the assignment of this new species to this genus is legitimate on the basis of the characteristics of ornamentation.

Comparison. *Acinosporites acanthomammillatus* Richardson, 1965 is single-layered and has ornamentation that is a more closely distributed than that *A. tristratus*. The three wall layers distinguish *A. tristratus* from all the other species of *Acinosporites* Richardson, 1965.

Occurrence. JNDL-1; Jauf (Muray Member) and Jubah Formation; *annulatus-protea* to *svalbardiae-eximius* zones.

Genus ALATISPORITES Ibrahim, 1933

Type species. *Alatisporites pustulatus* Ibrahim, 1933.

Alatisporites? *trisacculus* sp. nov.
Figure 8D–E

Derivation of name. From *trisacculus* (Latin) meaning three small sacci; refers to the size of the sacci.

Holotype. EFC W40/4 (Fig. 8E), slide 03CW195.

Paratype. EFC E52/4 (Fig. 8D), slide 03CW159; JNDL-3 core hole, sample 353.8 ft.

Type locality and horizon. JNDL-4 core hole, sample 87.2 ft; Jauf Formation, Domat Al-Jandal, Saudi Arabia.

Diagnosis. An *Alatisporites* bearing three individual proximo-equatorial sacci opposite the laesurae. Sacci thin and often folded.

Description. Amb is circular to sub-circular. Laesurae are distinct, straight, simple and equal one-half to two-thirds of the amb radius. Curvaturae are sometimes visible. Central body diameter equals commonly three-fifths to fourth-fifths of the total amb diameter. Exine of the central body is 1.5–3 μm thick equatorially. Zona is divided entirely or partially into three individual proximo-equatorial sacci, the maximum width (commonly 14–25 μm) of which is opposite the laesurae. Sacci are thin and often folded radially and can be folded back on the spore body. Thin generally sinuous attachment lines of the sacci on the central body can be distinguished on the proximal face. Proximal and distal surfaces are entirely laevigate.

Dimensions. 64(88)103 μm; five specimens measured.

Remarks. There is some doubt about the allocation of this species to the genus *Alatisporites* Ibrahim, 1933 because the sacci are opposite the laesurae sacci, whereas they are interradially arranged in the other species. Besides, representatives of *Alatisporites* usually first occur in Carboniferous age strata.

Comparison. *Alatisporites trialatus* Kosanke, 1950 has interradial granulate sacci and a sub-triangular to triangular central body. *A. pustulatus* Ibrahim, 1933 and *A. punctatus* Kosanke, 1950 differ by having a distal central body densely sculptured with fine rugulae and verrucae which may impart a microreticulate appearance. *A. hoffmeisteri* Morgan, 1955 possesses a verrucate ornament on the distal central body. *Zonotriletes* sp. 1 has a more sub-triangular amb is only zonate and thus possess an entire flange, which can appear tri-lobed, surrounding the central body.

Occurrence. JNDL-3 and JNDL-4; Jauf Formation (Hammamiyat Member); *lindlarensis-sextantii* Zone. MG-1; Ouan-Kasa and Awaynat Wanin I formations; *annulatus-protea* to *rugulata-libyensis* zones.

Genus AMBITISPORITES Hoffmeister, 1959

Type species. *Ambitisporites avitus* Hoffmeister, 1959.

Remarks. The genus *Archaicusporites* Rodriguez, 1983 is considered here as a junior synonym of *Ambitisporites* Hoffmeister, 1959. Indeed the diagnosis of *Archaicusporites* describes cingulate spores characterized by folds that are disposed parallel and concentrically with respect to the curvaturae. As folds are probably due to polar compression, this genus can be included in the circumscription of *Ambitisporites*.

Ambitisporites (*Archaicusporites*) *asturicus* (Rodriguez)
comb. nov.
Figure 9A–D

1978*b* *Retusotriletes rotundus* (Streel) Streel emend. Lele and Streel; Rodriguez, p. 420, pl. 4, fig. 2.

1978b *Retusotriletes triangulatus* (Streel) Streel; Rodriguez, p. 421, pl. 3, fig. 1.
1983 *Archaicusporites asturicus* Rodriguez, p. 32, pl. 6, fig. 5; pl. 7, fig. 3; text-fig. 3: 54.

Dimensions. 30(39)47 µm; 14 specimens measured.

Comparison. *Retusotriletes tenerimedium* Chibrikova, 1959 has a more pronounced wider darker sub-triangular apical zone and does not have elevated curvaturae.

Occurrence. BAQA-1, BAQA-2, JNDL-3, JNDL-4 and WELL-3; Jauf Formation (Sha'iba to Hammamiyat members); *ovalis-biornatus* to *lindlarensis-sextantii* zones. MG-1; Ouan-Kasa Formation; *svalbardiae-eximius* Zone but occurrences are probably reworked.

Previous record. From Přídolí–lower Pragian of Spain (Rodriguez 1978b).

Ambitisporites avitus Hoffmeister, 1959
Figure 9E–F

1959 *Ambitisporites avitus* Hoffmeister, p. 332, pl. 1, figs 1–8.
1969 *Ambitisporites* cf. *avitus* Hoffmeister; Richardson and Lister, p. 228, pl. 40, fig. 2.

Dimensions. 31(38)67 µm; 13 specimens measured.

Comparison. As *A. avitus* and *A. dilutus* (Hoffmeister) Richardson and Lister, 1969 may intergrade, Steemans *et al.* (1996) defined the *A. avitus* Morphon to include them. *A. avitus* may be distinguished from *A. dilutus* by its larger diameter and thicker cingulum.

Occurrence. BAQA-1, BAQA-2, JNDL-1, JNDL-3, JNDL-4, WELL-3 and WELL-7; Jauf and Jubah formations; *papillensis-baqaensis* to *svalbardiae-eximius* zones. A1-69; Ouan-Kasa Formation; *lindlarensis-sextantii* Zone. MG-1; Ouan-Kasa and Awaynat Wanin I formations; *lindlarensis-sextantii* to *rugulata-libyensis* zones.

Previous record. *Ambitisporites avitus* has a worldwide distribution and has been widely reported from Early Silurian through Early Devonian assemblages.

Ambitisporites eslae (Cramer and Díez)
Richardson *et al.*, 2001
Figure 9G–I

? 1968 Spore no. 2513 Magloire, pl. 1, fig. 13.
1973 *Ambitisporites* sp. B Richardson and Ioannides, p. 277 (*pars*), pl. 6, fig. 8 (*non* figs 7, 9).
1975 *Retusotriletes eslae* Cramer and Díez, p. 343, pl. 1, figs 11–12.
1976 *Ambitisporites tripapillatus* Moreau-Benoit, p. 37, pl. 7, fig. 2–4.
2001 *Ambitisporites eslae* (Cramer and Díez) Richardson *et al.*, p. 142, pl. 5, fig. 1.

Dimensions. 28(44)62 µm; 13 specimens measured.

Comparison. *Ambitisporites tripapillatus* Moreau-Benoit, 1976 is considered as a junior synonym of *A. eslae* (Cramer and Díez) Richardson *et al.*, 2001. *Scylaspora elegans* Richardson *et al.*, 2001 has a large darkened apical area and proximal rugulate sculpture. *Retusotriletes maculatus* McGregor and Camfield, 1976 and, in part, *Ambitisporites* sp. B in McGregor and Camfield (1976) appear to have an equatorially thinner exine, but are otherwise similar to *A. eslae. Synorisporites papillensis* McGregor, 1973 has smaller proximal papillae. Its distal face is commonly verrucate and its cingulum is irregular.

Occurrence. BAQA-1, BAQA-2, JNDL-4 and WELL-2; Jauf Formation (Sha'iba to Subbat members); *papillensis-baqaensis* to *ovalis-biornatus* zones.

Previous records. From lower Lochkovian – lower Pragian of Spain (Cramer and Díez 1975; Rodriguez 1978b; Richardson *et al.* 2001); Pragian of Armorican Massif, France (Le Hérissé 1983); middle Přídolí of Libya (Rubinstein and Steemans 2002); upper Pragian – lower Emsian of Argentina (Rubinstein and Steemans 2007); and lower Lochkovian of Brazil (Steemans *et al.* 2008).

Genus AMICOSPORITES Cramer, 1966a

Type species. *Amicosporites splendidus* Cramer, 1966a.

Amicosporites jonkeri (Riegel) Steemans, 1989
Figure 9J–L

1989 *Amicosporites jonkeri* (Riegel) Steemans, p. 91 (*cum syn.*), pl. 19, figs 9–11.

FIG. 8. Each figured specimen is identified by borehole, sample, slide number and England Finder Co-ordinate location. All figured specimens are at magnification ×1000 except where mentioned otherwise. A–C, *Acinosporites tristratus* sp. nov. A, Holotype, JNDL-1, 177.0 ft, 60849, N43/2. B, JNDL-1, 177.0 ft, 60849, F35/1. C, Paratype, JNDL-1, 172.7 ft, 60845, J30/1. D–F, *Alatisporites? trisacculus* sp. nov. D, Paratype, JNDL-3, 353.8 ft, 03CW159, E52/4. E, Holotype, JNDL-4, 87.2 ft, 03CW195, W40/4. F, MG-1, 2713 m, 62810, M34.

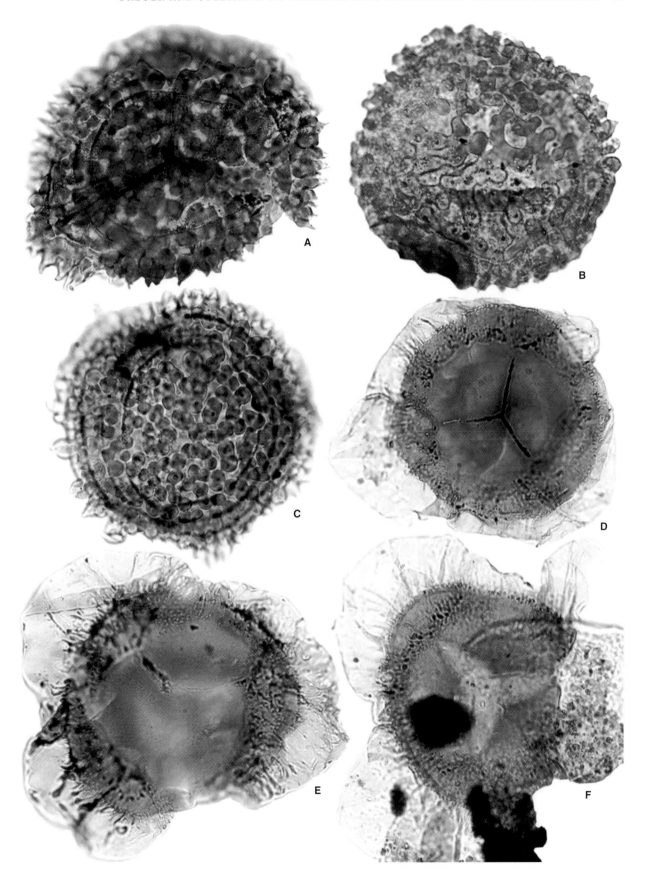

Dimensions. 32(40)46 µm; 11 specimens measured.

Comparison. Amicosporites splendidus Cramer, 1966a has a narrower annulus close to the equator. *A. streelii* Steemans, 1989 has an annulus formed by verrucae resulting in an irregular appearance. Moreover, other verrucae are present in the centre of the annulus. *Concentricosisporites barbulatus* (Rodriguez) Rodriguez, 1983 has a similar appearance except that the author interpreted the annulus to be in a proximal position. The smooth specimens of *Coronaspora* sp. *in* Le Hérissé (1983) are synonymous with *A. jonkeri* whereas those sculptured with verrucae are synonymous with *A. streelii*.

Occurrence. BAQA-1, BAQA-2, JNDL-3 and JNDL-4, Jauf Formation (Sha'iba to Hammamiyat members); *papillensis-baqaensis* to *lindlarensis-sextantii* zones. A1-69; Ouan-Kasa Formation; *lindlarensis-sextantii* Zone. MG-1; Ouan-Kasa and Awaynat Wanin I formations; *lindlarensis-sextantii* to *svalbardiae-eximius* zones but the youngest occurrences are probably due to reworking.

Previous records. From upper Lochkovian–Emsian of Belgium (Steemans 1989); Pragian–lower Emsian of Armorican Massif, France (Le Hérissé 1983); upper Lochkovian – lower Eifelian of Germany (Riegel 1973; Steemans 1989); and Pragian–Emsian of Poland (Turnau 1986; Turnau *et al.* 2005).

Amicosporites streelii Steemans, 1989
Figure 9M–O

1967 *Cirratriradites* sp. F Streel, pl. 5, fig. 59.
1981 *Coronaspora mariae* Rodriguez; Streel *et al.*, p. 184, pl. 3, figs 1–4.
1983 *Coronaspora* sp. Le Hérissé, p. 41 (*pars*), pl. 7, fig. 16–17; pl. 8, figs 1–2 (*non* fig. 3).
1989 *Amicosporites streelii* Steemans, p. 92, pl. 19, figs 15–17; pl. 20, figs 1–2.
2006 *Amicosporites* spp. Wellman, pl. 19, fig. m.

Dimensions. 30(38)45 µm; 18 specimens measured.

Remarks. In Steemans (1989), *A. streelii* does not often show a single sub-circular verruca in the centre of the

annulus; the specimens are generally sculptured with irregularly distributed verrucae. The varied verrucae organization in the centre of the annulus could represent different varieties (PS, pers. obs.).

Comparison. Although some specimens of *Coronospora* sp. *in* Le Hérissé (1983, pl. 7, figs 16–17; pl. 8, figs 1–2) are synonymous with *A. streelii*, others (pl. 8, fig. 3) have no ornamentation other than the annulus. The latter are consequently placed in synonymy with *A. jonkeri* (Riegel) Steemans, 1989.

Occurrence. BAQA-1, BAQA-2, JNDL-3, JNDL-4, WELL-2, WELL-3, WELL-4 and WELL-7; Jauf Formation; *papillensis-baqaensis* to *annulatus-protea* zones. MG-1; Ouan-Kasa Formation; *svalbardiae-eximius* Zone but occurrences are probably reworked.

Previous records. From upper Lochkovian – ?Emsian of Belgium (Streel 1967; Steemans 1989); lower Lochkovian of Amazon Basin, Brazil (Steemans *et al.* 2008); Pragian–lower Emsian of Armorican Massif, France (Le Hérissé 1983); Lochkovian–Pragian of Saudi Arabia (Steemans, 1995); and upper Pragian – ? lowermost Emsian of Scotland (Wellman 2006).

Genus ANCYROSPORA Richardson, 1960 emend. Richardson, 1962

Type species. Ancyrospora grandispinosa Richardson emend. Richardson, 1962.

Remarks. Morphology and wall ultrastructure of *Ancyrospora* are discussed in details in Wellman (2002).

Ancyrospora langii (Taugourdeau-Lantz) Allen, 1965
Figure 9P

? 1953 *Hymenozonotriletes incisus* Naumova, p. 68, pl. 9, fig. 11.
1960 *Archaeotriletes langii* Taugourdeau-Lantz, p. 145, pl. 3, fig. 33–34, 39.
1965 *Ancyrospora langii* (Taugourdeau-Lantz) Allen, p. 743, pl. 106, figs 5–7.

FIG. 9. Each figured specimen is identified by borehole, sample, slide number and England Finder Co-ordinate location. All figured specimens are at magnification ×1000 except where mentioned otherwise. A–D, *Ambitisporites (Archaicusporites) asturicus* (Rodriguez) comb. nov. A, JNDL-4, 495.2 ft, 68702, G37. B, BAQA-1, 308.3 ft, 66791, Q41. C, MG-1, 2631.2 m, 62553, V49/1. D, MG-1, 2631.2 m, 62552, K39/4. E–F, *Ambitisporites avitus* Hoffmeister, 1959. E, BAQA-2, 54.8 ft, 03CW129, N28. F, BAQA-2, 52.0 ft, 03CW128, O27/1. G–I, *Ambitisporites eslae* (Cramer and Díez) Richardson *et al.*, 2001. G, BAQA-2, 57.2 ft, 66817, F31. H, BAQA-1, 395.2 ft, 66807, P26. I, BAQA-2, 50.8 ft, 66813, L23/1. J–L, *Amicosporites jonkeri* (Riegel) Steemans, 1989. J, BAQA-2, 134.4 ft, 66826, U38/3. K, JNDL-4, 120.0 ft, 68612, E53/2. L, MG-1, 2741.4 m, 62611, L41. M–O, *Amicosporites streelii* Steemans, 1989. M, BAQA-1, 371.1 ft, 03CW118, F28/2. N, WELL-3, 14195.3 ft, 66838, F62/2. O, BAQA-2, 133.0 ft, 03CW136, F40/4. P, *Ancyrospora langii* (Taugourdeau-Lantz) Allen, 1965, magnification ×500. S-462, 1470–1475 ft, 63212, U30/4. Q, *Ancyrospora nettersheimensis* Riegel, 1973, magnification ×500. A1-69, 1540 ft, 26988, K35. R–V, *Aneurospora* cf. *A. bollandensis* Steemans, 1989. R, WELL-7, 13689.7 ft, 62319, Q43. S, WELL-7, 13614.6 ft, 62377, L27. T, BAQA-1, 285.5 ft, 03CW111, E31. U, WELL-7, 13689.7 ft, 62316, C37. V, WELL-7, 13689.7 ft, 62317, J36/2.

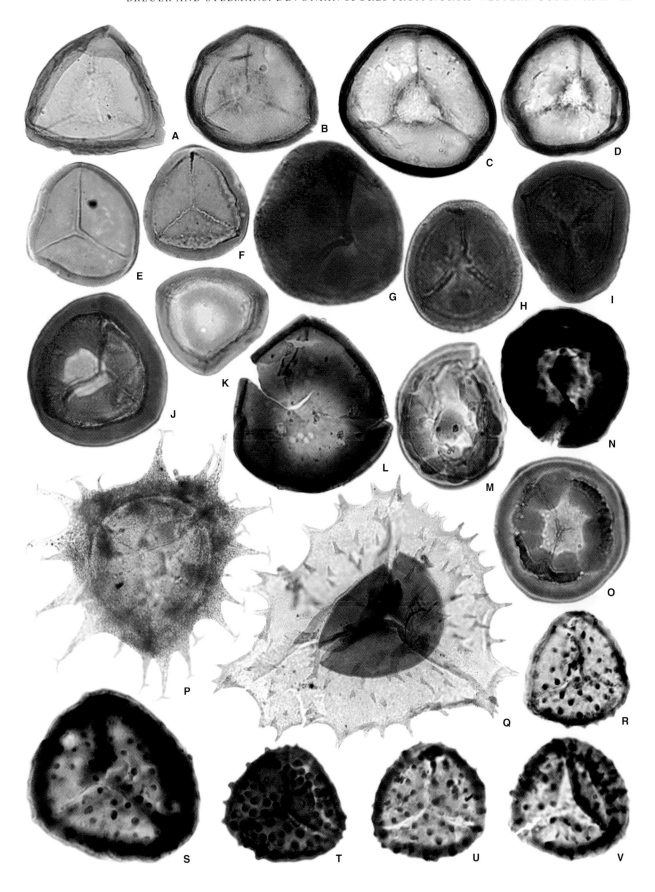

1966 *Ancyrospora* sp.; de Jersey, p. 21, pl. 9, figs 5, 7.
1968 *Ancyrospora amadei* Hodgson, p. 74, pl. 8, fig. 8.
1968 *Ancyrospora* cf. *A. simplex* Guennel; Hodgson, p. 75, pl. 8, figs 9–11; text-fig. 1.
1975 *Ancyrospora* sp. A; Grey, fig. 61b.

Dimensions. 83(103)128 μm; 10 specimens measured.

Occurrence. S-462; Jubah Formation; *langii-concinna* Zone but some specimens may be caved in older strata. A1-69; Awaynat Wanin II Formation; *langii-concinna* Zone. MG-1; Awaynat Wanin III Formation; *langii-concinna* Zone.

Previous records. From Frasnian–Famennian of Algeria (Boumendjel *et al.* 1988; Moreau-Benoit *et al.* 1993); upper Givetian – lower Frasnian of Argentina (Ottone 1996); middle Givetian – lower Frasnian of Australia (Balme 1988; Grey 1991; Hashemi and Playford 2005); ?Givetian–Famennian of Belgium (Becker *et al.* 1974; Steel and Loboziak 1987); upper Givetian–Famennian of Bolivia (Perez-Leyton 1990); Givetian of China (Gao Lianda 1981); upper Givetian – upper Frasnian of France (Brice *et al.* 1979; Loboziak and Steel 1980, 1988); uppermost Eifelian – Givetian of Germany (Loboziak *et al.* 1990); middle Givetian – ?early Famennian of Libya (Paris *et al.* 1985; Coquel and Moreau-Benoit 1986; Steel *et al.* 1988; Moreau-Benoit 1989); Givetian of Spitsbergen, Norway (Allen 1965); and uppermost Givetian – lower Frasnian (Marshall *et al.* 1996).

Ancyrospora nettersheimensis Riegel, 1973
Figure 9Q

1973 *Ancyrospora nettersheimensis* Riegel, p. 100, pl. 17, figs 6–8.

Dimensions. 128(140)155 μm; nine specimens measured.

Occurrence. A1-69; Awaynat Wanin I and Awaynat Wanin II formations; *svalbardiae-eximius* to *rugulata-libyensis* zones.

Previous records. From upper Emsian – lower Eifelian of Algeria (Moreau-Benoit *et al.* 1993); upper Emsian–Eifelian of Germany (Riegel 1973; Loboziak *et al.* 1990); lower Eifelian – lower Givetian of Libya (Paris *et al.* 1985; Steel *et al.* 1988; Moreau-Benoit 1989); uppermost Emsian – ?lowermost Eifelian of Poland (Turnau 1986; Turnau *et al.* 2005); and ?late Emsian–Eifelian of Saudi Arabia (PB, pers. obs.).

Genus ANEUROSPORA Steel, 1964 emend. Richardson *et al.*, 1982

Type species. *Aneurospora goensis* Steel, 1964.

Comparison. *Synorisporites* Richardson and Lister, 1969 has distal sculpture of verrucae and/or muri. *Ambitisporites* Hoffmeister, 1959 has a uniform cingulum and is very often laevigate.

Aneurospora cf. *A. bollandensis* Steemans, 1989
Figure 9R–V

1983 *Cymbosporites echinatus* Richardson and Lister; Le Hérissé, p. 50, pl. 7, fig. 10.
cf. 1989 *Aneurospora bollandensis* Steemans, p. 96, pl. 20, figs 14–19, 46.
2008 *Aneurospora* cf. *A. bollandensis* Steemans; Steemans *et al.*, pl. 6, fig. 6.

Description. Amb is sub-circular to sub-triangular. Laesurae are straight, accompanied by labra, 1–3 μm wide individually, approximately three-quarters of the amb radius in length. Equatorial crassitude is 2–5 μm wide. Proximal region is laevigate. Distal and equatorial regions are sculptured with evenly distributed conical bacula, 1–2.5 μm wide and high, 0.5–3 μm apart. The tops of elements are flat or slightly rounded and can be slightly flared.

Dimensions. 30(36)49 μm; nine specimens measured.

Comparison. *Aneurospora bollandensis* Steemans, 1989 is distinguished by the presence of proximal inspissations or papillae. *Cymbosporites echinatus* Richardson and Lister, 1969 *in* Le Hérissé (1983) is similar to *A. bollandensis* except for the absence of proximal inspissations or papillae. The specimen figured by Le Hérissé (1983, pl. 7, fig. 10) is thus considered to be synonymous with the specimens described here. *C. echinatus* is patinate and bears larger, not always parallel-sided, sculptural elements. *C. dammamensis* Steemans, 1995 is also patinate, with simple laesurae and the tops of elements are generally bifurcate.

Occurrence. BAQA-1, JNDL-4 and WELL-7; Jauf Formation (Subbat and Hammamiyat members); *ovalis-biornatus* to *lindlarensis-sextantii* zones.

Previous records. From Pragian of Armorican Massif (France; Le Hérissé 1983); and lower Lochkovian of Brazil (Steemans *et al.* 2008).

Genus APICULIRETUSISPORA (Steel) Steel, 1967

Type species. *Apiculiretusispora brandtii* Steel, 1964.

Apiculiretusispora arabiensis Al-Ghazi, 2009
Figure 10A–B

? 1986 Unidentified; Turnau, pl. 8, fig. 4.
2007 *Apiculiretusispora densa* Al-Ghazi, p. 68, pl. 1, figs 1–6; text-fig. 5 (*nom. nud.*).
2009 *Apiculiretusispora arabiensis* Al-Ghazi, p. 193.

Dimensions. 35(53)70 μm; 21 specimens measured.

Remarks. Apiculiretusispora densa Al-Ghazi, 2007 was renamed as *A. arabiensis*, because the first species name was preoccupied by *A. densa* Lu Lichang, 1988.

Comparison. Apiculiretusispora arabiensis, differs from all published species of the genus *Apiculiretusispora* (Streel) Streel, 1967 by its characteristic dark-coloured, rounded interradial thickenings on the proximal face. *A. brandtii* Streel, 1964 possesses a more densely distributed conate sculpture. Turnau (1986) illustrated an unnamed specimen resembling *A. arabiensis* but no description was given.

Occurrence. BAQA-1, BAQA-2, JNDL-1, JNDL-3, JNDL-4 and WELL-4; Jauf and Jubah formations; *papillensis-baqaensis* to *svalbardiae-eximius* zones.

Previous record. From Emsian of Saudi Arabia (Al-Ghazi 2007).

Apiculiretusispora brandtii Streel, 1964
Figure 10C–D

1963 *Cyclogranisporites* sp. Chaloner, fig. 8.
1964 *Apiculiretusispora brandtii* Streel, p. 8, pl. 1, figs 6–10.
1966 *Cyclogranisporites* sp. McGregor and Owens, pl. 2, fig. 7, 17–19.
1966 ?*Perotrilites* sp. McGregor and Owens, pl. 5, fig. 13.
1967 *Cyclogranisporites* sp. McGregor, pl. 1, fig. 1.

Dimensions. 50(69)95 μm, 15 specimens measured.

Remarks. The specimens of *Apiculiretusispora brandtii* Streel, 1964 which show local detachment of sexine could be transitional with those of *Rhabdosporites minutus* Tiwari and Schaarschmidt, 1975 that has a completely detached sexine. The two species have a very similar fine ornamentation. In addition, they appear to intergrade with *Cymbosporites asymmetricus* Breuer *et al.*, 2007c that has the same type of ornamentation but is patinate and with an oval shape. To accommodate the morphological intergradation between these genera and species, the *A. brandtii* Morphon is defined here (Table 1). Wellman (2009) concluded that there is a smooth evolutionary transition between spores belonging to *Apiculiretusispora* (Streel) Streel, 1967 and those related to *Rhabdosporites* Richardson emend. Marshall and Allen, 1982.

Comparison. Cymbosporites asymmetricus Breuer *et al.*, 2007c is patinate. *A. plicata* (Allen) Streel, 1967 is smaller and does not show partial detachment of the sexine.

Occurrence. BAQA-1, BAQA-2, JNDL-1, JNDL-3, JNDL-4, WELL-1, WELL-3, WELL-4, WELL-6 and WELL-7; Jauf and Jubah formations; *papillensis-baqaensis* to *lemurata-langii* zones. A1-69; Ouan-Kasa Formation; *annulatus-protea* Zone. MG-1; Ouan-Kasa Formation; *annulatus-protea* Zone.

Previous records. Apiculiretusispora brandtii has an almost worldwide distribution and has been reported from Pragian through Givetian assemblages from many parts of the world; e.g. Emsian–lower Eifelian of Algeria (Moreau-Benoit *et al.* 1993); upper Emsian–upper Eifelian of Belgium (Streel 1964; Lele and Streel 1969; Lessuise *et al.*, 1979; Laloux *et al.* 1996); upper Pragian–lower Emsian of Paraná Basin, Brazil (Mendlowicz Mauller *et al.* 2007); Emsian–Eifelian of Canada (McGregor and Owens 1966; McGregor 1973); Pragian–lower Emsian of Armorican Massif, France (Le Hérissé 1983); upper Emsian–lower Givetian of Germany (Riegel, 1968; Tiwari and Schaarschmidt 1975); middle Emsian–lower Givetian of Libya (Paris *et al.* 1985; Streel *et al.* 1988; Moreau-Benoit 1989); upper Emsian of Luxembourg (Steemans *et al.* 2000a); upper Pragian–Emsian of Morocco (Rahmani-Antari and Lachkar 2001); Pragian–middle Eifelian of Poland (Turnau and Matyja 2001; Turnau *et al.* 2005); Emsian of Saudi Arabia (Al-Ghazi 2007); upper Pragian–?lowermost Emsian of Scotland (Wellman 2006); and ?Emsian–Eifelian of Georgia, USA (Ravn and Benson 1988).

Apiculiretusispora plicata (Allen) Streel, 1967
Figure 10E–F

1965 *Cyclogranisporites plicatus* Allen, p. 695, pl. 94, figs 6–9.
1966 ?*Perotrilites* sp.; McGregor and Owens, pl. 5, fig. 13.
1967 *Apiculiretusispora plicata* (Allen) Streel, p. 33, pl. 2, figs 31, 34.
1968 *Cyclogranisporites plicatus* Allen; Lanninger, p. 120, pl. 22, fig. 3.
1972 *Granulatisporites* sp.; Kemp, p. 110, pl. 52, figs 4–5, 7.
1974 ?*Apiculiretusispora plicata* (Allen) Streel; McGregor, pl. 1, fig. 39.

Description. Amb is sub-circular to more rarely sub-triangular. Laesurae are simple, straight, *c.* 1 μm wide and equals three-fifths or almost extends to the amb radius. Curvaturae are visible. Distal and equatorial regions are ornamented with small coni or spinae, up to 1 μm high, rarely up to 1 μm wide and 0.5–1 μm apart. proximal surface laevigate. Exine is 0.5–2 μm thick.

Dimensions. 41(49)73 μm; 39 specimens measured.

Comparison. Apiculiretusispora plicata is distinguished from *A. brandtii* Streel, 1964 by its smaller size and does not show partial detachment of the sexine. In addition, ornamentation of the former is more regular and slightly less packed. *A. microconus* (Richardson) Streel, 1967 has an ornamentation similar to *A. plicata* but is much larger.

Occurrence. BAQA-1, BAQA-2, JNDL-1, JNDL-3, JNDL-4 and WELL-3, Jauf Formation; *papillensis-baqaensis* to *annulatus-protea* zones.

Previous record. Apiculiretusispora plicata has a worldwide distribution and has been widely reported from Lower Devonian through Middle Devonian assemblages.

Genus ARCHAEOZONOTRILETES Naumova emend. Allen, 1965

Type species. Archaeozonotriletes variabilis Naumova emend. Allen, 1965.

Comparison. Cymbosporites Allen, 1965 has a variably sculptured patina. Cyrtospora Winslow, 1962 has a laevigate distal surface with a large irregular intumescent or glebous mass and tubercules.

Archaeozonotriletes chulus (Cramer) Richardson and Lister, 1969
Figure 10G–H

1966b *Retusotriletes chulus* Cramer, p. 74, pl. 2, fig. 14.
1969 *Archaeozonotriletes chulus* var. *chulus* (Cramer) Richardson and Lister, p. 234, pl. 43, figs 1–6; text-fig. 4.
1969 *Archaeozonotriletes chulus* (Cramer) var. *nanus* Richardson and Lister, p. 238, pl. 43, figs 10–11; text-fig. 4.
1973 *Tholisporites chulus* var. *chulus* (Cramer) McGregor, p. 56, pl. 7, figs 13–15.

Dimensions. 27(31)37 μm; 14 specimens measured.

Remarks. The specimens encountered here are included in the *A. chulus* Morphon introduced by Steemans *et al.* (1996). It includes two varieties: *chulus* and *nanus* Richardson and Lister (1969), which collectively extend over a large size range. As these varieties closely intergrade, they are not distinguished here. In polar compression, some *Archaeozonotriletes* Naumova emend. Allen, 1965 resemble spores of the genus *Ambitisporites* Hoffmeister, 1959

except that the former have very thin contact areas. However, some tetrads and obliquely compressed specimens show clearly that the exine is much thicker at the equator and over the distal surface (Richardson and Lister 1969).

Comparison. Ambitisporites avitus Hoffmeister, 1959 and *A. dilutus* (Hoffmeister) Richardson and Lister, 1969 appear to have a more or less similar equatorial thickening. They can be distinguished from *Archaeozonotriletes chulus* since these species show an equatorial thickening that clearly narrows opposite the laesurae. They lack the thin proximal face and thick distal wall. *Retusotriletes semizonalis* McGregor, 1964 does not have the pronounced differential thickening between the proximal and distal surfaces of the spores described above and has minute sculpture.

Occurrence. BAQA-1, BAQA-2, JNDL-1, JNDL-3, JNDL-4, S-462, WELL-1, WELL-2, WELL-3, WELL-4 and WELL-6; Jauf and Jubah formations; *papillensis-baqaensis* to *lemurata-langii* zones. A1-69; Ouan-Kasa, Awaynat Wanin I and Awaynat Wanin II formations; *lindlarensis-sextantii* to *lemurata-langii* zones. MG-1; Ouan-Kasa, Awaynat Wanin I and Awaynat Wanin II formations; *lindlarensis-sextantii* to *lemurata-langii* zones.

Previous records. Archaeozonotriletes chulus has a worldwide distribution and has been widely reported from Silurian through Lower Devonian palynofloras. Locally it seems to reach rarely the Eifelian (Ravn and Benson 1988).

Archaeozonotriletes variabilis Naumova emend. Allen, 1965
Figure 10I–J

1953 *Archaeozonotriletes variabilis* Naumova, p. 30, 83, pl. 2, figs 12–13; pl. 12, figs 8–11; pl. 13, figs 7–9.
1965 *Archaeozonotriletes variabilis* Naumova emend. Allen, p. 721, pl. 100, figs 3–6.

Dimensions. 46(58)71 μm; 12 specimens measured.

Remarks. As *A. variabilis* has a very thick distal patina, it is therefore frequently preserved in oblique compression, thus giving the impression of an irregular cingulum. The thickness of the patina may comprise as much as two-fifths of the amb diameter.

FIG. 10. Each figured specimen is identified by borehole, sample, slide number and England Finder Co-ordinate location. All figured specimens are at magnification ×1000 except where mentioned otherwise. A–B, *Apiculiretusispora arabiensis* Al-Ghazi, 2009. A, JNDL-1, 174.6 ft, 60848, V29. B, BAQA-1, 308.3 ft, 03CW112, E43. C–D, *Apiculiretusispora brandtii* Streel, 1964. C, JNDL-4, 182.5 ft, 03CW220, T31. D, BAQA-1, 227.1 ft, 03CW110, L-M25. E–F, *Apiculiretusispora plicata* (Allen) Streel, 1967. E, WELL-3, 14186.3 ft, 66833, W44. F, BAQA-1, 169.1 ft, 03CW103, V24. G–H, *Archaeozonotriletes chulus* (Cramer) Richardson and Lister, 1969. G, JNDL-1, 172.7 ft, PPM007, U40/1. H, JNDL-1, 172.7 ft, 60845, L47/4. I–J, *Archaeozonotriletes variabilis* Naumova emend. Allen, 1965. I, MG-1, 2292 m, 63025, H32. J, MG-1, 2241 m, 62964, F34/1. K, *Auroraspora macromanifesta* (Hacquebard) Richardson, 1960, magnification ×750. MG-1, 2241 m, 62966, U46. L–M, *Auroraspora minuta* Richardson, 1965, magnification ×750. L, MG-1, 2258 m, 62948, U28/4. M, MG-1, 2264 m, 62951, F34.

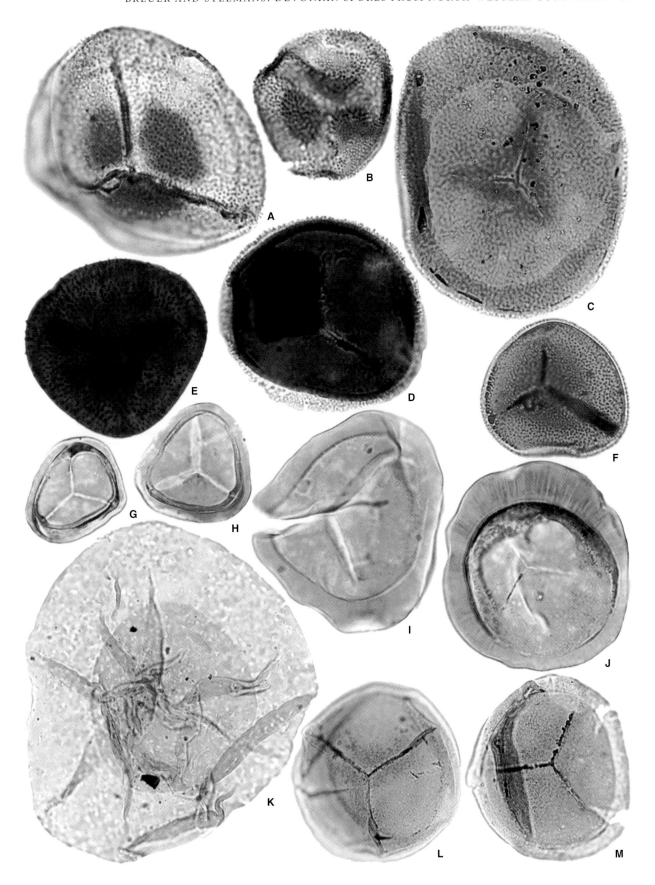

Comparison. Some extreme variants of *Lophozonotriletes media* Taugourdeau-Lantz, 1967, which show a similar exine and a much reduced ornamentation, could intergrade with *A. variabilis.* Consequently, the *A. variabilis* Morphon is defined here (Table 1).

Occurrence. S-462 and WELL-8; Jubah Formation; *lemurata-langii* to *triangulatus-catillus* zones. A1-69; Awaynat Wanin II Formation; *lemurata-langii* to *langii-concinna* zones. MG-1; Awaynat Wanin II and Awaynat Wanin III formations; *incognita* to *langii-concinna* zones.

Previous records. Archaeozonotriletes variabilis has been widely reported from Middle through Upper Devonian; e.g. upper Givetian – lower Frasnian of Argentina (Ottone 1996); middle Givetian – lower Frasnian (Grey 1991; Hashemi and Playford 2005); Frasnian of Belgium (Streel and Loboziak 1987); upper Givetian–Frasnian of Bolivia (Perez-Leyton 1990); lower Givetian – lowermost Famennian of Brazil (Loboziak *et al.* 1988; Loboziak *et al.* 1992b; Melo and Loboziak 2003; Breuer and Grahn 2011); Eifelian–lower Famennian of Canada (McGregor and Owens 1966; Owens 1971; McGregor and Uyeno 1972; McGregor and Camfield 1982); Givetian of China (Gao Lianda 1981), Poland (Turnau 1996; Turnau and Racki 1999) and Spitsbergen, Norway (Allen 1965); upper Givetian–upper Frasnian of France (Loboziak and Streel 1980, 1988; Loboziak *et al.* 1983); Eifelian–Givetian of Germany (Tiwari and Schaarschmidt 1975; Loboziak *et al.* 1990); middle Givetian of Greenland (Friend *et al.* 1983; Marshall and Hemsley 2003); middle Givetian – upper Famennian of Libya (Streel *et al.* 1988; Moreau-Benoit 1989); Givetian–Famennian of Morocco (Rahmani-Antari and Lachkar 2001); lower Frasnian of Russian Platform (Avkhimovitch *et al.* 1993); and uppermost Givetian – lower Frasnian of Scotland (Marshall *et al.* 1996).

Genus AURORASPORA Hoffmeister *et al.,* 1955

Type species. Auroraspora solisorta Hoffmeister *et al.,* 1955.

Auroraspora macromanifesta (Hacquebard) Richardson, 1960
Figure 10K

1925 Type A Lang, p. 255, pl. 1, figs 1–2.
1957 *Endosporites macromanifestus* Hacquebard, p. 317, pl. 3, figs 14–15.
1960 *Auroraspora macromanifestus* (Hacquebard) Richardson, p. 50, pl. 14, figs 1–2; text-fig. 6A.

Dimensions. 122–124 μm; two specimens measured.

Occurrence. MG-1; Awaynat Wanin II Formation; *lemurata-langii* to *triangulatus-catillus* zones.

Previous records. From middle Givetian–Frasnian of Algeria (Moreau-Benoit *et al.* 1993); Givetian–Mississippian of Canada (Hacquebard 1957; McGregor and Owens 1966; Owens 1971;

McGregor and Uyeno 1972); upper Emsian – lower Givetian of Germany (Tiwari and Schaarschmidt 1975); upper Givetian – lower Frasnian of France (Brice *et al.* 1979; Loboziak and Streel 1980); upper Eifelian–Givetian of Poland (Turnau 1996; Turnau and Racki 1999); and middle Eifelian – lower Frasnian (Richardson 1965; Marshall 1988, 2000; Marshall *et al.* 1996; Marshall and Fletcher 2002).

Auroraspora minuta Richardson, 1965
Figure 10L–M

1965 *Auroraspora minuta* Richardson, p. 586, pl. 93, fig. 2.

Dimensions. 59(72)92 μm; 14 specimens measured.

Occurrence. JNDL-1, S-462 and WELL-8; Jubah Formation; *svalbardiae-eximius* to *triangulatus-catillus* zones. A1-69; Awaynat Wanin II Formation; *rugulata-libyensis* to *lemurata-langii* zones. MG-1; Awaynat Wanin I, Awaynat Wanin II and Awaynat Wanin III formations; *rugulata-libyensis* to *langii-concinna* zones.

Previous records. From middle Givetian of Algeria (Moreau-Benoit *et al.* 1993); upper Eifelian–middle Givetian (Breuer and Grahn 2011); Emsian–?lower Eifelian of Canada (McGregor and Owens 1966); lower Eifelian of Libya (Moreau-Benoit 1989); and Givetian of Scotland (Richardson 1965).

Genus BIORNATISPORA Lele and Streel, 1969

Type species. Biornatispora dentata Lele and Streel, 1969.

Comparison. Acinosporites Richardson, 1965 has convoluted pattern of muri that bear characteristically biform ornaments. *Dictyotriletes* Naumova, 1939 ex Ishchenko, 1952 is ornamented with a perfectly closed reticulum.

Biornatispora dubia (McGregor) Steemans, 1989
Figure 11A–D

1973 *Camptotriletes dubius* McGregor, p. 42, pl. 5, figs 9–11, 13–14.
non 1989 *Biornatispora dubia* (McGregor) Steemans, p. 104, pl. 22, figs 21–25.

Description. Amb is sub-circular to broadly triangular. Laesurae are simple or with low labra individually up to 1 μm wide, two-thirds to nine-tenths of the amb radius in length. Contact areas are laevigate. Distal and equatorial regions are ornamented with coni, truncate coni, bacula or biform tubercules generally 1–2 μm wide at base, 1–2.5 μm long, 1–4 μm apart, commonly interconnected by ridges 0.5–2.5 μm wide and high that form an incomplete reticulum. Muri thicken where the ornamentation is rooted. Lumina of the reticulum 2–8 μm in greatest diameter. Exine is up to 2 μm thick at equator.

Dimensions. 29(37)51 μm; 30 specimens measured.

Remarks. Camptotriletes dubius McGregor, 1973 was transferred to the genus *Biornatispora* Lele and Streel, 1969 by Steemans (1989), who, however, as is evident from his description, erroneously applied the new combination *B. dubia* to another species of *Biornatispora*, namely *B. elegantula* sp. nov. The combination *B. dubia* (McGregor) Steemans, 1989 must not be applied to *B. elegantula* but must be retained for *C. dubius*.

Biornatispora dubia is a morphologically variable species. All gradations are observed between specimens characterized by small muri and ornament to ones showing thicker ornamentation. The appearance of the reticulum varies from an almost complete reticulum to an incomplete reticulum.

Comparison. Biornatispora elegantula sp. nov. is proximally granulate, finely sculptured and possesses a typical darker apical sub-triangular band. *B. dubia* is smaller than *B. dentata* (Streel) Lele and Streel, 1969 and has a more variable, less regularly disposed sculpture. It is distinguished from *Brochotriletes* sp. B *in* McGregor (1973) by its smaller size and low-ridged or incomplete foveo-reticulate sculpture.

Occurrence. BAQA-1, WELL-2, WELL-3, JNDL-3, JNDL-4, WELL-1, WELL-4, WELL-5, WELL-6 and WELL-7; Jauf Formation (Subbat to Murayr members); *ovalis-biornatus* to *annulatus-protea* zones. A1-69; Ouan-Kasa Formation; *lindlarensis-sextantii* Zone. MG-1; Ouan-Kasa Formation; *annulatus-protea* Zone.

Previous record. From Emsian of Eastern Canada (McGregor 1973).

Biornatispora elegantula sp. nov.
Figure 11E–I

1989 *Biornatispora dubia* (McGregor) Steemans, p. 104, pl. 22, figs 21–25.

Derivation of name. From *elegantulus* (Latin) meaning elegant, graceful; refers to the fine sculpture of the equatorial and distal regions.

Holotype. EFC P32/4 (Fig. 11G), slide 66812.

Paratype. EFC E41/2 (Fig. 11H), slide 03CW114; BAQA-1 core hole, sample 345.5 ft.

Type locality and horizon. BAQA-2 core hole, sample 50.2 ft; Jauf Formation at Baq'a, Saudi Arabia.

Diagnosis. A *Biornatispora* sculptured with densely distributed, small coni or spines interconnected by thin ridges. Contact areas granulate and characterized by a darker apical sub-triangular band, with concave to straight margins, which extends between the ends of the laesurae.

Description. Amb is sub-circular. Laesurae are accompanied with low labra, individually 1.5–2.5 μm wide, one-half to three-quarters of the amb radius in length. As the labrate laesurae are commonly open, it gives a darker apical sub-triangular band, with concave to straight margins, which extends between the ends of the laesurae. Curvaturae perfectae are commonly confluent with the equator for part of their length and invaginate proximally to join the extremities of laesurae. Contact areas are granulate. Distal and equatorial regions are ornamented with coni or spines *c.* 0.5 μm wide at base, *c.* 0.5–1.5 μm long, 0.5–2 μm apart, commonly interconnected by ridges up to 1 μm wide and high that form an incomplete reticulum. Muri thicken where the discrete sculptural elements are rooted. Lumina of the reticulum 1–4 μm in greatest diameter. Exine is 0.5–1.5 μm thick.

Dimensions. 30(34)39 μm; 18 specimens measured.

Remarks. Biornatispora dubia (McGregor) Steemans, 1989 *sensu* Steemans (1989) is a later homonym of *B. dubia sensu* McGregor (1973; see above). *B. elegantula* sp. nov. is consequently nominated (nomen novum) to encompass the description of the specimens described in Steemans (1989) and here. Holotype and paratype are also designated.

Comparison. Biornatispora dubia (McGregor) Steemans, 1989 is more coarsely ornamented and does not have a darker apical sub-triangular band.

Occurrence. BAQA-1, BAQA-2 and JNDL-4; Jauf Formation (Sha'iba to Subbat members); *papillensis-baqaensis* to *ovalis-biornatus* zones.

Previous record. From upper Lochkovian–Pragian of Belgium and upper Lochkovian–Emsian of Germany (Steemans 1989).

Biornatispora microclavata sp. nov.
Figure 11J–L

Derivation of name. From *microclavatus* (Latin), meaning sculptured with small club-shaped elements (clavae); refers to the sculpture of the equatorial and distal regions.

Holotype. EFC V36/4 (Fig. 11J), slide 03CW126.

Paratype. EFC E50/3 (Fig. 11L), slide 03CW126; BAQA-2 core hole, sample 50.2 ft.

Type locality and horizon. BAQA-2 core hole, sample 50.2 ft; Jauf Formation at Baq'a, Saudi Arabia.

Diagnosis. A *Biornatispora* bearing small clavae or pila interconnected by low ridges.

Description. Amb is sub-circular. Laesurae are straight, simple or up to 2 μm wide, commonly three-fifths to four-fifths of the amb radius in length. Curvaturae are visible and sometimes bounded by a single continuous thin and translucent murus, up to 2.5 μm high. Contact areas are granular. Distal and equatorial regions are ornamented with clavae or pila (elements with a swollen tip), or more rarely bacula and parallel-sided spinae. Sculptural elements are 0.5–1 μm wide at base, 1–3 μm long, 1–3 μm apart, commonly interconnected by low ridges, generally *c.* 0.5 μm wide and high that form an incomplete reticulum. Irregular lumina of the reticulum are 3–8 μm in greatest diameter. Exine is 1–1.5 μm thick at equator.

Dimensions. 41(54)72 μm; 13 specimens measured.

Comparison. This species is easily recognizable because it is the only species of *Biornatispora* Lele and Streel, 1969 that is ornamented with pila.

Occurrence. BAQA-1 and BAQA-2; Jauf Formation (Sha'iba and Qasr members); *ovalis* Zone.

Genus BROCHOTRILETES Naumova, 1939 ex Ishchenko, 1952

Type species. *Brochotriletes magnus* Ishchenko, 1952.

Brochotriletes crameri sp. nov.
Figure 11M–R

Derivation of name. In honour of the Dutch palynologist, Fritz H. Cramer, for his pioneering work on Gondwanan Lower Palaeozoic spores and acritarchs.

Holotype. EFC O54 (Fig. 11N), slide 68659.

Paratype. EFC Q38/3 (Fig. 11M), slide 68567; JNDL-3 core hole, sample 413.2 ft.

Type locality and horizon. JNDL-4 core hole, sample 277.6 ft; Jauf Formation at Domat Al-Jandal, Saudi Arabia.

Diagnosis. A *Brochotriletes* foveolate proximally and distally.

Description. Amb is sub-circular to sub-triangular. Laesurae are often open, straight, simple, extending to or almost to the equator. Exine is 1.5–3.5 μm thick. Proximal and distal regions are foveolate. Foveolae from both regions are round to oval in plan view, 3–6 μm in diameter, up to 2 μm deep and 2–4 μm apart. Number of foveolae on the equator are around 15. Exine between foveolae is laevigate.

Dimensions. 33(36)41 μm; nine specimens measured.

Comparison. *Coronospora reticulata* Richardson *et al.*, 2001 has kyrtome on the contact areas and is not reticulate proximally. *Brochotriletes* sp. B *in* Tekbali and Wood (1991) may correspond to specimens described above but there are only illustrations in the paper.

Occurrence. BAQA-1, JNDL-1, JNDL-3 and JNDL-4, Jauf (Subbat to Murayr members) and Jubah formations, *asymmetricus* to *svalbardiae-eximius* zones. A1-69; Ouan-Kasa and Awaynat Wanin I formations; *lindlarensis-sextantii* to *svalbardiae-eximius* zones.

Brochotriletes foveolatus Naumova, 1953
Figure 11S

1953 *Brochotriletes foveolatus* var. *major* Naumova, p. 58 (*cum syn.*), pl. 7, fig. 23–24.

Dimensions. 47(63)77 μm; 17 specimens measured.

Comparison. The absence of sculpture between foveolae distinguishes this species from *B. bellatulus* Steemans, 1989 and *B. robustus* (Scott and Rouse) McGregor, 1973. *Perforosporites*

FIG. 11. Each figured specimen is identified by borehole, sample, slide number and England Finder Co-ordinate location. All figured specimens are at magnification ×1000 except where mentioned otherwise. A–D, *Biornatispora dubia* (McGregor) Steemans, 1989. A, WELL-7, 13614.1 ft, 62373, E34. B, JNDL-3, 499.5 ft, 03CW183, V37/3. C, WELL-3, 14159.5 ft, 60521, H34. D, JNDL-4, 331.9 ft, 03CW246, K38. E–I, *Biornatispora elegantula* sp. nov. E, BAQA-2, 50.8 ft, 66813, G29/4. F, BAQA-1, 345.5 ft, 03CW114, E41/2. G, Holotype, BAQA-2, 50.2 ft, 66812, P32/4. H, Paratype, BAQA-1, 395.2 ft, 66807, Y39/2. I, BAQA-1, 223.5 ft, 66783, P37. J–L, *Biornatispora microclavata* sp. nov. J, Holotype, BAQA-2, 50.2 ft, 03CW126, V36/4. K, BAQA-2, 52.0 ft, 03CW128, U44/1. L, Paratype, BAQA-2, 50.2 ft, 03CW126, H50/3. M–R, *Brochotriletes crameri* sp. nov. M, Paratype, JNDL-3, 413.2 ft, 68567, Q38/3. N, Holotype, JNDL-4, 277.6 ft, 68659, O54. O, JNDL-4, 471.6 ft, 68697, M42. P, JNDL-4, 316.4 ft, 68667, K44. Q, JNDL-1, 156.0 ft, 60840, F47/3. R, JNDL-4, 471.6 ft, 68697, N53. S, *Brochotriletes foveolatus* Naumova, 1953. BAQA-1, 345.5 ft, 03CW114, H54/4. T, *Brochotriletes hudsonii* McGregor and Camfield, 1976. JNDL-3, 462.0 ft, 68576, K54. U–V, *Brochotriletes robustus* (Scott and Rouse) McGregor, 1973. U, WELL-8, 16642.3 ft, 62407, S32. V, WELL-8, 16642.3 ft, 62406, Q45.

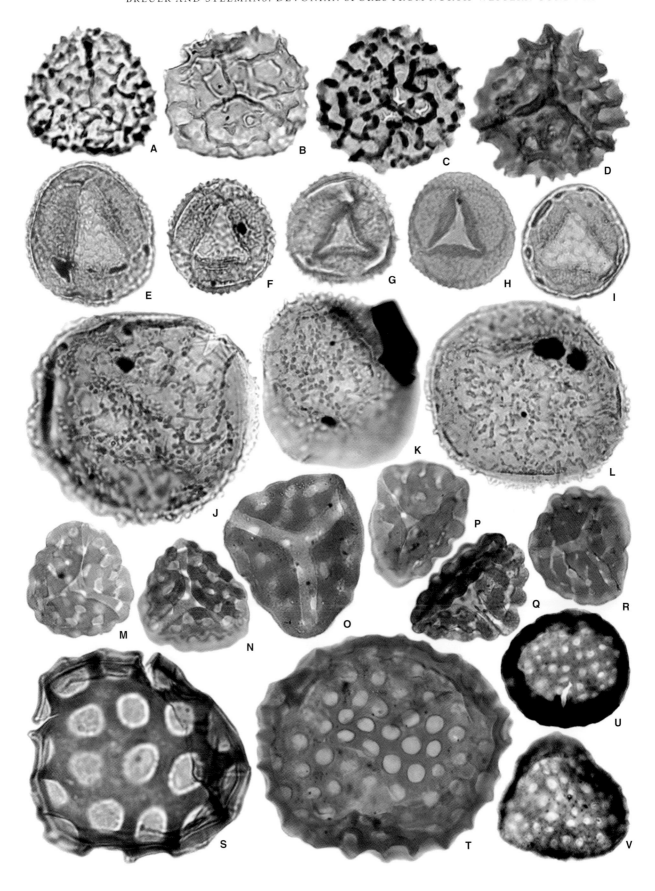

sp. *in* Allen (1965) is much larger and has a thicker exine. *B. hudsonii* McGregor and Camfield, 1976 is ornamented with more foveolae which are moreover smaller. *Chelinospora carnosa* sp. nov. is distinguished by the presence of constriction of the muri between the lumina.

Occurrence. BAQA-1, BAQA-2, JNDL-3, JNDL-4, WELL-1, S-462, WELL-2, WELL-3, WELL-4, WELL-6 and WELL-7; Jauf and Jubah formations; *papillensis-baqaensis* to *triangulatus-catillus* zones. A1-69; Ouan-Kasa and Awaynat Wanin II formations; *lindlarensis-sextantii* to *lemurata-langii* zones. MG-1; Ouan-Kasa, Awaynat Wanin II and Awaynat Wanin III formations; *lindlarensis-sextantii* to *langii-concinna* zones.

Previous records. *Brochotriletes foveolatus* has been widely reported in Early through Middle Devonian palynofloras; e.g. Algeria (Boumendjel *et al.* 1988), Brazil (Melo and Loboziak 2003; Grahn *et al.* 2005; Mendlowicz Mauller *et al.* 2007; Steemans *et al.* 2008), Belgium (Steemans 1989), Canada (McGregor and Owens 1966; McGregor 1973; McGregor and Camfield 1976), France (Le Hérissé 1983; Steemans 1989), Germany (Steemans 1989), Libya (Moreau-Benoit 1989), Poland (Turnau *et al.* 2005) and Romania (Steemans 1989).

Brochotriletes hudsonii McGregor and Camfield, 1976
Figure 11T

1966 *Brochotriletes* sp. McGregor and Owens, pl. 1, figs 8–9.
? 1968 Spore no. 380 Magloire, pl. 1, fig. 9 pl. 2, fig. 2.
1970 *Brochotriletes* sp. McGregor *et al.*, pl. 1, figs 14–15.
1975 *Brochotriletes* sp. *in* McGregor; Sanford and Norris, pl. 1, figs 14–15.
1976 *Brochotriletes hudsonii* McGregor and Camfield, p. 12, pl. 3, figs 1–2.

Dimensions. 52(69)80 μm; 10 specimens measured.

Occurrence. BAQA-1, JNDL-3 and JNDL-4; Jauf Formation (Subbat and Hammamiyat members); *ovalis-biornatus* to *lindlarensis-sextantii* zones. A1-69; Ouan-Kasa Formation; *lindlarensis-sextantii* Zone. MG-1; Ouan-Kasa Formation; *lindlarensis-sextantii* Zone.

Previous records. From upper Lochkovian–Pragian of Canada (McGregor and Owens 1966; McGregor and Camfield 1976);

upper Pragian of Germany (Steemans 1989); and Pragian–uppermost Emsian of Poland (Turnau 1986; Turnau *et al.* 2005).

Brochotriletes robustus (Scott and Rouse) McGregor, 1973
Figure 11U–V

1961 *Perforosporites robustus* Scott and Rouse, p. 978 (*pars*), pl. 113, figs 1–2.
1966 *Brochotriletes* spp. McGregor and Owens (*pars*), pl. 4, fig. 5.
1967 *Perforosporites robustus* Scott and Rouse; Beju, pl. 1, figs 23–24.
1973 *Brochotriletes robustus* (Scott and Rouse) McGregor, p. 40, pl. 5, figs 1, 6.
? 1983 *Brochotriletes* cf. *robustus* (Scott and Rouse) McGregor; Le Hérissé, p. 34, pl. 5, fig. 11.

Dimensions. 35(42)64 μm; eight specimens measured.

Comparison. This species is distinguished from *B. bellatulus* Steemans, 1989 by its smaller size and spinose sculpture. It is very close to it, and may represent an extreme member of the *B. bellatulus* population. Unfortunately, few specimens have been found.

Occurrence. JNDL-3, JNDL-4, WELL-3 and WELL-8; Jauf (Hammamiyat Member) and Jubah Formation; *lindlarensis-sextantii* Zone only as youngest specimens from the Jubah Formation are probably reworked.

Previous records. From upper Lochkovian–Emsian of Belgium (Steemans 1989); Emsian of Canada (McGregor and Owens, 1966; McGregor 1973; McGregor and Camfield 1976); and upper Lochkovian – lower Pragian of Germany (Steemans 1989).

Brochotriletes tenellus sp. nov.
Figure 12A–D

Derivation of name. From *tenellus* (Latin), meaning delicate, wispy; refers to the size of the foveolae and the spore.

Holotype. EFC T32/2 (Fig. 12D), slide 68617.

Paratype. EFC T36/3 (Fig. 12B), slide 66782; BAQA-1 core hole, sample 222.5 ft.

FIG. 12. Each figured specimen is identified by borehole, sample, slide number and England Finder Co-ordinate location. All figured specimens are at magnification ×1000 except where mentioned otherwise. A–D, *Brochotriletes tenellus* sp. nov. A, JNDL-4, 316.4 ft, 68667, L50. B, Paratype, BAQA-1, 222.5 ft, 66782, T36/3. C, JNDL-4, 484.1 ft, 68699, S24. D, Holotype, JNDL-4, 135.8 ft, 68617, T32/2. E–H, *Brochotriletes tripapillatus* sp. nov. E, Paratype, A1-69, 1109 ft, 27274, Q46/4. F–G, A1-69, 1334 ft, 27127, O51. H, A1-69, 1109 ft, 27273, P53/3. I–M, *Camarozonotriletes asperulus* sp. nov. I, MG-1, 2247 m, 62941, O40/1. J, MG-1, 2278 m, 62936, G43. K, MG-1, 2295 m, 63007, M38/3. L, Paratype, MG-1, 2264 m, 62950, J47/2. M, Holotype, MG-1, 2160.6 m, 62747, X46/2. N–O, *Camarozonotriletes filatoffii* Breuer *et al.*, 2007c. N, BAQA-1, 227.1 ft, 03CW110, B25. O, BAQA-1, 346.8 ft, 66796, E37/3.

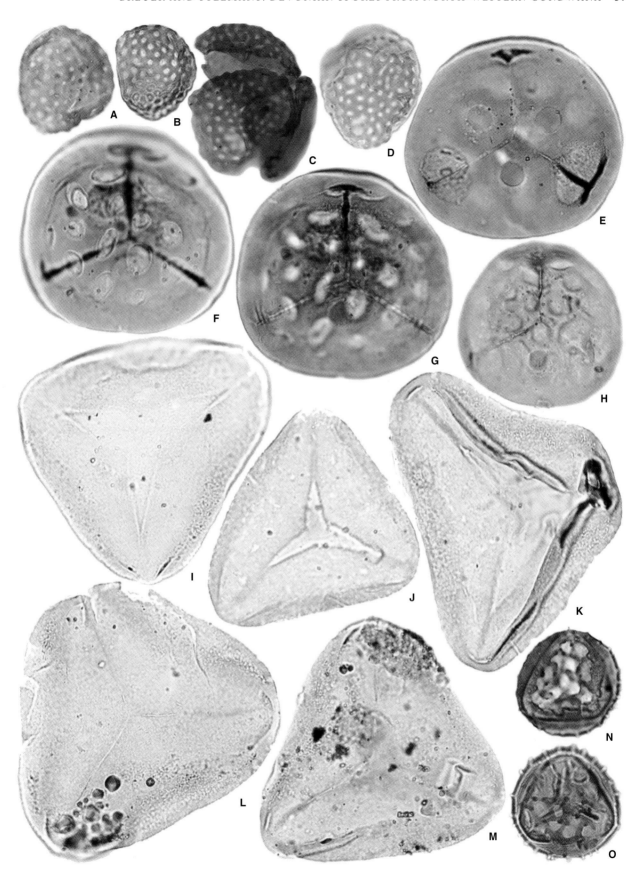

Type locality and horizon. JNDL-4 core hole, sample 135.8 ft; Jauf Formation at Domat Al-Jandal, Saudi Arabia.

Diagnosis. A small *Brochotriletes* sculptured with small regular circular to sub-circular, evenly distributed foveolae. Exine laevigate.

Description. Amb is rounded sub-triangular to sub-circular. Laesurae are not always perceptible, simple, straight and extending to equator. Exine is 0.5–1.5 μm thick equatorially and distally, thinner on contact areas of proximal face. Contact areas are laevigate. Distal and proximo-equatorial regions are foveolate. Foveolae are circular or sub-circular in plan view, U-shaped in profile, 0.5–2 μm in diameter, *c.* 0.5 μm deep, and 1.5–2 μm apart. Exine between foveolae is laevigate.

Dimensions. 25(30)40 μm; 20 specimens measured.

Comparison. Brochotriletes rarus Arkhangelskaya, 1978 is larger and shows small foveolae, usually less than 1 μm in diameter. *B. robustus* (Scott and Rouse) McGregor, 1973 is ornamented between the foveae with discrete elements, which are interconnected by barely perceptible fine muri.

Occurrence. BAQA-1, JNDL-3 and JNDL-4, Jauf Formation (Subbat and Hammamiyat members); *milleri* to *lindlarensis-sextantii* zones.

Brochotriletes tripapillatus sp. nov.
Figure 12E–H

Derivation of name. From *tripapillatus* (Latin), meaning sculptured with three papillae; refers to the proximal papillae.

Holotype. EFC O51 (Fig. 12F–G), slide 27127.

Paratype. EFC Q46/4 (Fig. 12E), slide 27274; A1-69 borehole, sample 1109 ft.

Type locality and horizon. A1-69 borehole, sample 1109 ft; Awaynat Wanin II Formation in A1-69, Libya.

Diagnosis. A *Brochotriletes* bearing three proximal sub-circular papillae in the interradial areas.

Description. Amb is sub-circular. Laesurae, simple, straight, three-quarters to nine-tenths of the amb radius in length. Patinate exine is 2–6.5 μm thick equatorially and distally, thinner on contact areas. Contact areas are laevigate to infragranular. Sub-circular papillae, 4–9 μm in diameter, are developed on each interradial region, one-third to one-half of the distance from apical pole to equator. Distal surface is foveolate. Foveolae are circular to elongate or roughly polygonal in plan view, 4–9 μm in diameter and 1.5–5 μm apart. Exine between foveolae is laevigate.

Dimensions. 45(55)61 μm, three specimens measured.

Comparisons. Although the specimen figured as *Brochotriletes* sp. cf. *B. foveolatus* Naumova, 1953 *in* McGregor and Playford (1992, pl. 4, fig. 11) seems to have proximal sculpture, the species described above is the only representative of *Brochotriletes* Naumova, 1939 ex Ishchenko, 1952 known that shows proximal papillae.

Occurrence. A1-69; Awaynat Wanin II Formation; *undulatus* to *catillus* zones.

Genus CAMAROZONOTRILETES Naumova, 1939 ex Naumova, 1953

Type species. Camarozonotriletes devonicus Naumova, 1953.

Camarozonotriletes asperulus sp. nov.
Figure 12I–M

Derivation of name. From *asperulus* (Latin), meaning slightly rough; refers to the sculpture of the proximo-equatorial and distal regions.

Holotype. EFC X46/2 (Fig. 12M), slide 62747.

Paratype. EFC J47/2 (Fig. 12L), slide 62950; MG-1 borehole, sample 2264 m.

Type locality and horizon. MG-1 borehole, sample 2160.6 m; Awaynat Wanin III Formation at Mechiguig, Tunisia.

Diagnosis. A large triangular *Camarozonotriletes* sculptured with minute, closely spaced grana or coni.

Description. Amb is triangular. The corners are rounded, while the margins are slightly convex, straight or sometimes slightly concave. Exine is thin. Laesurae are sometimes open, distinct, simple, straight and extending to the inner margin of the cingulum. Curvaturae are not easily distinguishable. Cingulum is generally 1–3 μm wide equatorially opposite the laesurae and commonly 4–10 μm interradially. Contact faces are laevigate. Proximo-equatorial and distal regions are infragranular, sculptured with minute closely spaced grana or coni, less than 0.5 μm in diameter and sometimes barely visible. Cingulum is slightly darker than central area.

Dimensions. 42(69)85 μm; 17 specimens measured.

Comparison. Camarozonotriletes minutus Naumova ex Chibrikova, 1959 and *C. antiquus* Kedo, 1955 are smaller and are described as shagreenate. *C. parvus* Owens, 1971 is smaller and more rounded. *C. laevigatus* McGregor and Camfield, 1982 is unsculptured and also smaller. *C. rugulosus* Breuer *et al.*, 2007*c* is commonly smaller and finely rugulate distally, but ornamentation is sometimes barely visible. *Camarozonotriletes? concavus* Loboziak and Streel, 1989 is smaller and the width of cingulum is barely reduced opposite the laesurae, which calls its allocation to the genus *Camarozonotriletes* Naumova, 1939 ex Naumova, 1953 into question. *Leiotriletes bonitus* Cramer, 1966*b* has the same thickening of the interradial margins but shows proximal thickenings along the laesurae. In addition, it is laevigate and smaller (*c.* 50 μm).

Occurrence. A1-69; Awaynat Wanin II Formation; *incognita* to *lemurata* zones. MG-1; Awaynat Wanin I, Awaynat Wanin II and Awaynat Wanin III formations; *rugulata-libyensis* to *langii-concinna* zones.

Previous record. From lower–middle Givetian of Parnaíba Basin, Brazil (Breuer and Grahn 2011).

Camarozonotriletes filatoffii Breuer *et al.*, 2007c
Figures 12N–O, 13A–B

2007c *Camarozonotriletes filatoffii* Breuer *et al.*, p. 49, pl. 4, figs 14–23; pl. 5, fig. 1.

Dimensions. 24(30)35 μm; 36 specimens measured.

Comparisons. Camarozonotriletes (*Rotaspora*) *retiformis* (Hashemi and Playford) comb. nov. is distally reticulate but not ornamented with spines. In contrast, *Rotaspora rara* (Raskatova) Hashemi and Playford, 2005 has the same type of spines but is not distally reticulate.

Occurrence. BAQA-1, BAQA-2, JNDL-3, JNDL-4, WELL-2, WELL-3, WELL-4 and WELL-7; Jauf Formation (Sha'iba to Hammamiyat members); *papillensis-baqaensis* to *lindlarensis-sextantii* zones.

Previous record. From upper Pragian – lower Emsian of Paraná Basin, Brazil (Mendlowicz Mauller *et al.* 2007).

Camarozonotriletes parvus Owens, 1971
Figure 13C–F

1966 *Camarozonotriletes* sp. cf. *C. breviculus* Ishchenko; McGregor and Owens, pl. 9, fig. 5.
1971 *Camarozonotriletes parvus* Owens, p. 40, pl. 11, figs 1–4.
1972 *Camarozonotriletes* n. sp. McGregor and Uyeno, pl. 2, fig. 2.
non 1989 *Camarozonotriletes parvus* Owens; Steemans, p. 112, pl. 26, figs 4–8, 56.
non 2007 *Camarozonotriletes parvus* Owens; Mendlowicz Mauller *et al.*, pl. 5, fig. 7.

Dimensions. 28(36)43 μm; 15 specimens measured.

Comparison. The specimens described as *C. parvus* Owens, 1971 *in* Steemans (1989) are misidentified; they show higher pila and bacula. This misidentified species needs to be redefined because it is the key species of the Pa Interval Zone of Streel *et al.* (1987). This biozone remains valid but not its name. *C. minutus* Naumova ex Chibrikova, 1959 and *C. antiquus* Kedo, 1955 are described as shagreenate. In all other respects, they appear identical to *C. parvus. C. laevigatus* McGregor and Camfield, 1982 strongly resembles the latter but is unsculptured.

Occurrence. S-462 and WELL-8; Jubah Formation; *lemurata-langii* Zone but some specimens from S-462 may be caved in older strata. A1-69; Awaynat Wanin II Formation; *undulatus* Zone. MG-1; Awaynat Wanin III Formation; *langii-concinna* Zone.

Previous records. From middle Givetian of Algeria (Moreau-Benoit *et al.* 1993) and Parnaíba Basin, Brazil (Breuer and Grahn 2011); upper Eifelian–Frasnian of Canada (McGregor and Owens 1966; Owens 1971; McGregor and Uyeno 1972; McGregor and Camfield 1982); lower Eifelian – upper Frasnian of Libya (Moreau-Benoit 1989); and Givetian of Morocco (Rahmani-Antari and Lachkar 2001).

Camarozonotriletes (*Rotaspora*) *retiformis* (Hashemi and Playford) comb. nov.
Figure 13G

1972 ?*Reticulatisporites* sp. Kemp, p. 115, pl. 55, fig. 9.
1992 *Camarozonotriletes* spp. McGregor and Playford (*pars*), pl. 18, fig. 4 (*non* figs 1–3, 5).
2005 *Rotaspora retiformis* Hashemi and Playford, p. 362, pl. 7, figs 13–17.

Dimensions. 25–27 μm; two specimens measured.

Remarks. Although *Camarozonotriletes* Naumova, 1939 ex Naumova, 1953 is considered by Hashemi and Playford (2005) as a junior synonym of *Rotaspora* Schemel, 1950, the present species is transferred to the genus *Camarozonotriletes*. The criterion to differentiate the two genera is the differing body and equatorial crassitude colour densities, but it seems unconvincing as a taxonomic distinction according to Hashemi and Playford (2005).

Comparison. ?*Reticulatisporites* sp. *in* Kemp (1972), which is distally reticulate but not ornamented with spines, is synonymous with *Camarozonotriletes* (*Rotaspora*) *retiformis* (Hashemi and Playford) comb. nov.

Occurrence. JNDL-3 and JNDL-4; Jauf Formation (Hammamiyat Member); *lindlarensis-sextantii* Zone.

Previous records. From Pragian of Antarctica (Kemp 1972; Troth *et al.* 2011); and Emsian of Adavale Basin, Australia (Hashemi and Playford 2005).

Camarozonotriletes rugulosus Breuer et al., 2007c
Figure 13H–I

2007c *Camarozonotriletes rugulosus* Breuer *et al.*, p. 49, pl. 5, figs 2–9.

Dimensions. 37(46)59 µm; 30 specimens measured.

Occurrence. JNDL-1. Jubah Formation; *svalbardiae-eximius* Zone. A1-69; Awaynat Wanin II Formation; *triangulatus* Zone.

Camarozonotriletes sextantii McGregor and Camfield, 1976
Figure 13J–K

1976 *Camarozonotriletes sextantii* McGregor and Camfield, p. 12 (*cum syn.*), pl. 4, figs 13–14, 16–18.
1982 *Craspedispora arctica* McGregor and Camfield, p. 28, pl. 5, figs 5–9; text-fig. 38.

Dimensions. 31(39)59 µm; 21 specimens measured.

Remarks. Two populations can be distinguished in our material. Size ranges of North African and Saudi Arabian specimens are 31–36 and 37–59 µm, respectively.

Comparison. *Craspedispora arctica* McGregor and Camfield, 1982 is herein considered as synonymous of *Camarozonotriletes sextantii*. *Craspedispora arctica* has notably straight to convex interradial margins of the contact area, while those of *Camarozonotriletes sextantii* are convex to more commonly concave. This feature does not constitute a discriminatory criterion to erect two different species as the shape of interradial margins may be similar in both species. *Craspedispora arctica* strongly resembles the specimens from the North African population.

Occurrence. JNDL-1, JNDL-3, JNDL-4, WELL-4, WELL-5, WELL-6 and WELL-7; Jauf (Hammamiyat and Murayr members) and Jubah formations; *lindlarensis-sextantii* to *svalbardiae-eximius* zones. A1-69; Ouan-Kasa and Awaynat Wanin I formations; *lindlarensis-sextantii* to *annulatus-protea* zones. MG-1; Ouan-Kasa and Awaynat Wanin I formations; *annulatus-protea* to *svalbardiae-eximius* zones.

Previous records. *Camarozonotriletes sextantii* is eponymous for the Emsian *annulatus-sextantii* Assemblage Zone of the Old Red Sandstone Continent and adjacent regions (Richardson and McGregor 1986). *C. sextantii* has an almost worldwide distribution extending from Emsian into the lower Eifelian. It has been reported from many parts of the world; e.g. Algeria (Moreau-Benoit *et al.* 1993), Belgium (Steemans 1989), Brazil (Mendlowicz Mauller *et al.* 2007), Canada (McGregor and Camfield 1976), Germany (Steemans 1989), Libya (Moreau-Benoit, 1989), Morocco (Rahmani-Antari and Lachkar 2001), Poland (Turnau *et al.* 2005) and Saudi Arabia (Steemans 1995; Al-Ghazi 2007).

Camarozonotriletes? *concavus* Loboziak and Streel, 1989
Figure 13L–P

1989 *Camarozonotriletes?* *concavus* Loboziak and Streel, p. 175, pl. 1, figs 13–15.

Description. Amb is sub-triangular to triangular with rounded corners and generally concave to almost straight interradial margins. Laesurae are simple, straight and extend to the inner margin of cingulum. Cingulum, 2–5 µm wide, is slightly reduced at corners, slightly darker than central area of the spore. Exine is proximally laevigate, equatorially and distally infragranulate to granulate giving a spongy appearance. Sculptural elements are less than 1 µm wide and high, often barely perceptible and closely distributed.

Dimensions. 32(40)48 µm; 23 specimens measured.

Remarks. Sometimes, two slightly separated walls can be detected. Reduction in the cingulum width at corners is not often very conspicuous in this species, and attribution to *Camarozonotriletes* Naumova, 1939 ex Naumova, 1953 is therefore questionable (Loboziak and Streel 1989).

FIG. 13. Each figured specimen is identified by borehole, sample, slide number and England Finder Co-ordinate location. All figured specimens are at magnification ×1000 except where mentioned otherwise. A–B, *Camarozonotriletes filatoffii* Breuer *et al.*, 2007c. A, BAQA-1, 308.3 ft, 03CW112, U26. B, BAQA-1, 223.5 ft, 03CW109, F30. C–F, *Camarozonotriletes parvus* Owens, 1971. C, MG-1, 2178 m, 62997, Y33/2. D, MG-1, 2160.6 m, 62747, J31. E, MG-1, 2180 m, 62971, M42/1. F, MG-1, 2178 m, 62997, P38/3. G, *Camarozonotriletes (Rotaspora) retiformis* (Hashemi and Playford) comb. nov. JNDL-4, 163.3 ft, 68625, U60. H–I, *Camarozonotriletes rugulosus* Breuer *et al.*, 2007c. H, JNDL-1, 172.7 ft, PPM007, R28/1. I, JNDL-1, 156.0 ft, 60840, V47/4. J–K, *Camarozonotriletes sextantii* McGregor, 1973. J, JNDL-4, 37.1 ft, 03CW184, R42. K, MG-1, 2639 m, 62779, M36/3. L–P, *Camarozonotriletes?* *concavus* Loboziak and Streel, 1989. L, A1-69, 1486 ft, 26977, R46. M, A1-69, 1483 ft, 26995, G38. N, A1-69, 1483 ft, 26995, B54. O, A1-69, 1486 ft, 26977, P40/3. P, WELL-1, 16354.0 ft, 61959, M36/1. Q–V, *Chelinospora carnosa* sp. nov. Q, BAQA-1, 395.2 ft, 03CW121, R38. R, Paratype, BAQA-2, 134.4 ft, 03CW137, P40/1. S, Holotype, BAQA-1, 395.2 ft, 62277, K48/2. T, BAQA-1, 395.2 ft, 62274, M49/4. U, BAQA-2, 133.0 ft, 03CW136, E30/1. V, BAQA-2, 133.0 ft, 03CW136, V42.

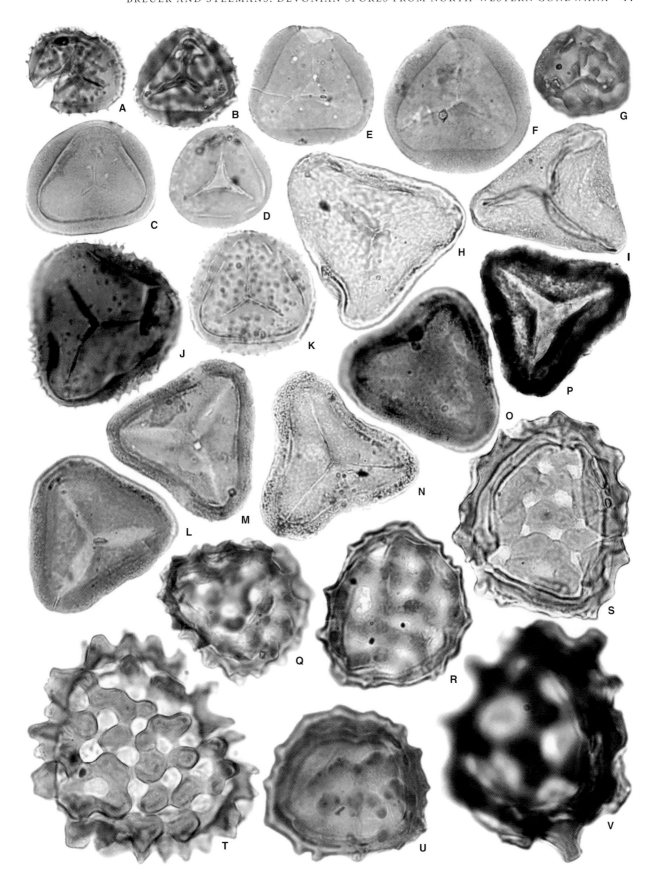

Comparison. Among sculptured species, *C. antiquus* Kedo, 1955 has convex interradial margins. *C. parvus* Owens, 1971 has the cingulum clearly reduced in front of the laesurae and a more rounded general amb. *C. pusillus* Naumova ex Chibrikova, 1959 has ornamentation up to 1.5 μm high.

Occurrence. S-462, WELL-1 and WELL-8; Jubah Formation; *lemurata-langii* to *triangulatus-catillus* zones. A1-69; Awaynat Wanin II formations; *incognita* to *triangulatus* zones. MG-1; Awaynat Wanin I, Awaynat Wanin II and Awaynat Wanin III formations; *rugulata-libyensis* to *langii-concinna* zones.

Previous records. From upper Eifelian–Frasnian of Brazil (Loboziak *et al.* 1988; Melo and Loboziak 2003; Breuer and Grahn 2011).

Genus CHELINOSPORA Allen, 1965

Type species. *Chelinospora concinna* Allen, 1965.

Chelinospora carnosa sp. nov.
Figure 13Q–V

Derivation of name. From *carnosus* (Latin), meaning fleshy; refers to the distal sculpture.

Holotype. EFC K48/2 (Fig. 13S), slide 62277.

Paratype. EFC P40/1 (Fig. 13R), slide 03CW137; BAQA-2 core hole, sample 134.4 ft.

Type locality and horizon. BAQA-1 core hole, sample 395.2 ft; Jauf Formation at Baq'a, Saudi Arabia.

Diagnosis. A thick-walled *Chelinospora* sculptured with broad reticulum and large verrucae showing constrictions between each pair of junctions.

Description. Amb is sub-triangular to triangular. Laesurae are straight and simple, but often not observed because of the thinness of proximal exine, frequently torn. Exine is laevigate to infragranulate, 2–7 μm equatorially thick, thinner proximally. Patina is sculptured with broad reticulum. Muri are 2–7 μm wide and 1–4 μm high. At junctions, muri commonly widen into large rounded or flat-topped verrucae, 4–9 μm wide, up to 8 μm high, sometimes fused together. Lumina, polygonal or irregular in plan view, are 2–10 μm in greatest diameter. Muri show constrictions between each pair of junctions.

Dimensions. 37(55)69 μm; 16 specimens measured.

Comparison. The broad muri with distinct constrictions and more or less polygonal lumina distinguish this species from other species of *Chelinospora* Allen, 1965.

Occurrence. BAQA-1, BAQA-2 and WELL-7; Jauf Formation (Sha'iba to Subbat members); *papillensis-baqaensis* to *ovalis* zones.

Chelinospora concinna Allen, 1965
Figure 14A–C

? 1964 *Knoxisporites reticulatus* Vigran, p. 22, pl. 1, figs 10–12; pl. 2, figs 8–9.
  1965 *Chelinospora concinna* Allen, p. 728, pl. 101, figs 12–20.

Dimensions. 36(51)65 μm; 22 specimens measured.

Remarks. There is considerable variation in width of the patina, thickness of muri and size of lumina.

Occurrence. S-462; Jubah Formation; *langii-concinna* Zone. A1-69; Awaynat Wanin II Formation; *langii-concinna* Zone. MG-1; Awaynat Wanin II and Awaynat Wanin III formations; *langii-concinna* Zone.

Previous records. *Chelinospora concinna* is eponymous for the upper Givetian – lower Frasnian TCo Oppel Zone of Western Europe (Streel *et al.* 1987). *C. concinna* has an almost world-wide distribution extending from Givetian into Frasnian and has been reported from many parts of the world; e.g. Bolivia (Perez-Leyton 1990), Brazil (Loboziak *et al.* 1988; Breuer and Grahn 2011), Canada (McGregor and Uyeno 1972), France (Brice *et al.* 1979; Loboziak and Streel 1980, 1988), Greenland (Friend *et al.* 1983; Marshall and Hemsley 2003), Spitsbergen, Norway (Vigran 1964; Allen 1965), Poland (Turnau 1996; Turnau and Racki 1999), Portugal (Lake *et al.* 1988), Russian Platform (Avkhimovitch *et al.* 1993; Arkhangelskaya and

FIG. 14. Each figured specimen is identified by borehole, sample, slide number and England Finder Co-ordinate location. All figured specimens are at magnification ×1000 except where mentioned otherwise. A–C, *Chelinospora concinna* Allen, 1965. A–B, S-462, 1810–1815 ft, 63256, S50/1. C, MG-1, 2205 m, 62597, M43. D–G, *Chelinospora condensata* sp. nov. D, Holotype, BAQA-1, 371.1 ft, 03CW118, R48. E, BAQA-1, 371.1 ft, 03CW118, H42/1. F, Paratype, BAQA-1, 371.1 ft, 03CW118, H27/3. G, BAQA-1, 219.2 ft, 03CW107, G31/2. H–K, *Chelinospora densa* sp. nov. H, Paratype, BAQA-2, 50.8 ft, 03CW127, R23/4. I, BAQA-1, 408.3 ft, 03CW124, F29/2. J, BAQA-1, 416.6 ft, 03CW125, P30. K, Holotype, BAQA-2, 54.8 ft, 03CW129, P-Q36. L–P, *Chelinospora laxa* sp. nov. L, Holotype, JNDL-4, 499.1 ft, 68704, D27/1. M, BAQA-2, 50.8 ft, 66813, G54/4. N, Paratype, BAQA-2, 64.5 ft, 03CW132, X29. O, BAQA-2, 64.5 ft, 66818, V53/1. P, BAQA-2, 54.8 ft, 03CW129, U28/4. Q, *Chelinospora retorrida* Turnau, 1986. BAQA-2, 134.4 ft, 66826, H41/3.

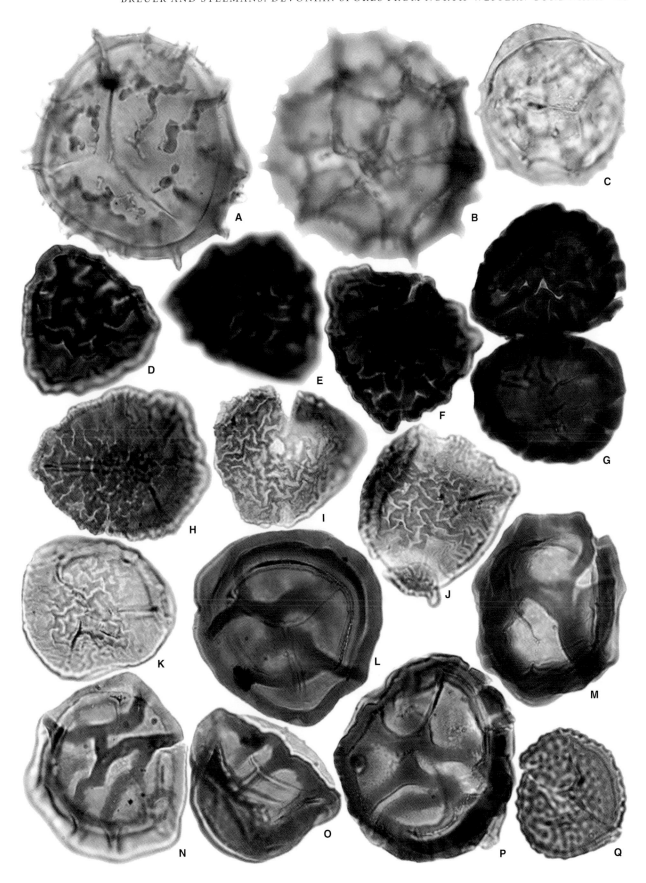

Turnau 2003) and Scotland (Marshall and Allen 1982; Marshall *et al.* 1996).

Chelinospora condensata sp. nov.
Figure 14D–G

Derivation of name. From *condensatus* (Latin), meaning dense; refers to the colour of the body and the distal sculpture.

Holotype. EFC R48 (Fig. 14D), slide 03CW118.

Paratype. EFC H27/3 (Fig. 14F), slide 03CW118; BAQA-1 core hole, sample 371.1 ft.

Type locality and horizon. BAQA-1 core hole, sample 371.1 ft; Jauf Formation at Baq'a, Saudi Arabia.

Diagnosis. A thick-walled *Chelinospora* sculptured with brain-like convoluted muri almost adjoining.

Description. Amb is sub-circular to triangular. Laesurae are straight, simple and extending to the inner edge of patina. Exine is laevigate to infragranulate, 3–5 μm thick equatorially, thinner proximally. Contact areas are laevigate. Patina is sculptured with brain-like convoluted muri almost adjoining, 1.5–4 μm wide and commonly less than 1 μm apart (rarely up to 2 μm). Muri become radially oriented over the equatorial and subequatorial regions.

Dimensions. 36(47)68 μm; 16 specimens measured.

Comparison. *Chelinospora densa* sp. nov. and *C. hemiesferica* (Cramer and Díez) Richardson *et al.*, 2001 have narrower distal muri. *C. vulgata* sp. nov. and *Chelinospora* cf. *hemiesferica* (Cramer and Díez) Richardson *et al.*, 2001 have the same muri, but these are loosely spaced. All these convolute forms of *Chelinospora* Allen, 1965 are not always distinguished easily. They may derived from a group of closely related plants and can be grouped into the *C. vulgata* Morphon defined here (Table 1).

Occurrence. BAQA-1 and BAQA-2; Jauf Formation (Sha'iba to Subbat members); *papillensis-baqaensis* to *ovalis-biornatus* zones.

Chelinospora densa sp. nov.
Figure 14H–K

Derivation of name. From *densus* (Latin) meaning dense; refers to the distal sculpture.

Holotype. EFC P-Q36 (Fig. 14K), slide 03CW129.

Paratype. EFC R23/4 (Fig. 14H), slide 03CW127; BAQA-2 core hole, sample 50.8 ft.

Type locality and horizon. BAQA-2 core hole, sample 54.8 ft; Jauf Formation at Baq'a, Saudi Arabia.

Diagnosis. A *Chelinospora* sculptured with numerous, closely spaced, narrow brain-like convoluted muri.

Description. Amb is sub-circular to sub-triangular. Laesurae are straight, simple or labrate (up to 2 μm wide) and extend to the inner edge of patina. Exine is laevigate or infragranulate, 3–7 μm equatorially thick, thinner proximally. Contact areas are laevigate, sometimes torn. Patina is sculptured with brain-like convoluted muri, 0.5–2 μm wide and up to 1.5 μm apart. Muri become radially oriented over the equatorial and subequatorial regions.

Dimensions. 37(47)62 μm; 10 specimens measured.

Comparison. The spores may be two-layered, and on some specimens some localized detachment of the outer layer is apparent (Fig. 14J) as in *C. hemiesferica* (Cramer and Díez) Richardson *et al.*, 2001. The latter, however, show a membranous curvatural zone bearing well-pronounced radial extensions of the distal muri. *C. vulgata* sp. nov. and *Chelinospora* cf. *hemiesferica* (Cramer and Díez) Richardson *et al.*, 2001 have wider muri.

Occurrence. BAQA-1 and BAQA-2; Jauf Formation (Sha'iba to Subbat members); *ovalis* Zone.

Chelinospora laxa sp. nov.
Figure 14L–P

Derivation of name. From *laxus* (Latin), meaning loose; refers to the distal sculpture.

Holotype. EFC D27/1 (Fig. 14L), slide 68704.

Paratype. EFC X29 (Fig. 14N), slide 03CW132; BAQA-1 core hole, sample 64.5 ft.

Type locality and horizon. JNDL-4 core hole, sample 499.1 ft; Jauf Formation at Domat Al-Jandal, Saudi Arabia.

Diagnosis. A *Chelinospora* sculptured with loosely spaced broad muri, forming an irregular reticulum with broad lumina.

Description. Amb is sub-circular to sub-triangular. Laesurae are straight, simple and extend to the inner edge of patina. Exine is laevigate to infragranulate, 3–6 μm equatorially thick, thinner proximally. Contact areas are laevigate. Patina is sculptured with convoluted muri widely distributed, 3–5 μm wide and from 1 to more than 10 μm apart. They form an irregular reticulate pattern.

Dimensions. 40(54)70 μm; 12 specimens measured.

Remarks. Depending on compression, specimens may give the appearance of having a zona (Fig. 14N).

Comparison. Chelinospora cassicula Richardson and Lister, 1969 and *C. lavidensis* Richardson *et al.*, 2001 show the same kind of irregular reticulate pattern but the former has narrower, high and fold-like muri and the latter is sculptured with narrower, low muri (1 μm or less wide). *C. vulgata* sp. nov. and *Chelinospora* cf. *hemiesferica* (Cramer and Díez) Richardson *et al.*, 2001 have narrower distal muri which are closely spaced and more numerous. *Chelinospora laxa* sp. nov. represents an end-member of the *C. vulgata* Morphon (Table 1).

Occurrence. BAQA-1, BAQA-2 and JNDL-4; Jauf Formation (Sha'iba to Subbat members); *papillensis-baqaensis* to *ovalis-biornatus* zones.

Chelinospora retorrida Turnau, 1986
Figure 14Q

1969 ?*Chelinospora* sp. A Richardson and Lister, p. 243, pl. 41, fig. 15.
? 1983 *Synorisporites* cf. *dittonensis*; Rodriguez, pl. 1, fig. 13.
1986 *Chelinospora retorrida* Turnau, p. 339, pl. 1, figs 1–4.
1989 *Chelinospora retorrida* Turnau; Steemans, p. 118, pl. 29, figs 12–17.

Dimensions. 33–35 μm; two specimens measured.

Remarks. Steemans (1989) noted the presence of inspissations, rarely evident, on the interradial margins of the proximal face on 50 per cent of the specimens of *C. retorrida*. These could represent different varieties of *C. retorrida* (Steemans 1989), although this feature was not mentioned by Turnau (1986).

Comparison. ?*Chelinospora* sp. A *in* Richardson and Lister (1969) is similar to *C. retorrida* Turnau, 1986. ?*Archaeozonotriletes dubius* Richardson and Lister, 1969 has muri that represent an internal structure. *C. hemiesferica* (Cramer and Díez) Richardson *et al.*, 2001 is sculptured with closely spaced narrow muri, geniculate in plan, becoming radially oriented over the equatorial and subequatorial regions. *Chelinospora densa* sp. nov. is larger and has a thicker exine.

Occurrence. BAQA-2; Jauf Formation (Sha'iba Member); *papillensis-baqaensis* to *ovalis* zones.

Previous records. From lower Lochkovian – upper Pragian of Belgium (Steemans 1989); Lochkovian of Canada (Burden *et al.* 2002); upper Lochkovian of France, Germany and Romania (Steemans 1989); Lochkovian of Iran (Ghavidel-Syooki 2003), Saudi Arabia (Steemans 1995) and Wales (Richardson and Lister 1969);

middle Přídolí of Libya (Rubinstein and Steemans 2002); and Lochkovian–Pragian of Poland (Turnau 1986; Turnau *et al.* 2005).

Chelinospora timanica (Naumova) Loboziak and Streel, 1989
Figure 15A–B

1953 *Archaeozonotriletes timanicus* Naumova, p. 81, pl. 12, fig. 14.
? 1962 *Convolutispora fromensis* Balme and Hassell, p. 8, pl. 1, figs 14–16.
? 1959 *Archaeozonotriletes polymorphus* Naumova var. *takatinicus* Chibrikova, p. 58, pl. 7, figs 2–3.
? 1962 *Archaeozonotriletes timanicus* Naumova var. *radiatus* Chibrikova, p. 412, pl. 7, fig. 1.
? 1965 *Convolutispora tegula* Allen, p. 705, pl. 97, figs 4–8.
1965 *Archaeozonotriletes ignoratus* Naumova; Hemer, pl. 2.
1965 *Archaeozonotriletes timanicus* Naumova var. no 1; Nazarenko, pl. 1, figs 49–50.
? 1965 *Tholisporites ancylus* Allen, p. 724 (*pars*), pl. 101, fig. 5 (*non* figs 1–4, 6–7).
? 1966 *Archaeozonotriletes laticolaris* Mikhailova, p. 209, pl. 3, fig. 4.
1982 *Archaeozonotriletes timanicus* Naumova; McGregor and Camfield, p. 20, pl. 3, figs 13–15.
1989 *Chelinospora timanica* (Naumova) Loboziak and Streel, p. 175, pl. 2., figs 8–9.
1992 *Archaeozonotriletes timanicus* Naumova; McGregor and Playford, pl. 4, figs 3–4.

Dimensions. 43(54)74 μm; 54 specimens measured.

Remarks. Numerous specimens are allocated to this species which may include several species defined in the literature and probably belonging to genera other than *Chelinospora* Allen, 1965. These forms are very variable and are not easily distinguishable. In such forms, it is difficult to determine whether the character of sculptural elements are positive or negative because variation between the two configurations seems to be continuous. Intergradations from *C. timanica* to other species of *Convolutispora* Hoffmeister *et al.*, 1955 genus most likely exist.

Comparison. The diagnosis of *Convolutispora fromensis* Balme and Hassell, 1962 is similar to that of *Chelinospora timanica*, but illustrations do not allow a direct comparison. *C. timanica* is difficult to distinguish from *Convolutispora tegula* Allen, 1965. The main difference is that the patina is dissected into elements (negative) in *Chelinospora timanica* while exine of *Convolutispora tegula* is sculptured with positive elements. The two characters are distinguishable with difficulty (see above). *C. tegula* may thus be included in the present taxon. *C. florida* Hoffmeister *et al.*, 1955 has a more extensively anastomosing muroid pattern, and wider lumina. *C. uistatas* Playford, 1962 has similar sculpture, but is much larger. *Convolutispora crassata*? (Naumova) McGregor and Camfield, 1982 is not patinate. *C. subtilis* Owens, 1971 possesses narrower convolutae. *Archaeozonotriletes asymmetricus* Panshina, 1971 appears to be of similar basic construction but its large, flat, irregular tubercules are larger towards the equator where they are commonly fused with one another. Extreme forms of spores with convolute and verrucose sculpture, including notably species of *Convolutispora* Hoffmeister *et al.*, 1955, *Dibolisporites uncatus* (Naumova) McGregor and Camfield, 1982 and *Verrucosisporites scurrus* (Naumova) McGregor and Camfield, 1982, may intergrade with *Chelinospora timanica*. Indeed, all these species are morphologically very close and occur mostly in the same Givetian strata.

Occurrence. S-462; Jubah Formation; *lemurata-langii* to *langii-concinna* zones. A1-69; Awaynat Wanin II Formation; *lemurata-langii* to *langii-concinna* zones. MG-1; Awaynat Wanin I, Awaynat Wanin II and Awaynat Wanin III formations; *rugulata-libyensis* to *langii-concinna* zones.

Previous records. From middle Givetian – lower Frasnian of Australia (Grey 1991; Hashemi and Playford 2005); upper Givetian –Frasnian of Bolivia (Perez-Leyton 1990); lower Givetian – lower-most Famennian of Brazil (Loboziak *et al.* 1988; Loboziak *et al.* 1992*b*; Melo and Loboziak 2003); Eifelian–lower Givetian of Canada (McGregor and Camfield 1982); upper Eifelian–Givetian of Germany (Loboziak *et al.* 1990); middle Givetian of Greenland (Friend *et al.* 1983; Marshall and Hemsley 2003); Givetian of Poland (Turnau 1996; Turnau and Racki 1999); Givetian–lower Frasnian of Russian Platform (Avkhimovitch *et al.* 1993); and uppermost Givetian–lower Frasnian of Scotland (Marshall *et al.* 1996).

Chelinospora vulgata sp. nov.
Figure 15C–G

Derivation of name. From *vulgatus* (Latin), meaning common; refers to its abundance in the lower Jauf Formation.

Holotype. EFC M-N41 (Fig. 15E), slide 03CW118.

Paratype. EFC W39/3 (Fig. 15D), slide 03CW118; BAQA-1 core hole, sample 371.1 ft.

Type locality and horizon. BAQA-1 core hole, sample 371.1 ft; Jauf Formation at Baq'a, Saudi Arabia.

Diagnosis. A large *Chelinospora* sculptured with brain-like well-defined convoluted muri. Exine infragranulate.

Description. Amb is circular to sub-triangular. Laesurae are straight, simple or accompanied by labra, up to 3 µm in overall width, extending to the inner edge of patina. Exine is infragranulate, 2–8 µm thick equatorially, thinner proximally. Contact areas are laevigate, sometimes torn. Patina is sculptured with brain-like convoluted muri, 1–4 µm wide and 1–3 µm apart. Muri become radially oriented over the equatorial and subequatorial regions.

Dimensions. 43(55)71 µm; 17 specimens measured.

Comparison. *Chelinospora laxa* sp. nov. has fewer muri and *C. condensata* sp. nov. has the same type of muri, but these are more densely spaced. All these convolute forms of *Chelinospora* Allen, 1965 may be included in the *C. vulgata* Morphon (Table 1). They were probably derived from a group of closely related plants. *C. densa* sp. nov. and *C. hemiesferica* (Cramer and Díez) Richardson *et al.*, 2001 have narrower distal muri which are closely spaced. Nevertheless, the two species may intergrade within the same morphon.

Occurrence. BAQA-1, BAQA-2, JNDL-4 and WELL-3; Jauf Formation (Sha'iba to Subbat members); *papillensis-baqaensis* to *ovalis-biornatus* zones.

Chelinospora? sp. 1
Figure 15H–I

Description. Amb is circular to sub-circular. Laesurae are simple, straight, three-quarters to full central area radius in length. Contact areas are thinner and support a sparse ornament of broad rugulae and muri 2–4 µm wide. Patina, *c.* 3 µm thick equatorially and distally, is sculptured with irregularly distributed broad rounded muri, 2–5 µm wide and high, forming an imperfect to perfect reticulum. Lumina are irregular in plan view, commonly 3–10 µm in greatest diameter.

Dimensions. 49(52)56 µm; three specimens measured.

Remarks. A doubt remains about the assignment of this form to the genus *Chelinospora* Allen, 1965 as the contact areas show a sparse ornamentation of rugulae.

Comparison. *Chelinospora concinna* Allen, 1965 has narrower muri, larger lumina and commonly a thicker patina. *C. timanica* (Naumova) Loboziak and Streel, 1989 has also a thicker patina with a very irregular ornament pattern.

FIG. 15. Each figured specimen is identified by borehole, sample, slide number and England Finder Co-ordinate location. All figured specimens are at magnification ×1000 except where mentioned otherwise. A–B, *Chelinospora timanica* (Naumova) Loboziak and Streel, 1989. A, MG-1, 2476 m, 63016, U36. B, MG-1, 2178 m, 62996, Q43. C–G, *Chelinospora vulgata* sp. nov. C, BAQA-2, 133.0 ft, 03CW136, J52. D, Paratype, BAQA-1, 371.1 ft, 03CW118, W39/3. E, Holotype, BAQA-1, 371.1 ft, 03CW118, M-N41. F, BAQA-1, 376.4 ft, 03CW119, J49. G, BAQA-1, 366.9 ft, 03CW117, J38/4. H–I, *Chelinospora?* sp. 1. H, A1-69, 971 ft, 62369, P44. I, A1-69, 971 ft, 62641, U52. J, *Cirratriradites? diaphanus* Steemans, 1989. BAQA-2, 50.2 ft, 03CW126, S26/2. K, *Clivosispora verrucata* McGregor, 1973 var. *convoluta* McGregor and Camfield, 1976. JNDL-4, 316.4 ft, 03CW244, K43/2. L–M, *Clivosispora verrucata* McGregor, 1973 var. *verrucata* McGregor and Camfield, 1976. L, BAQA-1, 395.2 ft, 66807, G39. M, JNDL-4, 87.2 ft, 03CW195, F34/1.

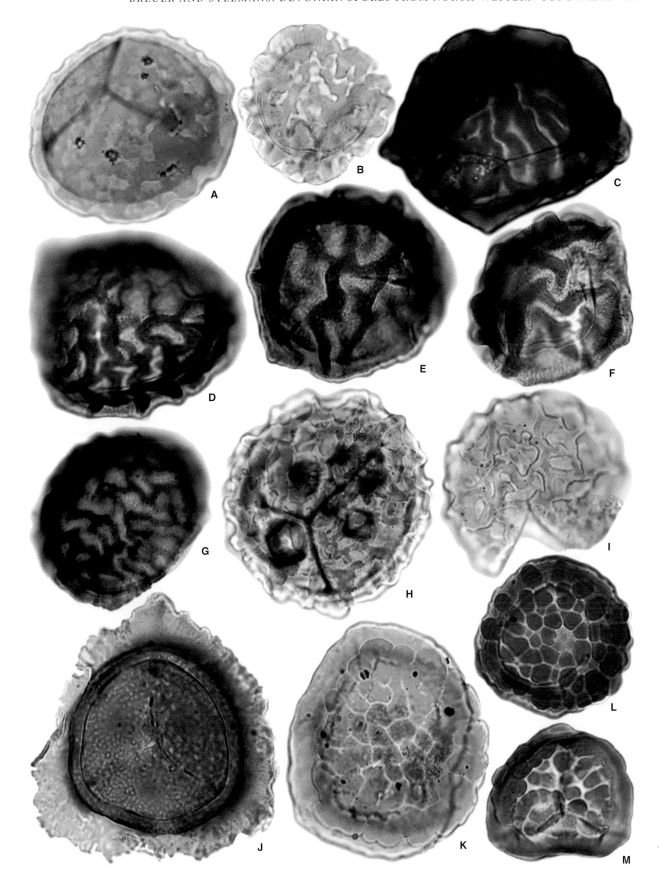

Occurrence. A1-69; Awaynat Wanin II Formation; *langii-concinna* Zone.

Genus CIRRATRIRADITES Wilson and Coe, 1940

Type species. Cirratriradites maculatus Wilson and Coe, 1940.

Cirratriradites? diaphanus Steemans, 1989
Figure 15J

1981   *Cirratriradites* sp. A Steemans, p. 53.
1981   *Cirratriradites* sp. A *in* Steemans; Streel *et al.*, pl. 3, figs 10–11.
1989   *Cirratriradites diaphanus* Steemans, p. 119, pl. 30, figs 4–9.
2006   Unidentified zonate spores Wellman, pl. 19, fig. d.

Dimensions. 46(58)71 µm; 10 specimens measured.

Remarks. There is some doubt about the allocation of this species to the genus *Cirratriradites* Wilson and Coe, 1940 because no distal foveolae are observed in the present population.

Occurrence. BAQA-1, BAQA-2, JNDL-3, JNDL-4, WELL-2, WELL-3, WELL-4, WELL-5, WELL-6 and WELL-7; Jauf Formation (Sha'iba to Hammamiyat members); *papillensis-baqaensis* to *lindlarensis-sextantii* zones. MG-1; Ouan-Kasa Formation; *svalbardiae-eximius* Zone but occurrences are probably reworked.

Previous records. From upper Lochkovian – upper Pragian of Belgium (Steemans 1989); Lochkovian–lower Emsian of Brazil (Melo and Loboziak 2003; Grahn *et al.* 2005; Mendlowicz Mauller *et al.* 2007); upper Lochkovian – upper Pragian of Germany (Steemans 1989); middle Pragian – middle Emsian of Luxembourg (Steemans *et al.* 2000a); upper Lochkovian (Steemans 1989); and upper Pragian – ?lowermost Emsian of Scotland (Wellman 2006).

Genus CLIVOSISPORA Staplin and Jansonius, 1964

Type species. Clivosispora variabilis Staplin and Jansonius, 1964.

Clivosispora verrucata McGregor, 1973 var. convoluta
McGregor and Camfield, 1976
Figure 15K

1976   *Clivosispora verrucata* McGregor var. *convoluta* McGregor and Camfield, p. 15, pl. 2, figs 13–21.

Dimensions. 36(44)63 µm; 11 specimens measured.

Comparison. Although ornamentation of *Chelinospora poecilomorpha* (Richardson and Ioannides) Richardson *et al.*, 2001 could be comparable, it exhibits muri produced by the terminal fusion of verrucae, circular, sub-circular, sub-polygonal to irregular in plan view. This species is also distinguished by its smaller size. *Clivosispora verrucata* var. *convoluta* seems to intergrade with *Clivosispora verrucata* McGregor, 1973 var. *verrucata* McGregor and Camfield, 1976.

Occurrence. BAQA-1, BAQA-2, JNDL-4, WELL-2, WELL-4 and WELL-7; Jauf Formation (Sha'iba to Subbat members); *papillensis-baqaensis* to *ovalis-biornatus* zones. A1-69; Ouan-Kasa Formation; *lindlarensis-sextantii* Zone.

Previous records. From lower Přídolí – lower Emsian of Amazon and Paraná basins, Brazil (Mendlowicz Mauller *et al.* 2007; Steemans *et al.* 2008); Pragian–Emsian of Canada (McGregor and Camfield 1976) and Iran (Ghavidel-Syooki 2003); upper Pragian of Armorican Massif, France (Le Hérissé 1983); Ludlow or Přídolí of Libya (Rubinstein and Steemans 2002); upper Pragian – Emsian of Morocco (Rahmani-Antari and Lachkar 2001); and upper Pragian – ?lowermost Emsian of Scotland (Wellman 2006).

Clivosispora verrucata McGregor, 1973 var. verrucata
McGregor and Camfield, 1976
Figure 15L–M

1954   Spore type C6 Radforth and McGregor, pl. 1, fig. 35.
1966   cf. *Clivosispora* McGregor and Owens, pl. 3, figs 16–17.
? 1968   Trilete verruquée sp. 1 Jardiné and Yapaudjian, pl. 1, fig. 23.
1970   *Clivosispora* sp. McGregor *et al.*, pl. 1, figs 28–29.
1973   *Clivosispora verrucata* McGregor, p. 54, pl. 7, figs 4–5, 10.
1976   *Clivosispora verrucata* McGregor var. *verrucata*; McGregor and Camfield, p. 15, pl. 3, figs 11–14.
non 1981   *Clivosispora verrucata* McGregor var. *verrucata*; Gao Lianda, pl. 2, fig. 3.

Dimensions. 36(43)55 µm; 16 specimens measured.

Comparison. This variety differs from *C. verrucata* McGregor, 1973 var. *convoluta* McGregor and Camfield, 1976 in that the distal sculpture consists of convolute muri. The two varieties are otherwise alike, and intergrade. *Chelinospora poecilomorpha* (Richardson and Ioannides) Richardson *et al.*, 2001 is distinguished by the terminal fusion of comparable verrucae that produce muri sometimes anastomosing in places. It is also smaller and the simple sutures are often barely perceptible because of its very thin proximal face. *Synorisporites verrucatus* Richardson and Lister, 1969 has a narrower cingulum and smaller verrucae.

Occurrence. BAQA-1, BAQA-2, JNDL-1, JNDL-4, WELL-2, WELL-3, WELL-4, WELL-5, WELL-6 and WELL-7; Jauf Formation; *papillensis-baqaensis* to *annulatus-protea* zones.

Previous records. From lower Přídolí of Amazon Basin, Brazil (Steemans *et al.* 2008); Pragian–Emsian of Canada (McGregor and Owens 1966; McGregor 1973; McGregor and Camfield 1976) and Iran (Ghavidel-Syooki 2003); and uppermost Pragian – ?lowermost Emsian of Scotland (Wellman 2006).

Genus CONCENTRICOSISPORITES Rodriguez, 1983

Type species. *Concentricosisporites sagittarius* (Rodriguez) Rodriguez, 1983.

Concentricosisporites sagittarius (Rodriguez) Rodriguez, 1983
Figure 16A–C

1978a *Stenozonotriletes sagittarius* Rodriguez, p. 219, pl. 1, fig. 7.
1983 *Concentricosisporites sagittarius* (Rodriguez); Rodriguez, p. 36, pl. 3, fig. 15.

Dimensions. 28(38)47 μm; 10 specimens measured.

Occurrence. BAQA-1, BAQA-2, JNDL-4 and WELL-7; Jauf Formation (Sha'iba to Subbat members); *papillensis-baqaensis* to *lindlarensis-sextantii* zones. MG-1; Ouan-Kasa Formation; *lindlarensis-sextantii* Zone.

Previous records. From upper Lochkovian of Solimões Basin, Brazil (Rubinstein *et al.* 2005); middle Přídolí of Libya (Rubinstein and Steemans 2002); upper Ludfordian – lower Lochkovian (Rodriguez 1978a, b; Richardson *et al.* 2001); and middle–upper Ludfordian of Pennsylvania, USA (Beck and Strother 2008).

Genus CONTAGISPORITES Owens, 1971

Type species. *Contagisporites optivus* (Chibrikova) Owens, 1971.

Contagisporites optivus (Chibrikova) Owens, 1971
Figures 16D, 47A–C

1959 *Archaeozonotriletes optivus* Chibrikova, p. 60, pl. 7, fig. 9.
1960 *Retusotriletes* sp. Taugourdeau-Lantz, p. 145, pl. 1, fig. 5.
1962 *Archaeozonotriletes optivus* var. *vorobjevensis* Chibrikova, p. 430, pl. 2, fig. 6.
1964 *Biharisporites spitsbergensis* Vigran, p. 12, pl. 2, figs 1–4.
1965 *Calyptosporites optivus* (Chibrikova) Allen, p. 736, pl. 104, figs 1–4.
1966 *Archaeozonotriletes* cf. *A. optivus* var. *vorobjevensis* Chibrikova; McGregor and Owens, pl. 16, figs 3–4.
1966 *Archaeozonotriletes optivus* Chibrikova; McGregor and Owens, pl. 17, fig. 6.
1967 *Rhabdosporites cuvillieri* Taugourdeau-Lantz, p. 54, pl. 3, figs 1–6.
1971 *Contagisporites optivus* (Chibrikova) var. *optivus* Owens, p. 52, pl. 16, figs 1–3.
1971 *Contagisporites optivus* var. *vorobjevensis* (Chibrikova) Owens, p. 53, pl. 16, figs 4–6.
1987 Megaspore (*Biharisporites*) of *Tanaitis furchihasta* Krassilov *et al.*, p. 173, pl. 4, figs 1–2; pl. 7, figs 1–2.

Dimensions. 200(223)250 μm; seven specimens measured.

Remarks. Although *C. optivus* var. *vorobjevensis* (Chibrikova) Owens, 1971 differs mainly from *C. optivus* (Chibrikova) var. *optivus* Owens, 1971 by its coarser, low verrucose or blunt pointed conate elements, both are grouped here together because they appear to intergrade and the difference is often impossible to discern under a transmitted light microscope. It is the first time that this megaspore is observed on the Gondwana (de Ville de Goyet *et al.* 2007; Steemans *et al.* 2011b).

Comparison. *Contagisporites optivus* (Chibrikova) Owens, 1971 differs from *Rhabdosporites langii* (Eisenack) Richardson, 1960 by its larger size, well-developed curvaturae and elevated labra.

Occurrence. S-462; Jubah Formation; *lemurata-langii* to *triangulatus-catillus* zones, although some specimens may be slightly caved. A1-69; Awaynat Wanin II Formation; *undulatus* to *triangulatus-catillus* zones.

Previous records. *Contagisporites optivus* is eponymous for the upper Givetian – lower Frasnian optivus-triangulatus Assemblage Zone of the Old Red Sandstone Continent and adjacent regions (Richardson and McGregor 1986). *C. optivus* has been mainly recorded from Givetian–Frasnian from Euramerica; e.g. Canada (McGregor and Owens 1966; Owens 1971; McGregor and Uyeno 1972), France (Brice *et al.* 1979; Loboziak and Streel 1980, 1988; Loboziak *et al.* 1983), Greenland (Friend *et al.* 1983; Marshall and Hemsley 2003), Spistsbergen, Norway (Vigran 1964; Allen 1965), Poland (Turnau 1996; Turnau and Racki 1999), Russian Platform (Avkhimovitch *et al.* 1993; Arkhangelskaya and Turnau 2003) and Scotland (Marshall *et al.* 1996; Marshall 2000). Outside Euramerica, it has only been reported from Givetian assemblages of Spain (Cramer 1969), Libya (de Ville de Goyet 2007; Steemans *et al.* 2011b) and China, which was originally claimed as Eifelian by Gao Lianda (1981).

Genus CONVOLUTISPORA Hoffmeister *et al.*, 1955

Type species. *Convolutispora florida* Hoffmeister *et al.*, 1955.

Convolutispora subtilis Owens, 1971
Figure 16E–G

1971 *Convolutispora subtilis* Owens, p. 35, pl. 9, figs 3–6.
1987 *Chelinospora* sp. Burjack *et al.*, pl. 2, fig. 1.
1988 *Chelinospora paravermiculata* Loboziak *et al.*,
    p. 355, pl. 3, figs 7–13.

Dimensions. 37(49)62 μm; seven specimens measured.

Comparison. *Chelinospora paravermiculata* Loboziak *et al.*, 1988 is herein considered as synonymous with *Convolutispora subtilis*.

Occurrence. MG-1; Awaynat Wanin II and Awaynat Wanin III formations; *undulatus* to *langii-concinna* zones.

Previous records. From Givetian–lower Frasnian of Paraná Basin, Brazil (Loboziak *et al.* 1988); upper Eifelian–Frasnian of Canada (Owens 1971; McGregor and Camfield 1982); Frasnian of Iran (Ghavidel-Syooki 2003); Givetian of Poland (Turnau 1996; Turnau and Racki 1999); lower Givetian of Russian Platform (Avkhimovitch *et al.* 1993); ?Eifelian–Givetian of Saudi Arabia (PB, pers. obs.); and uppermost Givetian – lower Frasnian of Scotland (Marshall *et al.* 1996).

Genus CORONASPORA Rodriguez emend. Richardson *et al.*, 2001

Type species. *Coronaspora mariae* Rodriguez, 1978*a*.

Coronaspora inornata sp. nov.
Figure 16H–M

Derivation of name. From *inornatus* (Latin), meaning without ornament; refers to the absence of distal sculptural elements.

Holotype. EFC V27 (Fig. 16I), slide 68704.

Paratype. EFC E34/4 (Fig. 16J), slide 68697; JNDL-4 core hole, sample 471.6 ft.

Type locality and horizon. JNDL-4 core hole, sample 499.1 ft; Jauf Formation at Domat Al-Jandal, Saudi Arabia.

Diagnosis. A *Coronaspora* with a broad kyrtome and a laevigate distal surface.

Description. Amb is circular to sub-triangular. Laesurae are straight, simple and extending to the inner edge of crassitude, which is invaginated at the radial apices. Equatorial crassitude is smooth to irregularly thickened, 3–6 μm wide. Proximal region is laevigate and bears a broad kyrtome on each interradial area. Kyrtome is distinct, formed by raised ridges (more or less semicircular in profile) paralleling the laesurae and increasing in width (up to 8.5 μm wide) towards the spore apex in the interradial areas. Distal surface is laevigate.

Dimensions. 32(37)45 μm; 18 specimens measured.

Comparison. Differs from other members of the genus, with the exception of *C. primordiale* (Rodriguez) Rodriguez, 1983 in having a laevigate distal surface. *C. primodiale* has thick labra.

Occurrence. BAQA-1, BAQA-2, JNDL-3, JNDL-4 and WELL-7; Jauf Formation (Sha'iba to Hammamiyat members); *papillensis-baqaensis* to *lindlarensis-sextantii* zones.

Genus CORYSTISPORITES Richardson, 1965

Type species. *Corystisporites multispinosus* Richardson, 1965.

Corystisporites collaris Tiwari and Schaarschmidt, 1975
Figures 16N, 47D–F

1975 *Corystisporites collaris* Tiwari and Schaarschmidt,
    p. 28, pl. 6, figs 2–5; text-fig. 18.

Dimensions. 72(74)77 μm; three specimens measured.

Comparison. *Corystisporites multispinosus* Richardson, 1965 has smaller regular spines, which do not posses any collars.

FIG. 16. Each figured specimen is identified by borehole, sample, slide number and England Finder Co-ordinate location. All figured specimens are at magnification ×1000 except where mentioned otherwise. A–C, *Concentricosisporites sagittarius* (Rodriguez) Rodriguez, 1983. A, BAQA-1, 395.2 ft, 03CW121, H48. B, BAQA-2, 57.2 ft, 66817, X42/4. C, BAQA-1, 346.8 ft, 66795, V47/2. D, *Contagisporites optivus* (Chibrikova) Owens, 1971, magnification ×500. S-462, 2260–2265 ft, 63281, Q28/2. E–G, *Convolutispora subtilis* Owens, 1971. E, MG-1, 2264 m, 62950, J35. F, MG-1, 2160.6 m, 62747, X26. G, MG-1, 2181.2 m, 62525, V42. H–M, *Coronaspora inornata* sp. nov. H, JNDL-4, 495.2 ft, 68702, D26. I, Holotype, JNDL-4, 499.1 ft, 68704, V27. J, Paratype, JNDL-4, 471.6 ft, 68697, E34/4. K, JNDL-4, 306.3 ft, 68665, F26/4. L, BAQA-1, 406.0 ft, 66809, F39. M, WELL-7, 13738.5 ft, 62322, S37/2. N, *Corystisporites collaris* Tiwari and Schaarschmidt, 1975. A1-69, 1109 ft, 27274, O45/1. O, *Corystisporites undulatus* Turnau, 1996, magnification ×500. A1-69, 1277 ft, 62636, E35/3.

Occurrence. S-462; Jubah Formation; *langii-concinna* Zone. A1-69; Awaynat Wanin II Formation; *undulatus* to *triangulatus-catillus* zones.

Previous records. From lower Eifelian – lower Givetian of Germany (Tiwari and Schaarschmidt 1975); and upper Eifelian–Givetian of Poland (Turnau 1996; Turnau and Racki 1999).

Corystisporites undulatus Turnau, 1996
Figures 16O, 17A, 47G–L

1989 *Hystricosporites mitratus* Allen; Loboziak and Streel, pl. 8, figs 3–4.
1996 *Corystisporites undulatus* Turnau, p. 117, pl. 1, fig. 1.

Dimensions. 75(110)156 μm; 15 specimens measured.

Remarks. The megaspore *Heliotriletes longispinosus* Fuglewicz and Prejbisz, 1981, which strongly resembles the microspore *C. undulatus*, is present in North Africa in same samples but was not studied here (de Ville de Goyet *et al.* 2007; Steemans *et al.* 2011b). These two species also co-occur in Poland.

Comparison. Loboziak and Streel (1989) misidentified specimens of *C. undulatus* from North Africa as *Hystricosporites mitratus* Allen, 1965. No typical grapnel-tipped *Hystricosporites* ornamentation was recognized on specimens in the restudied slides of Loboziak and Streel (1989). The figured specimen in Loboziak and Streel (1989, pl. 8, fig. 3) shows the characteristic morphology of *C. undulatus*.

Occurrence. WELL-8; Jubah Formation; *undulatus* Zone. A1-69; Awaynat Wanin II Formation; *undulatus* to *langii-concinna* zones. MG-1; Awaynat Wanin II and Awaynat Wanin III formations; *undulatus* to *langii-concinna* zones.

Previous record. From upper Eifelian of Poland (Turnau 1996).

Genus CRASPEDISPORA Allen, 1965

Type species. *Craspedispora craspeda* Allen, 1965.

Craspedispora ghadamesensis Loboziak and Streel, 1989
Figures 17B, 47M–O

1989 *Craspedispora ghadamesensis* Loboziak and Streel, p. 177, pl. 2, figs 1–4; pl. 9, fig. 4.

Dimensions. 71(80)95 μm; nine specimens measured.

Comparison. *Craspedispora craspeda* Allen, 1965 is smaller and has a laevigate or sparsely sculptured zona. *Samarisporites eximius* (Allen) Loboziak and Streel, 1989 has the same type of ornament but has a larger amb and the zona is as wide interradially as radially.

Occurrence. S-462; Jubah Formation; *triangulatus-catillus* Zone. A1-69; Awaynat Wanin I and Awaynat Wanin II formations; *svalbardiae-eximius* to *triangulatus-catillus* zones. MG-1; Ouan-Kasa, Awaynat Wanin I and Awaynat Wanin II formations; *annulatus-protea* to *triangulatus-catillus* zones.

Previous records. From Eifelian–Givetian of Brazil (Loboziak *et al.* 1988; Melo and Loboziak 2003; Breuer and Grahn 2011).

Craspedispora paranaensis Loboziak et al., 1988
Figures 17C, 47P–R

1988 *Craspedispora paranaensis* Loboziak *et al.*, p. 355, pl. 2, figs 5–10.

Dimensions. 70(88)120 μm; nine specimens measured.

Comparison. This species differs from *C. ghadamesensis* Loboziak and Streel, 1989 by possessing a larger and somewhat coalescent ornamentation on the zona.

Occurrence. A1-69; Awaynat Wanin I and Awaynat Wanin II formations; *svalbardiae-eximius* to *triangulatus-catillus* zones.

Previous records. From upper Eifelian–Givetian of Brazil (Loboziak *et al.* 1988, 1992b; Melo and Loboziak 2003; Breuer and Grahn 2011) and Saudi Arabia (PB, pers. obs.); and Givetian–Frasnian of Tunisia (Loboziak *et al.* 1992a).

FIG. 17. Each figured specimen is identified by borehole, sample, slide number and England Finder Co-ordinate location. All figured specimens are at magnification ×1000 except where mentioned otherwise. A, *Corystisporites undulatus* Turnau, 1996, magnification ×500. A1-69, 1277 ft, 62636, R29/2. B, *Craspedispora ghadamesensis* Loboziak and Streel, 1989. A1-69, 1596 ft, 26990, D44. C, *Craspedispora paranaensis* Loboziak *et al.*, 1988. A1-69, 1700 ft, 62632, R50. D–E, *Craspedispora* sp. *in* Paris *et al.* (1985). D, A1-69, 2108–2111 ft, 26913, L57-58. E, A1-69, 2039–2040 ft, 27279, M37/1. F–H, *Cristatisporites* (*Calyptosporites*) *reticulatus* (Tiwari and Schaarschmidt) comb. nov., magnification ×750. F, MG-1, 2264 m, 62951, L42/3. G, A1-69, 1109 ft, 27273, J35. H, MG-1, 2264 m, 62951, V31/3. I, *Cristatisporites streelii* sp. nov., magnification ×750. MG-1, 2241 m, 62964, T30/1.

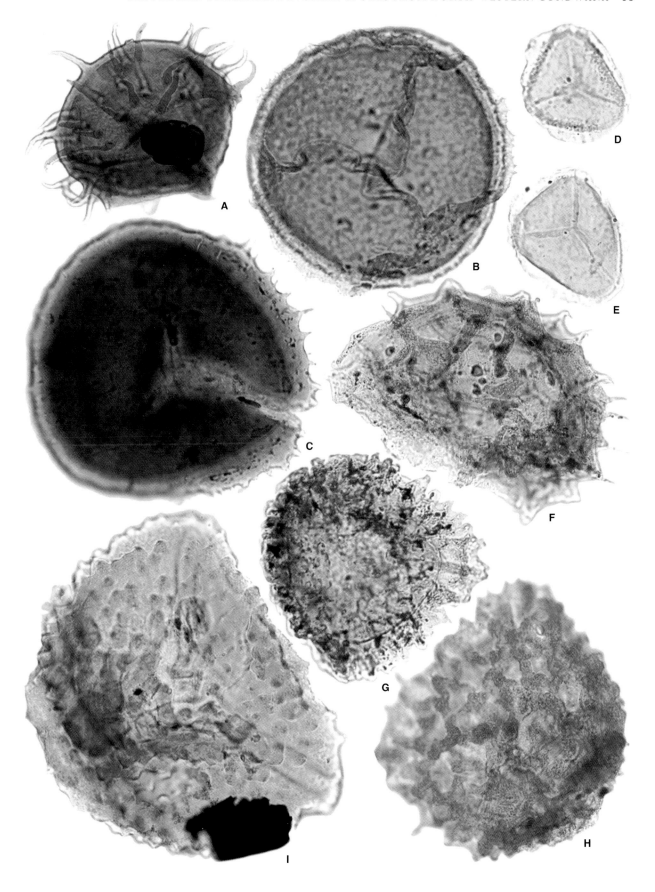

Craspedispora sp. Paris *et al.,* 1985
Figure 17D–E

? 1966a *Perotrilites gordianus* Cramer, p. 266, pl. 3, fig. 64.
1985 *Craspedispora* sp. Paris *et al.,* pl. 18, fig. 9.

Description. Amb is sub-triangular to triangular with rounded corners. Laesurae are distinct, straight to slightly sinuous, up to 1.5 µm high, extending to the inner margin of the zona. Curvaturae not visible. Central body radius equals more or less nine-tenths of the amb radius. Exine of the central body is thin equatorially. The thin proximo-equatorial flange is commonly 2–4.5 µm wide interradially. The flange is generally narrower opposite the laesurae. Thin transverse attachment lines of the flange on the central body often can be distinguished on the proximal face. Proximal and distal surfaces are entirely laevigate.

Dimensions. 32(35)39 µm; three specimens measured.

Comparison. Perotrilites gordianus Cramer, 1966a may be the same species. *Craspedispora* sp. *in* Paris *et al.* (1985) is the same species and was found from equivalent Libyan material.

Occurrence. A1-69; Ouan-Kasa and Awaynat Wanin II formations; *lindlarensis-sextantii* to *annulatus-protea,* the single specimen from the Awaynat Wanin II Formation (*annulatus-protea* Zone) may be reworked.

Previous record. From lower or middle Emsian of Libya (Paris *et al.* 1985).

Genus CRISTATISPORITES Potonié and Kremp, 1954

Type species. Cristatisporites indignabundus (Loose) Potonié and Kremp, 1954.

Comparison. Samarisporites Richardson, 1965 is considered as a junior synonym of *Cristatisporites* by Playford (1971). *Samarisporites* includes forms with a wide variety of distal sculpture (e.g. coni, cristae and verrucae), which cannot be accommodated within *Cristatisporites.* The species described from this study in *Samarisporites* show a more flimsy and better individualized zona than the species described in *Cristatisporites.*

Cristatisporites (Calyptosporites) reticulatus (Tiwari and Schaarschmidt) comb. nov.
Figures 17F–H, 47S–U

1975 *Calyptosporites reticulatus* Tiwari and Schaarschmidt, p. 45., pl. 27, figs 2–4; pl. 28, fig. 1; text-fig. 35.
? 1989 *Acinosporites acanthomammillatus* Richardson; Loboziak and Streel, pl. 1, fig. 4.

Dimensions. 57(85)113 µm; 10 specimens measured.

Remarks. The genus *Calyptosporites* Richardson, 1962, which is considered as a junior synonym of *Grandispora* Hoffmeister *et al.* emend. Neves and Owens, 1966, is not the most appropriate genus for the species described here. *Calyptosporites reticulatus* Tiwari and Schaarschmidt, 1975 is transferred to the genus *Cristatisporites* Potonié and Kremp, 1954 as it is sculptured with ridges supporting spines.

Comparison. Acinosporites acanthomammillatus Richardson, 1965 illustrated in Loboziak and Streel (1989) was probably misidentified because it does not bear contorted anastomosing ridges as described by Richardson (1965) but rather ridges forming a subreticulate pattern. The specimen figured in Loboziak and Streel (1989) may be similar to *C. reticulatus. C. streelii* sp. nov., which is morphologically very similar, larger and does not possess a discernible inner body. These two taxa seem to intergrade in the *C. reticulatus* Morphon defined here (Table 1).

Occurrence. S-462, WELL-1 and WELL-8; Jubah Formation; *rugulata-libyensis* to *triangulatus-catillus* zones, although some specimens from S-462 may be caved. A1-69; Awaynat Wanin I and Awaynat Wanin II formations; *rugulata-libyensis* to *triangulatus-catillus* zones. MG-1; Ouan-Kasa and Awaynat Wanin II formations; *annulatus-protea* to *lemurata-langii* zones.

Previous record. From Eifelian of Germany (Tiwari and Schaarschmidt 1975).

Cristatisporites streelii sp. nov.
Figures 17I, 18A–B, 47V–X, 48A–C

? 1985 *Samarisporites* sp. B Paris *et al.,* pl. 20, fig. 6.
? 1988 *Acinosporites acanthomammillatus* Richardson; Loboziak *et al.,* pl. 1, fig. 12.

Derivation of name. In honour of the Belgian palynologist, Prof. Maurice Streel, for his pioneering Devonian palynology.

Holotype. EFC R42/3 (Figs 18A, 47W), slide 62849.

Type locality and horizon. MG-1 borehole, sample 2270 m; Awaynat Wanin II Formation at Mechiguig, Tunisia.

Paratype. EFC H44/3 (Figs 18B, 47V), slide 62964; MG-1 borehole, sample 2241 m.

Diagnosis. A *Cristatisporites* sculptured with cristae closely distributed in a subconcentric, sinuous or subreticulate pattern. Laesurae extending to equatorial margin commonly obscured by thick triradiate fold-like labra. Central body not well differentiated.

Description. Amb is sub-triangular. Laesurae are straight or sinuous, extending to equatorial margin and commonly obscured by triradiate fold-like labra *c.* 4–9 μm thick in total width. A central area can sometimes be delimited by thick folds, but an inner body is not clearly present. Sexine is laevigate, infragranular or shagreenate, sculptured distally and equatorially with fold-like ridges, up to 6 μm thick and high, bearing spines or biform elements (bulbous coni supporting an accuminate apical spine), commonly 1–5 μm wide at their base and 3–7 μm high. Cristae thus formed are sometimes closely distributed and constitute a subconcentric, sinuous or subreticulate pattern. On some specimens, the ridges may be barely visible.

Dimensions. 87(104)130 μm; 12 specimens measured.

Comparison. *Acinosporites acanthomammillatus* Richardson, 1965 illustrated in Loboziak *et al.* (1988) was probably misidentified because the specimen does not bear contorted anastomosing ridges as in the diagnosis of Richardson (1965) but rather ridges forming a subconcentric, sinuous or subreticulate pattern. The specimen figured in Loboziak and Streel, 1989 may be similar to *C. streelii*. *C. reticulatus* sp. nov. is somewhat smaller and shows very often a thin inner body, and cristae are instead distributed in a subreticulate pattern. Extreme variants could intergrade with *C. reticulatus* and form a morphon (Table 1).

Occurrence. S-462, WELL-1 and WELL-8; Jubah Formation; *lemurata-langii* to *triangulatus-catillus* zones. A1-69; Awaynat Wanin II Formation; *lemurata-langii* Zone. MG-1; Awaynat Wanin I, Awaynat Wanin II and Awaynat Wanin III formations; *incognita* to *langii-concinna* zones.

Previous record. From lower–middle Givetian of Parnaíba Basin, Brazil (Breuer and Grahn 2011).

Genus CYMBOSPORITES Allen, 1965

Type species. *Cymbosporites cyathus* Allen, 1965.

Cymbosporites asymmetricus Breuer *et al.*, 2007c
Figure 18C–D

2007c *Cymbosporites asymmetricus* Breuer *et al.*, p. 49, pl. 5, figs 15–19; pl. 6, figs 1–2.

Dimensions. 43(53)69 μm; 29 specimens measured.

Remarks. It is possible that the sexine, which is sometimes slightly locally detached, may be completely removed and the resulting spores would resemble specimens of *Retusotriletes* Naumova emend. Streel, 1964.

Comparison. *Apiculiretusispora brandtii* Streel, 1964 has a similar size and ornamentation, and sometimes also has asymmetrically

placed laesurae, but differs in not being patinate. *Rhabdosporites minutus* Tiwari and Schaarschmidt, 1975 also possesses a similar ornamentation, but the sexine is totally detached from the nexine at the equator. These three species seem related and are included in the same morphon (Table 1).

Occurrence. BAQA-1, JNDL-1, JNDL-3, JNDL-4, WELL-1, WELL-3, WELL-5, WELL-6 and WELL-7; Jauf (Subbat to Murayr members) and Jubah formations; *asymmetricus* to *svalbardiae-eximius* zones. A1-69; Ouan-Kasa, Awaynat Wanin I and Awaynat Wanin II formations; *lindlarensis-sextantii* to *triangulatus-catillus* zones. MG-1; Awaynat Wanin I and Awaynat Wanin II formations; *svalbardiae-eximius* to *triangulatus-catillus* zones.

Previous record. From upper Pragian – lower Emsian of Paraná Basin, Brazil (Mendlowicz Mauller *et al.* 2007).

Cymbosporites catillus Allen, 1965
Figure 18E–F

1965 *Cymbosporites catillus* Allen, p. 727, pl. 100, figs 11–12.
non 1978b *Cymbosporites catillus* Allen; Rodriguez, p. 416, pl. 3, figs 17, 21.

Dimensions. 34(48)63 μm; 37 specimens measured.

Remarks. One monolete specimen of *C. catillus* was recorded.

Comparison. *Cymbosporites cyathus* Allen, 1965 has an ornamentation of larger coni. *C. cyathus* and *C. catillus*, which are generally found together in the same samples from the studied Saudi Arabian material, intergrade and consequently represent a morphon. The *C. catillus* Morphon is defined here (Table 1).

Occurrence. S-462 and WELL-8; Jubah Formation; *triangulatus-catillus* to *langii-concinna* zones, some specimens may be caved. A1-69; Awaynat Wanin II Formation; *catillus* to *langii-concinna* zones. MG-1; Awaynat Wanin II and Awaynat Wanin III formations; *catillus* to *langii-concinna* zones.

Previous records. From upper Givetian – lower Frasnian of Argentina (Ottone 1996); upper Eifelian – upper Givetian of Bolivia (Perez-Leyton 1990); lower Givetian – ?middle Famennian of Brazil (Loboziak *et al.* 1988; Loboziak *et al.* 1992b; Melo and Loboziak 2003); Givetian of Iran (Ghavidel-Syooki 2003) and Spisbergen, Norway (Allen 1965).

Cymbosporites cyathus Allen, 1965
Figure 18G–H

1965 *Cymbosporites cyathus* Allen, p. 725, pl. 101, figs 8–11.

Dimensions. 37(48)70 μm; 97 specimens measured.

Remarks. Some specimens show local detachments of sexine.

Comparison. Cymbosporites magnificus (McGregor) McGregor and Camfield, 1982 is larger, and may have fusion of ornament bases. *C. catillus* Allen, 1965 has a less developed ornamentation but some specimens intergrade with *C. cyathus* within the *C. catillus* Morphon (Table 1). *C. echinatus* Richardson and Lister, 1969 has a thinner patina and differs in the character of ornamentation.

Occurrence. S-462 and WELL-8; Jubah Formation; *triangulatus-catillus* to *langii-concinna* zones, some specimens may be caved. A1-69; Awaynat Wanin II Formation; *catillus* to *langii-concinna* zones. MG-1; Awaynat Wanin III Formation; *langii-concinna* Zone.

Previous records. From middle Givetian of Algeria (Moreau-Benoit *et al.* 1993); lower Frasnian–Famennian of Bolivia (Perez-Leyton 1990); lower Givetian – ?middle Famennian of Amazon and Paraná basins, Brazil (Loboziak *et al.* 1988; Melo and Loboziak 2003); Eifelian (but likely Givetian) of China (Gao Lianda 1981); lower Eifelian – lower Givetian of Germany (Tiwari and Schaarschmidt 1975); middle Givetian – upper Frasnian of Libya (Moreau-Benoit 1989); and Givetian of Spitsbergen, Norway (Allen 1965).

Cymbosporites dammamensis Steemans, 1995
Figure 18I–J

 ? 1973 *Raistrickia* sp. McGregor, p. 35–36, pl. 4, figs 9–10.
 1983 *Raistrickia* sp. A Le Hérissé, p. 24, pl. 4, figs 2, 8a–b.
 1983 *Raistrickia* sp. B Le Hérissé, p. 25, pl. 4, fig. 3.
 1983 *Raistrickia* sp. D Le Hérissé, p. 25, pl. 4, figs 6–7.
 1995 *Cymbosporites dammamensis* Steemans, p. 101 (*pars*), pl. 2, figs 10–12 (only).

Dimensions. 27(34)43 μm; 20 specimens measured.

Comparison. The different specimens described by Le Hérissé (1983) are very similar to the species described by Steemans (1995). The ornamentation of *Raistrickia* sp. *in* McGregor (1973) is similar, but there is no dimensions for the thickness of the exine. *C. echinatus* Richardson and Lister, 1969 is larger and sculptured with biform spines. *Cymbohilates baqaensis* Breuer *et al.*, 2007c is similar but hilate. Some specimens, where the proximal face has been torn, are difficult to assign. Some specimens of *Cymbosporites dammamensis* not illustrated by Steemans (1995) were probably misidentified and should be reassigned to the hilate species because they have just slits as laesurae (PS, pers. obs.). *Raistrickia jaufensis* sp. nov. is more triangular and sculptured with widely distributed and generally larger bacula. *Verrucosisporites* sp. 1 is not patinate and bears verrucae that are generally wider at the base.

Occurrence. BAQA-1, BAQA-2, JNDL-3, JNDL-4, WELL-2, WELL-3, WELL-4 and WELL-7; Jauf Formation (Sha'iba to Hammamiyat members); *papillensis-baqaensis* to *lindlarensis-sextantii* zones. MG-1; Ouan-Kasa Formation; *svalbardiae-eximius* Zone but occurrences are probably reworked.

Previous records. From upper Pragian – lower Emsian of Paraná Basin, Brazil (Mendlowicz Mauller *et al.* 2007); upper Pragian of Armorican Massif, France (Le Hérissé 1983); Lochkovian of Iran (Ghavidel-Syooki 2003); and Lochkovian–Pragian of Saudi Arabia (Steemans 1995).

Cymbosporites dittonensis Richardson and Lister, 1969
Figure 18K–N

 1969 *Cymbosporites dittonensis* Richardson and Lister, p. 241, pl. 41, figs 10–13.

Dimensions. 29(33)41 μm; 10 specimens measured.

Occurrence. BAQA-1, BAQA-2, JNDL-3, JNDL-4, WELL-2 and WELL-7; Jauf Formation (Sha'iba to Hammamiyat members); *papillensis-baqaensis* to *lindlarensis-sextantii* zones. MG-1; Ouan-Kasa Formation; *svalbardiae-eximius* Zone but occurrences are probably reworked.

Previous records. From Lochkovian of Belgium (Steemans 1989), Armorican Massif, France (Steemans 1989), Poland (Turnau *et al.* 2005) and Wales (Richardson and Lister 1969); upper Lochkovian of Solimões Basin, Brazil (Rubinstein *et al.* 2005) and Germany (Steemans 1989); upper Pragian–Emsian of China (Lu Lichang and Ouyang Shu 1976; Gao Lianda

---

FIG. 18. Each figured specimen is identified by borehole, sample, slide number and England Finder Co-ordinate location. All figured specimens are at magnification ×1000 except where mentioned otherwise. A–B, *Cristatisporites streelii* sp. nov., magnification ×750. A, Holotype, MG-1, 2270 m, 62849, R42/3. B, Paratype, MG-1, 2241 m, 62964, H44/3. C–D, *Cymbosporites asymmetricus* Breuer *et al.*, 2007c. C, JNDL-3, 294.0 ft, 03CW152, M29/1. D, JNDL-3, 341.0 ft, 03CW157, O32. E–F, *Cymbosporites catillus* Allen, 1965. E, MG-1, 2180 m, 62972, T37. F, S-462, 2010–2015 ft, 63399, N47. G–H, *Cymbosporites cyathus* Allen, 1965. G, S-462, 2010–2015 ft, 63399, E34/4. H, S-462, 2010–2015 ft, 63399, S53/1. I–J, *Cymbosporites dammamensis* Steemans, 1995. I, BAQA-1, 345.5 ft, 03CW114, W41/1. J, BAQA-1, 366.9 ft, 03CW117, H31/2. K–N, *Cymbosporites dittonensis* Richardson and Lister, 1969. K, WELL-7, 13614.1 ft, 62374, F46. L, WELL-7, 13614.1 ft, 62372, O28-29. M, WELL-2, 15893.9 ft, 63105, O49/4. N, MG-1, 2631.2 m, 62552, M47/2. O, *Cymbosporites echinatus* Richardson and Lister, 1969. BAQA-1, 366.9 ft, 03CW117, Q29. P–Q, *Cymbosporites ocularis* (Raskatova) comb. nov. P, MG-1, 2160.6 m, 62747, P43/4. Q, MG-1, 2181.2 m, 62524, V49.

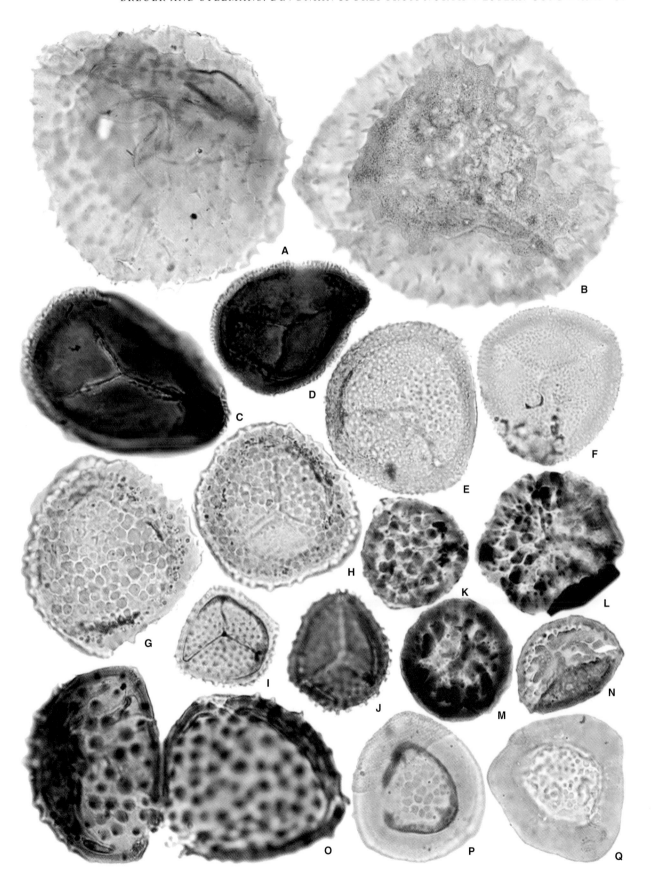

1981); middle Přídolí of Libya (Rubinstein and Steemans 2002); and upper Pragian of Morocco (Rahmani-Antari and Lachkar 2001).

Cymbosporites echinatus Richardson and Lister, 1969
Figure 18O

```
    1967  Cymbosporites Richardson, pl. 1, fig. f.
    1969  Cymbosporites echinatus Richardson and Lister,
            p. 239, pl. 42, figs 1–5.
non 1983  Cymbosporites echinatus Richardson and Lister;
            Le Hérissé, p. 50, pl. 7, figs 10.
```

Dimensions. 52–54 µm; two specimens measured.

Remarks. Cymbosporites echinatus frequently occurs in tetrads according to Richardson and Lister (1969).

Comparison. Cymbosporites cyathus Allen, 1965 has a thicker patina and spinose-tipped ornamentation is more densely packed.

Occurrence. BAQA-1; Jauf Formation (Subbat Member); *ovalis* Zone.

Previous records. Lochkovian of Belgium (Steemans 1989) and Romania (Steemans 1989); and Přídolí of Wales (Richardson and Lister 1969).

Cymbosporites ocularis (Raskatova) comb. nov.
Figure 18P–Q

```
    1993  Archaeozonotriletes ocularis Raskatova;
            Avkhimovitch et al., pl. 8, fig. 9.
```

Dimensions. 40(41)42 µm; three specimens measured.

Remarks. Archaeozonotriletes ocularis Raskatova, 1969 is herein transferred into the genus *Cymbosporites* Allen, 1965. The latter consists of ornamented patinate spores while *Archaeozonotriletes* has a laevigate or punctate patina.

Occurrence. MG-1; Awaynat Wanin III Formation; *langii-concinna* Zone.

Previous record. From middle Givetian of the Russian Platform (Avkhimovitch *et al.* 1993).

Cymbosporites rarispinosus Steemans, 1989
Figure 19A–B

```
    1981  Cymbosporites sp. G Steemans, p. 53,
            pl. 2, fig. 9.
    1984  Cymbosporites sp. 1 Steemans and Gerrienne, pl. 2,
            fig. 11.
    1989  Cymbosporites rarispinosus Steemans, p. 124, pl. 32,
            figs 26–28; pl. 33, figs 1–2.
```

Dimensions. 34(44)54 µm; 16 specimens measured.

Occurrence. BAQA-1, BAQA-2, JNDL-3, JNDL-4 and WELL-4; Jauf Formation (Sha'iba to Hammamiyat members); *papillensis-baqaensis* to *lindlarensis-sextantii* zones.

Previous records. From upper Lochkovian–Emsian of Belgium (Steemans 1989); uppermost Pragian – lowermost Emsian of Parnaíba Basin (Grahn *et al.* 2005); and upper Lochkovian of Germany (Steemans 1989).

Cymbosporites senex McGregor and Camfield, 1976
Figure 19C–D

```
    1970  New species McGregor et al., pl. 1, fig. 7.
    1976  Cymbosporites? senex McGregor and Camfield,
            p. 16, pl. 2, figs 1–4, text-fig. 14.
```

Dimensions. 47(60)72 µm; 28 specimens measured.

Remarks. Examination of the population presented here has proved that this form is clearly patinate and its attribution to the genus *Cymbosporites* is confirmed.

Comparison. Calyptosporites proteus McGregor and Camfield, 1976 is smaller, with sculpture consisting of grana and minute cones only. *Cymbohilates comptulus* Breuer *et al.,* 2007c is very similar and has the same size but is hilate.

Occurrence. BAQA-1, BAQA-2, JNDL-1, JNDL-3, JNDL-4, WELL-1, WELL-2, WELL-3, WELL-4, WELL-5, WELL-6 and

---

FIG. 19. Each figured specimen is identified by borehole, sample, slide number and England Finder Co-ordinate location. All figured specimens are at magnification ×1000 except where mentioned otherwise. A–B, *Cymbosporites rarispinosus* Steemans, 1989. A, BAQA-2, 56.0 ft, 03CW130, U29/3. B, BAQA-2, 134.4 ft, 03CW137, P28/1. C–D, *Cymbosporites senex* McGregor and Camfield, 1976. C, WELL-7, 13670.8 ft, 62382, S-T32. D, BAQA-1, 285.5 ft, 03CW111, O25. E–I, *Cymbosporites stellospinosus* var. *minor* var. nov. E, BAQA-2, 133.0 ft, 66825, K37/1. F, BAQA-1, 161.0 ft, 66775, J53. G, Paratype, BAQA-1, 161.0 ft, 66775, G40/4. H, BAQA-1, 161.0 ft, 66775, S36. I, Holotype, BAQA-1, 161.0 ft, 66773, G42. J–M, *Cymbosporites variabilis* var. *densus* sp. et var. nov. J, BAQA-1, 366.9 ft, 62256, H32. K, BAQA-1, 219.2 ft, 62237, K35. L, BAQA-1, 345.5 ft, 62253, P30. M, Paratype, BAQA-1, 366.9 ft, 62255, H34/4.

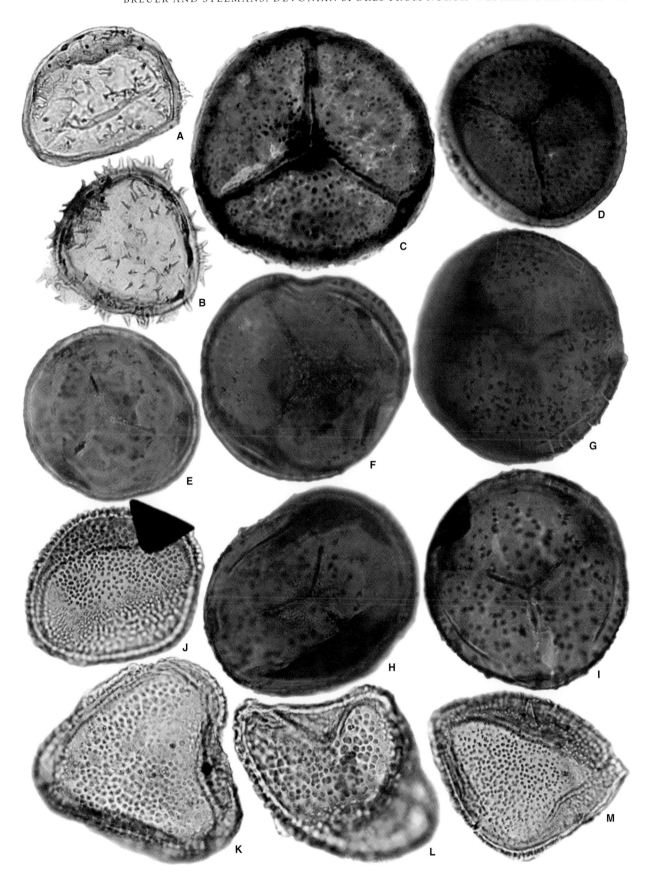

WELL-7; Jauf and Jubah formations; *papillensis-baqaensis* to *svalbardiae-eximius* zones.

Previous records. From Pragian–Emsian of Canada (McGregor and Camfield 1976); and Emsian of Saudi Arabia (Al-Ghazi 2007; Breuer *et al.* 2007*c*).

Cymbosporites stellospinosus Steemans, 1989 var. *minor* var. nov.
Figure 19E–I

Derivation of name. From *minor* (Latin), meaning smaller; refers to the size of distal sculptural elements.

Holotype. EFC G42 (Fig. 19I), slide 66773.

Paratype. EFC G40/4 (Fig. 19G), slide 66775; BAQA-1 core hole, sample 161.0 ft.

Type locality and horizon. BAQA-1 core hole, sample 161.0 ft; Jauf Formation at Baq'a, Saudi Arabia.

Diagnosis. A large *Cymbosporites stellospinosus* distally sculptured with irregularly distributed, minute spines in star-shaped clusters.

Description. Amb is sub-circular to circular. Laesurae are straight, simple, about two-thirds of the amb radius in length. Exine is proximally thin, equatorially and distally patinate, commonly 2–4 μm thick. Proximal surface is laevigate, sometimes differentially thickened. A sub-circular to sub-triangular apical zone (diameter about one-third of the amb diameter) surrounded by a curved, thicker band, 3–4 μm, characterized by a sharp inner outline and a commonly diffuse outer outline sometimes occurs. Patina is sculptured, irregularly distributed, short spines, 0.5–2 μm high and in star-shaped clusters (one to five), 0.5–5 μm apart.

Dimensions. 38(54)62 μm; 19 specimens measured.

Remarks. This form described herein is similar to *C. stellospinosus* Steemans, 1989 by having the same type of ornamentation but in smaller size. The two populations seem to constitute two different varieties of a same species: *C. stellospinosus* vars *stellospinosus* Steemans, 1989, in Western Europe, and *minor* var. nov., in Saudi Arabia.

Comparison. Cymbosporites stellospinosus var. *stellospinosus* Steemans, 1989 is smaller (31–36 μm) and the spines are longer (2–3 μm). *Cymbohilates cymosus* Richardson, 1996 has the same diagnostic clusters of short spines, but it occurs mainly in tetrads and sometimes as hilate monads. Its exine is also thinner (*c.* 1 μm thick).

Occurrence. BAQA-1, BAQA-2, JNDL-3 and JNDL-4; Jauf Formation (Sha'iba to Hammamiyat members); *papillensis-baqaensis* to *lindlarensis-sextantii* zones.

Cymbosporites variabilis sp. nov.
Figures 19J–M, 20

Derivation of name. From *variabilis* (Latin), meaning changeable, variable; refers to the distribution of distal sculptural elements.

Holotype. EFC N29/2 (Fig. 20H), slide 03CW128.

Paratype. EFC H37 (Fig. 20I), slide 62250; BAQA-1 core hole, sample 345.5 ft.

Type locality and horizon. BAQA-2 core hole, sample 52.0 ft; Jauf Formation at Baq'a, Saudi Arabia.

Diagnosis. A *Cymbosporites* distally sculptured with grana, coni or small verrucae, variably distributed.

Remarks. Clivosispora variabilis and *Dictyotriletes biornatus* Breuer *et al.*, 2007*c* constitute the *D. biornatus* Morphon (Table 1). It includes *C. variabilis* vars *variabilis*, *densus* and *dispersus* sp. et var. nov., and *D. biornatus* vars *biornatus* and. *murinatus* var. nov. The ornament and its organization on the spore distal surface vary between the two end-members which correspond to two distinct genera: *Cymbosporites* Allen, 1965 and *Dictyotriletes* Naumova, 1939 ex Ishchenko, 1952. All intermediary forms between the two end-members co-occur in the assemblages. In the simplest form of the spore, sculptural elements are evenly distributed on the distal surface (*C. variabilis* var. *densus* sp. et var. nov.). In the intermediary forms (*C. variabilis* vars *dispersus* sp. et var. nov. and *C. variabilis* var. *variabilis* sp. et var. nov.), elements organize progressively and combine until they form a pseudoreticulum, the walls of which are constituted by lines

FIG. 20. Each figured specimen is identified by borehole, sample, slide number and England Finder Co-ordinate location. All figured specimens are at magnification ×1000 except where mentioned otherwise. A–B, *Cymbosporites variabilis* var. *densus* sp. et var. nov. A, BAQA-1, 366.9 ft, 62257, F34. B, Holotype, BAQA-1, 366.9 ft, 62256, K42/2. C–G, *Cymbosporites variabilis* var. *dispersus* sp. et var. nov. C, BAQA-1, 366.9 ft, 62257, M51/2. D, BAQA-1, 345.5 ft, 62249, U45. E, BAQA-1, 366.9 ft, 62257, F-G42. F, Paratype, BAQA-1, 345.5 ft, 62248, R42/1. G, Holotype, BAQA-1, 395.2 ft, 03CW121, X43/1. H–L, *Cymbosporites variabilis* var. *variabilis* sp. et var. nov. H, Holotype, BAQA-2, 52.0 ft, 03CW128, N29/2. I, Paratype, BAQA-1, 345.5 ft, 62250, H37. J, BAQA-1, 345.5 ft, 62250, X38. K, BAQA-1, 308.3 ft, 03CW112, B36. L, BAQA-1, 345.5 ft, 62251, D45/2.

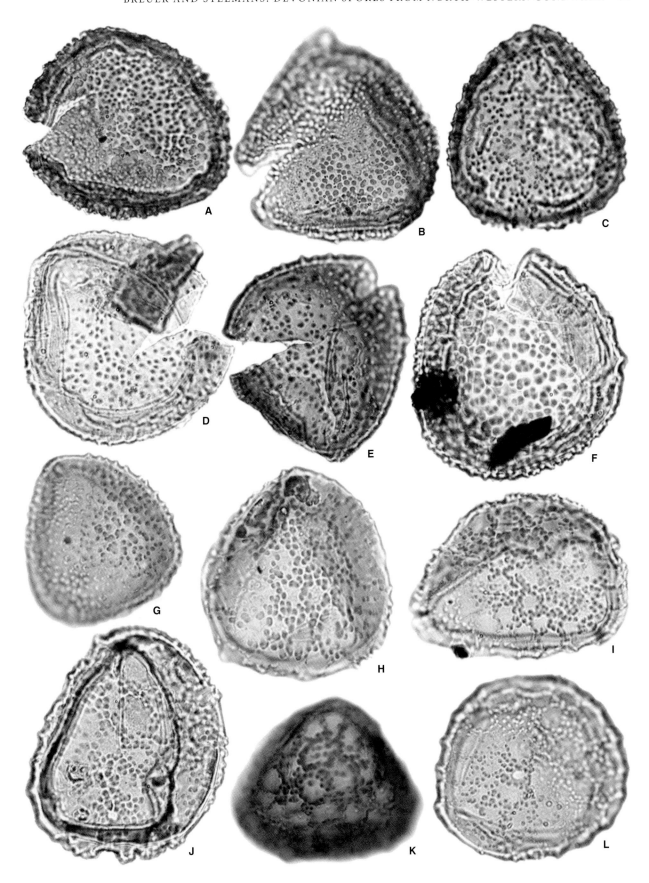

of discrete ornaments (*D. biornatus* Breuer *et al.*, 2007*c* var. *biornatus*). In the most complex spore form, ornaments merge to form elongated muri, which constitutes a perfectly closed reticulum (*D. biornatus* Breuer *et al.*, 2007*c* var. *murinatus* var. nov.). Thus, a progressive organization of the ornamentation appears from the simplest spores to the most complex ones.

Comparison. *Cymbosporites catillus* Allen, 1965 has a generally thicker patina and distinct laesurae straight accompanied by labra.

Cymbosporites variabilis var. *variabilis* sp. et var. nov.
Figure 20H–L

2007*a* *Cymbosporites*? sp. 3 Breuer *et al.*,
      text-fig. 13–C.
2007*b* Unnamed spore Breuer *et al.*, text-fig. 1–3.

Diagnosis. A *Cymbosporites variabilis* distally sculptured with grana, coni or small verrucae distributed into an irregular to almost imperfect reticulate pattern.

Description. Amb is sub-circular to roundly triangular. Laesurae are straight, simple, often indistinct, commonly three-fifths to four-fifths of the amb radius in length. Exine is proximally thin, equatorially and distally patinate, commonly 2–5 μm thick, homogeneous. Proximal surface is laevigate, often torn or collapsed. Patina is sculptured with grana, coni or small verrucae, 0.5–2 μm wide at base and high, 0.5–7 μm apart. Elements are distributed into an irregular to almost imperfect reticulate pattern. Elements are often merged at the base forming elongated elements or patches of several ornaments.

Dimensions. 47(57)68 μm; 34 specimens measured.

Comparison. *C. variabilis* var. *dispersus* sp. et var. nov. has more regularly distributed ornament. *Dictyotriletes biornatus* Breuer *et al.*, 2007*c* var. *biornatus* has all the ornament disposed in a perfect reticulate pattern. However, perfect lumina can occur locally in *C. variabilis* var. *variabilis*; others are incomplete or have isolated elements within the reticulum.

Occurrence. BAQA-1, BAQA-2 and JNDL-4; Jauf Formation (Sha'iba to Subbat members); *ovalis-biornatus* to *lindlarensis-sextantii* zones.

Cymbosporites variabilis var. *densus* sp. et var. nov.
Figures 19J–M, 20A–B

2007*a* *Cymbosporites* sp. 1 Breuer *et al.*,
      text-fig. 1: 3A.
2007*b* Unnamed spore Breuer *et al.*, text-fig. 1: 1.

Derivation of name. From *densus* (Latin), meaning dense; refers to the distribution of distal sculptural elements.

Holotype. EFC K42/2 (Fig. 20B), slide 62256.

Paratype. EFC H34/4 (Fig. 19M), slide 62255; BAQA-2 core hole, sample 366.9 ft.

Type locality and horizon. BAQA-1 core hole, sample 366.9 ft; Jauf Formation at Baq'a, Saudi Arabia.

Diagnosis. A *Cymbosporites variabilis* distally sculptured with grana, coni or small verrucae evenly distributed and sometimes locally merged at their bases.

Description. Amb is sub-circular to roundly triangular. Laesurae are straight, simple, often indistinct, commonly two-thirds to four-fifths of the amb radius in length. Exine is proximally thin, equatorially and distally patinate, commonly 2–5 μm thick, homogeneous. Proximal surface is laevigate, often torn or collapsed. Patina is sculptured with densely distributed grana, coni or small verrucae, 0.5–2 μm wide at base and high, 0.5–2 μm apart. Elements are evenly distributed and sometimes locally merged at the base.

Dimensions. 47(58)73 μm; 54 specimens measured.

Remarks. *C. variabilis* var. *densus* constitutes an end-member of the *Dictyotriletes biornatus* Morphon (Table 1).

Comparison. *C. variabilis* var. *dispersus* sp. et var. nov. has more irregularly distributed ornaments. As the two species intergrade (see discussion above), it is sometimes difficult to discriminate between them.

Occurrence. BAQA-1, BAQA-2 and JNDL-4; Jauf Formation (Sha'iba to Subbat members); *ovalis-biornatus* to *lindlarensis-sextantii* zones.

Cymbosporites variabilis var. *dispersus* sp. et var. nov.
Figure 20C–G

2007*a* *Cymbosporites* sp. 2 Breuer *et al.*,
      text-fig. 1: 3B.
2007*b* Unnamed spore Breuer *et al.*, text-fig. 1: 2.

Derivation of name. From *disperses* (Latin), meaning dispersed; refers to the distribution of distal sculptural elements.

Holotype. EFC X43/1 (Fig. 20G), slide 03CW121.

Paratype. EFC R42/1 (Fig. 20F), slide 62248; BAQA-1 core hole, sample 345.5 ft.

Type locality and horizon. BAQA-1 core hole, sample 395.2 ft; Jauf Formation at Baq'a, Saudi Arabia.

Diagnosis. A *Cymbosporites variabilis* distally sculptured with grana, coni or small verrucae irregularly distributed and often merged at the base forming patches of several elements.

Description. Amb is sub-circular to roundly triangular. Laesurae are straight, simple, often indistinct, commonly three-fifths to four-fifths of the amb radius in length. Exine is proximally thin, equatorially and distally patinate, commonly 2–5 μm thick, homogeneous. Proximal surface is laevigate, often torn or collapsed. Patina is sculptured with grana, coni or small verrucae, 0.5–2 μm wide at base and high, 0.5–3 μm apart. Elements are irregularly packed and often merged at the base forming patches of several elements.

Dimensions. 45(56)70 μm; 46 specimens measured.

Comparison. *C. variabilis* var. *densus* sp. et var. nov. has more regularly distributed ornament. *C. variabilis* var. *variabilis* sp. et var. nov. shows clearly a more advanced organization of elements.

Occurrence. BAQA-1, BAQA-2, JNDL-4 and WELL-4; Jauf Formation (Sha'iba to Subbat members); *ovalis-biornatus* to *lindlarensis-sextantii* zones.

Cymbosporites variegatus sp. nov.
Figure 21A–F

1972   cf. *Cymbosporites cyathus* Allen; Mortimer and Chaloner, p. 11, pl. 1, fig. 4.

1989   *Verrucosisporites bulliferus* Richardson and McGregor; Loboziak and Streel, pl. 1, fig. 6.

? 1992   *Cymbosporites* sp. cf. *C. magnificus* (McGregor) McGregor and Camfield; McGregor and Playford, pl. 5, figs 5–6.

? 1992a   *Geminospora piliformis* Loboziak *et al.*; Loboziak *et al.*, pl. 3, fig. 11.

2011   *Cymbosporites* sp. 1 Breuer and Grahn, pl. 2, fig. k.

Derivation of name. From *variegatus* (Latin), meaning variegated; refers to the size and type of distal sculptural elements.

Holotype. EFC F28/3 (Fig. 21E), slide 62940.

Paratype. EFC L51 (Fig. 21D), slide 62782; MG-1 borehole, sample 2315 m.

Type locality and horizon. MG-1 borehole, sample 2247 m; Awaynat Wanin II Formation at Mechiguig, Tunisia.

Diagnosis. A *Cymbosporites* sculptured with variable low verrucae or baculae, flat-topped or slightly rounded in profile, sub-circular, polygonal or irregular in plan view. Proximal surface infragranular to granular.

Description. Amb is sub-circular to sub-triangular. Laesurae are straight, simple and extending to, or almost to, the inner margin of patina. Exine is proximally thin, equatorially and distally patinate, commonly 2–5 μm thick, homogeneous to infragranulate. Proximal surface is infragranular to granular. Patina is sculptured with variable low verrucae or baculae, flat-topped or slightly rounded in profile, sub-circular, polygonal or irregular in plan view, 0.5–4 μm wide, up to 1.5 μm high, 0.5–3 μm apart. Elements are sometimes irregularly packed with some small areas without any elements.

Dimensions. 38(48)60 μm; 26 specimens measured.

Remarks. Some specimens appear to have a partly separate inner body similar to that of the genus *Geminospora* Balme, 1962. This type of separation is typical of a thick patinate structure of *Cymbosporites* Allen, 1965.

Comparison. The sculptural elements of cf. *C. cyathus* Allen, 1965 *in* Mortimer and Chaloner (1972) are flat-topped as those of the species described herein. The specimens figured as *Cymbosporites* sp. cf. *C. magnificus* (McGregor) McGregor and Camfield, 1982 *in* McGregor and Playford (1992) could be similar, but no description of the observed specimens is given. The specimen figured as *Geminospora piliformis* Loboziak *et al.*, 1988 *in* Loboziak *et al.* (1992a) could be misinterpreted since it seems to bear low verrucae and to be single-layered. Consequently, it is similar to the species described herein. *G. piliformis* is two-layered and bears pila. *C. magnificus* (McGregor) McGregor and Camfield, 1982 is larger and sculptured with verrucae, mammae and rounded coni, discrete or joined laterally into short irregularly-trending ridges while *C. cyathus* Allen, 1965 bears mainly coni. *Verrucosisporites bulliferus* Richardson and McGregor, 1986 has generally larger verrucae (2.5–5 μm), is proximally laevigate and not patinate. *V. bulliferus* illustrated by Loboziak and Streel (1989) has been re-examined and is exactly the same as *C. variegatus*. As no *V. bulliferus* Richardson and McGregor, 1986 *sensu stricto* has been found in the studied material, the specimens described here cannot be confused with the index species of Richardson and McGregor (1986).

Occurrence. A1-69; Awaynat Wanin II Formation; *lemurata-langii* to *triangulatus-catillus* zones. MG-1; Awaynat Wanin II and Awaynat Wanin III formations; *lemurata-langii* to *langii-concinna* zones.

Previous records. From middle Givetian of Parnaíba Basin, Brazil (Breuer and Grahn 2011); Frasnian of Libya (Loboziak and Streel 1989); Givetian–Frasnian of Saudi Arabia (PB, pers. obs.); and Givetian of England (Mortimer and Chaloner 1972).

Cymbosporites wellmanii sp. nov.
Figure 21G–K

? 2000*b* *Aneurospora* sp. A Wellman *et al.*, p. 171,
pl. 4, figs 8–9.

Derivation of name. In honour of the British palynologist, Dr. Charles H. Wellman, for his outstanding contribution to the understanding of the Silurian–Devonian spore assemblages, notably from Saudi Arabia.

Holotype. EFC K31 (Fig. 21J), slide 66825.

Paratype. Figure 21H, BAQA-2 core hole, sample 133.0 ft, slide 03CW136, EFC K31.

Type locality and horizon. BAQA-2 core hole, sample 133.0 ft; Jauf Formation at Baq'a, Saudi Arabia.

Diagnosis. A *Cymbosporites* sculptured with densely distributed, slender pointed spines.

Description. Amb is sub-circular to sub-triangular. Laesurae are straight, simple or labrate, usually 1–2 μm wide, three-fifths to four-fifths of the amb radius in length. Exine is proximally thin, distally and equatorially patinate, 2.5–6 μm thick equatorially. Proximal surface is laevigate to scabrate. Equatorial and distal regions sculptured with slender pointed spines, densely distributed, 2.5–5 μm high, 0.75–2 μm wide at base and 1–2 μm apart.

Dimensions. 40(48)61 μm; nine specimens measured.

Comparison. Although *Aneurospora* sp. A *in* Wellman *et al.* (2000*b*) is smaller (33–41 μm), slightly thinner equatorially and distally, it could be similar to *C. wellmanii* (C. H. Wellman, pers. comm. 2009). *Dibolisporites eifeliensis* (Lanninger) McGregor, 1973 has a thinner exine and more widely distributed sculptural elements. Moreover, their basal part is bulbous or slightly to strongly tapering, and their apical part is not accuminate.

Occurrence. BAQA-2; Jauf Formation (Sha'iba Member); *papillensis-baqaensis* Zone.

Genus CYRTOSPORA Winslow, 1962

Type species. *Cyrtospora cristifera* (Luber) Van der Zwan, 1979.

Cyrtospora tumida sp. nov.
Figures 21L–N, 22A–C

Derivation of name. From *tumidus* (Latin), meaning swollen up; refers to the general aspect of the spore body.

Holotype. EFC T36/2 (Fig. 21N), slide 62736.

Paratype. EFC E32 (Fig. 21L), slide 62936; MG-1 borehole, sample 2278 m.

Type locality and horizon. MG-1 borehole, sample 2421 m; Awaynat Wanin I Formation at Mechiguig, Tunisia.

Diagnosis. A *Cyrtospora* with a patina of irregular shape in plan view and dissected into irregularly distributed, large, broad based, rounded verrucae, coni, tubercules or protuberances.

Description. Amb is sub-circular to sub-triangular. Laesurae are simple, straight, two-thirds to full central area radius in length. Patina, 2–21 μm thick, is homogeneous sometimes with scattered infrapuncta. Contact areas thinner than distal exine and laevigate. Patina is of irregular shape in plan view and dissected into irregularly distributed, large, broad based, rounded verrucae, coni, tubercules or protuberances, 1.5–28 μm wide at base, 1.5–14 μm high.

Dimensions. 39(59)72 μm; 11 specimens measured.

Remarks. Spores in this population do not belong to the genus *Archaeozonotriletes* Naumova emend. Allen, 1965 because their patina is far from being uniform.

Comparison. *Archaeozonotriletes variabilis* Naumova emend. Allen, 1965 also has a laevigate or finely punctate and irregular patina, which is not dissected into ornamentation or protuberances. Some specimens of *A. variabilis* (Fig. 10I–J) may locally show slightly convex and concave zones on the patina. *A. variabilis* is considered to intergrade with *C. tumida* in the same morphon (Table 1). *Lophozonotriletes media* Taugourdeau-Lantz, 1967 is densely punctate and bears equatorially blunt, pointed or rounded verrucae. The latter is rather cingulate, but extreme variants could intergrade with *C. tumida*. *C. cristifera* (Luber) Van der Zwan, 1979 has a distal ornamentation mainly consisting of bacula of variable shape, size, and, to a minor degree, coni and verrucae (1–8 μm high, 1–7 μm wide).

FIG. 21. Each figured specimen is identified by borehole, sample, slide number and England Finder Co-ordinate location. All figured specimens are at magnification ×1000 except where mentioned otherwise. A–F, *Cymbosporites variegatus* sp. nov. A, MG-1, 2160.6 m, 62727, R40. B, MG-1, 2247 m, 62942, R40. C, MG-1, 2182.4 m, 62527, U-V40. D, Paratype, MG-1, 2315 m, 62782, L51. E, Holotype, MG-1, 2247 m, 62940, F28/3. F, MG-1, 2160.6 m, 62746, R28. G–K, *Cymbosporites wellmanii* sp. nov. G, BAQA-2, 134.4 ft, 66826, M29. H, Paratype, BAQA-2, 133.0 ft, 03CW136, O-P23. I, BAQA-2, 133.0 ft, 03CW136, G43/3. J, Holotype, BAQA-2, 133.0 ft, 66825, K31. K, BAQA-2, 133.0 ft, 66825, X33. L–N, *Cyrtospora tumida* sp. nov. L, Paratype, MG-1, 2278 m, 62936, E32. M, MG-1, 2456 m, 62737, H48. N, Holotype, MG-1, 2421 m, 62736, T36/2.

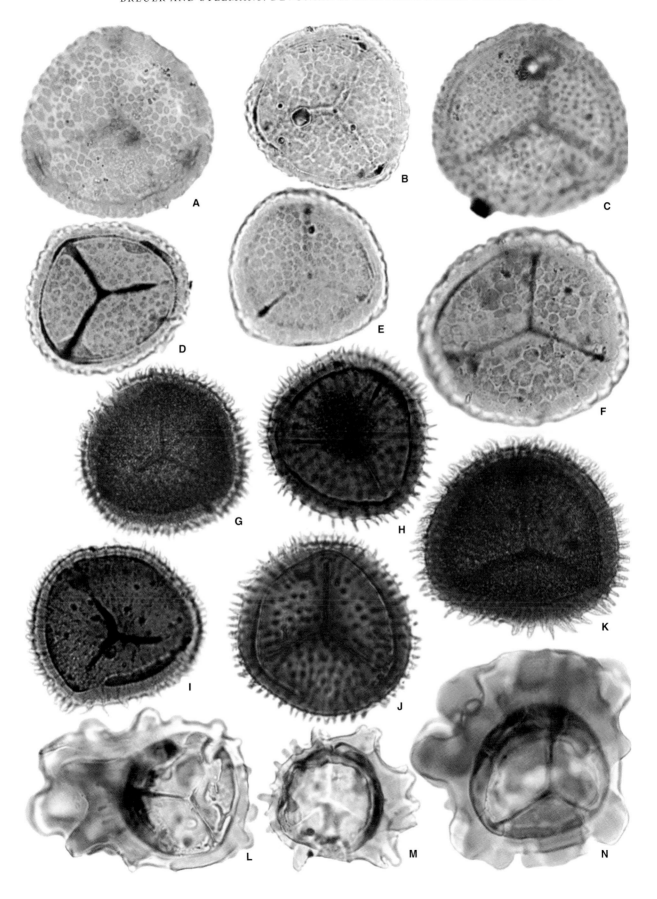

Occurrence. A1-69; Awaynat Wanin II Formation; *lemurata-langii* to *triangulatus-catillus* zones. MG-1; Awaynat Wanin I, Awaynat Wanin II and Awaynat Wanin III formations; *rugulata-libyensis* to *langii-concinna* zones.

Previous record. Cyrtospora tumida has also been recorded from Givetian of Saudi Arabia (PB, pers. obs.).

Genus DENSOSPORITES (Berry) Butterworth *et al.* in Staplin and Jansonius (1964)

Type species. Densosporites covensis Berry, 1937.

Densosporites devonicus Richardson, 1960
Figures 22D, 48D–I

1960 *Densosporites devonicus* Richardson, p. 57, pl. 14, figs 10–11; text-fig. 7.
1965 *Densosporites orcadensis* Richardson, p. 580, pl. 92, figs 1–2.
1976 *Hymenozonotriletes propolyacanthus* Arkhangelskaya, p. 55, pl. 10, figs 3–4.
1976 *Densosporites orcadensis* Richardson; McGregor and Camfield, p. 17, pl. 6, fig. 3.

Dimensions. 77(96)126 μm; 23 specimens measured.

Remarks. In lateral compression, proximal face flattened-pyramidal, distal region strongly rounded (McGregor and Camfield 1982).

Comparison. Richardson (1965) gave the sculptural details and the relative widths of the light and dark zones of the cingulum as criteria for distinguishing *D. devonicus* from *D. orcadensis* Richardson, 1960. These criteria were used by Marshall and Allen (1982) in an attempt to substantiate the differences between the two species, but no systematic variation of these characters, as alleged by Richardson (1965). Thus, McGregor and Camfield (1982) and Marshall and Allen (1982) consider that *D. devonicus* and *D. orcadensis* intergrade and it is impractical to separate them. *D. weatherallensis* McGregor and Camfield, 1982 differs from *D. devonicus* in having broader spines that rarely expand at the tip, in the close spacing or basal fusion of its sculpture towards the distal pole and its less conspicuous dark zone. *D. inaequus* (McGregor) McGregor and Camfield,

1982 has more prominent spines, which do not bifurcate but have papillate tips. *D. concinnus* (Owens) McGregor and Camfield, 1982 is smaller in size and less elongate, with a predominantly pointed sculpture but intergrades with extreme variants of *D. devonicus* (McGregor and Camfield, 1982).

Occurrence. MG-1; Ouan-Kasa, Awaynat Wanin I and Awaynat Wanin II formations; *annulatus-protea* to *langii-concinna* zones.

Previous records. Densosporites devonicus is eponymous for the late Eifelian – early Givetian *devonicus-naumovae* Assemblage Zone of the Old Red Sandstone Continent and adjacent regions (Richardson and McGregor 1986) and AD Oppel Zone of Western Europe (Streel *et al.* 1987). *D. devonicus* occurs from Eifelian into lower Frasnian and has been reported from Canada (McGregor and Uyeno 1972; McGregor and Camfield 1976; McGregor and Camfield 1982), Germany (Riegel 1973; Tiwari and Schaarschmidt 1975; Streel and Paproth 1982; Loboziak *et al.* 1990), Libya (Streel *et al.* 1988), Siptsbergen, Norway (Allen 1965), Poland (Turnau 1996; Turnau and Racki 1999), Russian Platform (Avkhimovitch *et al.* 1993), Saudi Arabia (PB, pers. obs.) and Scotland (Richardson 1965; Marshall and Allen 1982; Marshall *et al.* 1996; Marshall 2000; Marshall and Fletcher 2002).

Genus DIAPHANOSPORA Balme and Hassell, 1962

Type species. Diaphanospora riciniata Balme and Hassell, 1962.

Diaphanospora milleri sp. nov.
Figure 22E–K

2007*a* sp. 1 Breuer *et al.*, text-figs 1–4A–B.

Derivation of name. In honour of the American palynologist employed by Saudi Aramco, Merrell A. Miller, for his outstanding work on the acritarch and spore palynology from the Lower Palaeozoic.

Holotype. EFC N49/1 (Fig. 22J), slide 62317.

Paratype. EFC G27/2 (Fig. 22K), slide 03CW108; BAQA-1 core hole, sample 222.5 ft.

Type locality and horizon. WELL-7 well, sample 13689.7 ft; Jauf Formation at Uthmaniyah, Saudi Arabia.

FIG. 22. Each figured specimen is identified by borehole, sample, slide number and England Finder Co-ordinate location. All figured specimens are at magnification ×1000 except where mentioned otherwise. A–C, *Cyrtospora tumida* sp. nov. A, MG-1, 2264 m, 62951, E46/4. B, MG-1, 2536 m, 62742, T32. C, MG-1, 2435 m, 63018, R40/3. D, *Densosporites devonicus* Richardson, 1960, magnification ×750. MG-1, 2247 m, 62942, H39/3. E–K, *Diaphanospora milleri* sp. nov. E, WELL-7, 13689.7 ft, 62317, Y42/1. F, WELL-7, 13689.7 ft, 62317, E33/4. G, WELL-7, 13689.7 ft, 62319, F-G29. H, BAQA-1, 346.8 ft, 66797, H41/3. I, WELL-2, 15919.7 ft, 63108, O28. J, Holotype, WELL-7, 13689.7 ft, 62317, N49/1. K, Paratype, BAQA-1, 222.5 ft, 03CW108, G27/2. L, *Diatomozonotriletes franklinii* McGregor and Camfield, 1982. A1-69, 1830 ft, 26961, U30/4. M–N, *Dibolisporites bullatus* (Allen) Riegel, 1973. M, BAQA-1, 223.5 ft, 03CW109, K29/4. N, WELL-7, 13738.5 ft, 62323, W32/1.

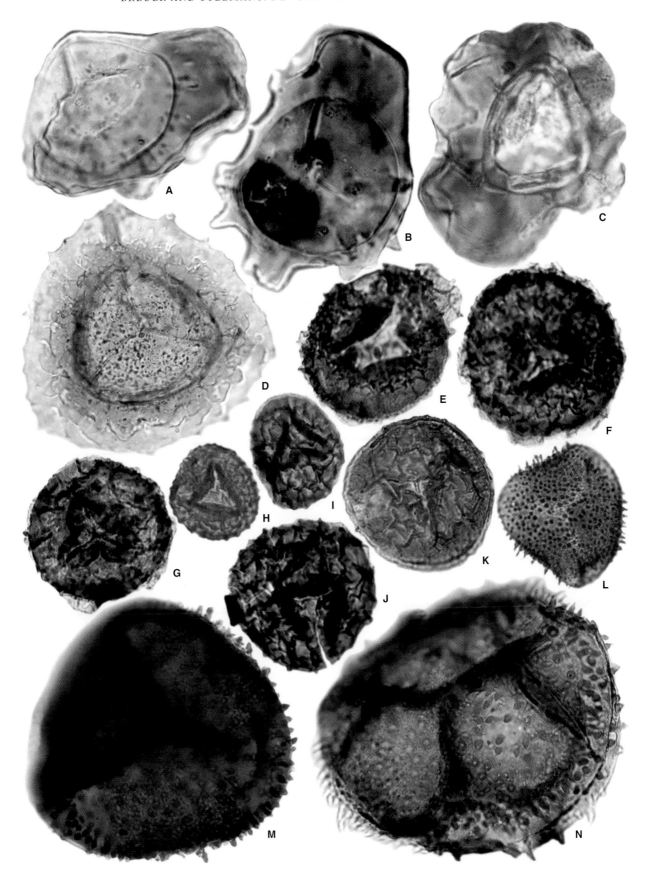

Diagnosis. A small *Diaphanospora* with a darker apical sub-triangular band on the spore body.

Description. Amb is sub-circular. Laesurae are straight, simple or rarely accompanied by narrow labra, *c.* 1 µm in overall width, three-fifths to three-quarters of the amb radius in length, connected by curvaturae perfectae not always well visible and often obscured by the folds of sexine. Nexine is laevigate, commonly 1 –2 µm thick. A darker apical sub-triangular band, with straight, slightly concave or convex sides, extends to, or almost to, the end of the laesurae. This thickened area is up 3–6 µm wide interradially. An inner lighter sub-triangular area (with a thinner nexine), generally with slightly concave sides is present proximally and surrounded by the darkened band. Sexine extremely thin, transparent, closely appressed to the spore body and attached proximally close to the curvaturae. Surface of sexine randomly and finely folded.

Dimensions. 26(41)51 µm; 17 specimens measured.

Remarks. Specimens in which the outer layer is locally detached are not uncommon (Fig. 22E, G). Specimens in which the outer layer is missing are common in the less well-preserved palynological assemblages and consequently are similar to *Retusotriletes celatus* sp. nov. described below. The very delicate sexine may have been torn off by sedimentary or taphonomic processes. The two form-species *D. milleri* and *R. celatus* thus represent a unique biological species with the different states of preservation between both. They are grouped into the *D. milleri* Morphon (Table 1). They sometimes co-occur and have similar stratigraphical ranges.

Comparison. *Diaphanospora riciniata* and *D. perplexa* Balme and Hassell, 1962 are labrate and do not have a thickened apical area on the spore body.

Occurrence. BAQA-1, JNDL-3, JNDL-4, WELL-2, WELL-4 and WELL-7; Jauf Formation (Subbat and Hammamiyat members); *milleri* to *lindlarensis-sextantii* zones. A1-69; Ouan-Kasa Formation; *lindlarensis-sextantii* Zone.

Genus DIATOMOZONOTRILETES Naumova emend. Playford, 1962

Type species. *Diatomozonotriletes saetosus* (Hacquebard and Barss) Hughes and Playford, 1961.

Diatomozonotriletes franklinii McGregor and Camfield, 1982
Figure 22L

1966 *Diatomozonotriletes devonicus* Naumova; Mikhailova, pl. 1, fig. 2.

1972 *Diatomozonotriletes devonicus* Naumova; McGregor and Uyeno, pl. 2, fig. 6.
1976 *Anapiculatisporites petilus* Richardson; Massa and Moreau-Benoit, pl. 3, fig. 4.
1979 *Anapiculatisporites petilus* Richardson; Moreau-Benoit, p. 32.
1985 *Anapiculatisporites petilus* Richardson; Massa and Moreau-Benoit, pl. 1, fig. 2.
1982 *Diatomozonotriletes franklinii* McGregor and Camfield, p. 36, pl. 7, figs 10–13.

Dimensions. 35(42)63 µm; 14 specimens measured.

Comparison. *Diatomozonotriletes oligodontus* Chibrikova, 1962 has predominantly wider, less closely spaced sculptural elements. *Diatomozonotriletes* sp. in Allen (1965) closely resembles *D. franklinii* McGregor and Camfield, 1982, except for a greater tendency towards concave-triangular amb. *Camarozonotriletes sextantii* McGregor and Camfield, 1976 has similar sculpture but has a prominent interradial cingulum.

Occurrence. BAQA-1, JNDL-3 and JNDL-4; Jauf Formation (Subbat and Hammamiyat members); *asymmetricus* to *lindlarensis-sextantii* zones. A1-69; Ouan-Kasa, Awaynat Wanin I and Awaynat Wanin II formations; *annulatus-protea* to *triangulatus-catillus* zones. MG-1; Ouan-Kasa, Awaynat Wanin I, Awaynat Wanin II and Awaynat Wanin III formations; *annulatus-protea* to *langii-concinna* zones.

Previous records. From upper Emsian – lower Eifelian of Algeria (Moreau-Benoit *et al.* 1993); upper Emsian–Givetian of Amazon and Paraná basins, Brazil (Loboziak *et al.* 1988; Melo and Loboziak 2003); upper Eifelian – lower Givetian of Canada (McGregor and Uyeno 1972); Emsian–lower Givetian of Libya (Moreau-Benoit 1989); Emsian–Eifelian of Morocco (Rahmani-Antari and Lachkar 2001); and upper Eifelian of Russian Platform (Avkhimovitch *et al.* 1993).

Genus DIBOLISPORITES Richardson, 1965

Type species. *Dibolisporites echinaceus* (Eisenack) Richardson, 1965.

Comparison. *Biharisporites* Potonié, 1956 contains megaspores with similar sculpture. *Apiculiretusispora* Streel, 1964 shows a variable sculpture of small grani, coni or spinae, which measure less than 1 µm high.

Dibolisporites bullatus (Allen) Riegel, 1973
Figure 22M–N

1965 *Bullatisporites bullatus* Allen, p. 703, pl. 96, figs 5–7.
1973 *Dibolisporites bullatus* Allen; Riegel, p. 84, pl. 10, figs 10–12; pl. 11, figs 1–2.

Dimensions. 54(75)84 µm; 14 specimens measured.

Remarks. Although McGregor and Camfield (1982) consider *D. bullatus* as synonymous with *D. echinaceus* (Eisenack) Richardson, 1965, these two species are easily differentiated here and the synonymy is rejected.

Comparison. McGregor (1973) observed an intergradation from spores represented by the holotype of *D. echinaceus* to those of the holotype of *D. bullatus*, which is figured in Allen (1965). The two species are easily distinguishable, in this study, because the specimens of *D. echinaceus* are more densely ornamented with elongate, more or less parallel-sided spinae.

Occurrence. BAQA-1, JNDL-4 and WELL-7; Jauf Formation (Subbat and Hammamiyat members); *asymmetricus* to *lindlarensis-sextantii* zones.

Previous records. From upper Givetian – lower Frasnian of France (Brice *et al.* 1979; Loboziak and Streel 1980); upper Emsian–Givetian of Germany (Riegel 1973; Loboziak *et al.* 1990); and Pragian–Eifelian of Spitsbergen, Norway (Allen 1965).

Dibolisporites echinaceus (Eisenack) Richardson, 1965
Figure 23A–B

```
       1944  Triletes echinaceus Eisenack, p. 113, pl. 2, fig. 5.
     ? 1953  Retusotriletes devonicus Naumova, pl. 22, fig. 108.
     ? 1962  Retusotriletes devonicus Naumova var. echinatus
              Chibrikova, p. 393, pl. 1, fig. 9.
       1965  Dibolisporites echinaceus Eisenack; Richardson,
              p. 568, P. 89, figs 5–6; text-figs 3B–D.
   non 1973  Dibolisporites bullatus Allen; Riegel, p. 84, pl. 10,
              figs 10–12; pl. 11, figs 1–2.
       1975  Dibolisporites triangulatus Tiwari and
              Schaarschmidt, p. 21, pl. 7, figs 3–4; pl. 8, figs 1–2;
              text-fig. 9.
```

Dimensions. 60(87)129 µm; 11 specimens measured.

Comparison. Dibolisporites triangulatus Tiwari and Schaarschmidt, 1975 is identical to *D. echinaceus* Richardson, 1965. *Retusotriletes devonicus* figured but not described by Naumova (1953) could be synonymous with *D. echinaceus*, but the details of the ornament cannot be seen. *R. devonicus* var. *echinatus* Chibrikova, 1962 could be also synonymous. *Dibolisporites* cf. *gibberosus* var. *major* (Kedo) Richardson, 1965 has shorter sculptural elements. *D. pseudoreticulatus* Tiwari and Schaarschmidt, 1975 has relatively high and wide curvatural ridges. Although *D. radiatus* Tiwari and Schaarschmidt, 1975 is considered as synonymous with *D. echinaceus* according to McGregor (1973), they are here distinguished because *D. radiatus* has larger sculptural elements disposed in a regular radial pattern. *D. varius* Tiwari and Schaarschmidt, 1975 has baculate ornamentation. Although *D. bullatus* (Allen) Riegel, 1973 is often

placed in synonymy with *D. echinaceus*, it has more bulbous, wider spinae. All these species listed display only minor differences with *D. echinaceus*, so their assignation may be sometimes difficult.

Occurrence. BAQA-1, JNDL-1, JNDL-3 and JNDL-4; Jauf (Subbat to Murayr members) and Jubah formations; *ovalis* to *svalbardiae-eximius* zones. A1-69; Awaynat Wanin I and Awaynat Wanin II formations; *svalbardiae-eximius* to *rugulata-libyensis* zones.

Previous records. Widely reported and often common in Early–Middle Devonian (particularly Middle Devonian) assemblages from many parts of the world; e.g. Algeria (Boumendjel *et al.* 1988; Moreau-Benoit *et al.* 1993), Belgium (Steemans 1989), Bolivia (Perez-Leyton 1990), Brazil (Mendlowicz Mauller *et al.* 2007), Canada (McGregor and Owens 1966; McGregor and Uyeno 1972; McGregor 1973; McGregor and Camfield 1976, 1982), China (Gao Lianda 1981), France (Brice *et al.* 1979; Loboziak and Streel 1980; Le Hérissé 1983), Germany (Lanninger 1968; Riegel 1968, 1973; Tiwari and Schaarschmidt 1975; Steemans 1989; Loboziak *et al.* 1990), Iran (Ghavidel-Syooki 2003), Libya (Paris *et al.* 1985; Streel *et al.* 1988; Moreau-Benoit 1989), Poland (Turnau 1986, 1996; Turnau and Racki 1999; Turnau *et al.* 2005), Saudi Arabia (Al-Ghazi 2007) and Scotland (Richardson 1965; Marshall 1988; Marshall and Fletcher 2002).

Dibolisporites eifeliensis (Lanninger) McGregor, 1973
Figure 23C–E

```
       1966  cf. Archaeotriletes setigerus Kedo; McGregor and
              Owens, pl. 3, fig. 12.
       1967  Acanthotriletes inferus Naumova; Beju, pl. 1, fig. 14.
       1968  Anapiculatisporites eifeliensis Lanninger, p. 124,
              pl. 22, fig. 11.
       1973  Dibolisporites eifeliensis (Lanninger) McGregor,
              p. 31, pl. 3, figs 17–22, 26.
   non 1988  Dibolisporites eifeliensis (Lanninger) McGregor;
              Ravn and Benson, pl. 3, figs 16–23.
```

Dimensions. 29(42)60 µm; 20 specimens measured.

Comparison. Dibolisporites quebecensis McGregor, 1973 has smaller sculptural elements consisting of bacula and verrucae as well as tubercules. *Anapiculatisporites petilus* Richardson, 1965 has more widely spaced thinner pointed elements; about 15 around the periphery.

Occurrence. BAQA-1, BAQA-2, JNDL-3, JNDL-4, WELL-2, WELL-3, WELL-4, WELL-5, WELL-6 and WELL-7; Jauf Formation (Sha'iba to Hammamiyat members); *papillensis-baqaensis* to *lindlarensis-sextantii* zones. A1-69; Ouan-Kasa and Awaynat Wanin I formations; *lindlarensis-sextantii* to *svalbardiae-eximius* zones. MG-1; Ouan-Kasa and Awaynat Wanin I formations; *lindlarensis-sextantii* to *svalbardiae-eximius* zones.

Previous records. From upper Lochkovian–Emsian of Belgium (Steemans 1989); upper Lochkovian – lower Emsian of Paraná and Solimões basins, Brazil (Rubinstein *et al.* 2005; Mendlowicz Mauller *et al.* 2007); Emsian–Eifelian of Canada (McGregor and Owens 1966; McGregor 1973; McGregor and Camfield 1976); upper Pragian – lower Emsian of China (Gao Lianda 1981); upper Lochkovian – ?lowermost Pragian of Germany (Steemans 1989); Emsian of Iran (Ghavidel-Syooki 2003); Pragian–lowermost Eifelian of Libya (Paris *et al.* 1985; Moreau-Benoit 1989); upper Pragian–Eifelian of Morocco (Rahmani-Antari and Lachkar 2001); Pragian–?lowermost Eifelian of Poland (Turnau 1986; Turnau *et al.* 2005); and Emsian of Saudi Arabia (Al-Ghazi 2007).

Dibolisporites farraginis McGregor and Camfield, 1982
Figure 23F–G

1982 *Dibolisporites farraginis* McGregor and Camfield, p. 38, pl. 8, figs 3–4; text-fig. 54.

Dimensions. 42(60)77 μm; 18 specimens measured.

Remarks. Specimens assigned to this species belong to a more or less intergrading series from those with predominantly conate and small verrucose sculpture (*D. farraginis* and *D. uncatus* (Naumova) McGregor and Camfield, 1982) to those with large verrucate sculptural elements, and thus conform rather closely to the diagnosis of *Verrucosisporites scurrus* (Naumova) McGregor and Camfield, 1982 and *V. premnus* Richardson, 1965. All these forms belong to the *V. scurrus* Morphon defined by McGregor and Playford (1992). Therefore, the *D. farraginis* Morphon also defined by McGregor and Playford (1992) is included in the *V. scurrus* Morphon (Table 1).

Comparison. *Dibolisporites farraginis* is distinguished from *D. vegrandis* McGregor and Camfield, 1982 by larger, more elongate sculptural elements, and from *Dibolisporites* cf. *correctus* (Naumova) Richardson, 1965 by relatively narrower base, more varied sculptural elements. *D. uncatus* (Naumova) McGregor and Camfield, 1982 has larger sculptural elements but intergrades with *D. farraginis* as it was noted as a component of the *D. farraginis* Morphon in McGregor and Playford (1992, table 4). This morphon includes sub-circular forms sculptured with a mixture of mostly discrete grana, coni, spinae, biform elements and verrucae of various sizes and simple laesurae.

Occurrence. S-462; Jubah Formation; *rugulata-libyensis* to *lemurata-langii* zones. A1-69; Awaynat Wanin I and Awaynat Wanin II formations; *rugulata-libyensis* to *triangulatus-catillus* zones. MG-1; Awaynat Wanin I, Awaynat Wanin II and Awaynat Wanin III formations; *rugulata-libyensis* to *langii-concinna* zones.

Previous records. From upper Eifelian – lower Givetian of Canada (McGregor and Camfield 1982); and lower Eifelian – lower Givetian of Libya (Moreau-Benoit 1989).

Dibolisporites (*Apiculiretusispora*) *gaspiensis* (McGregor) comb. nov.
Figure 23H–J

1966 cf. *Geminospora svalbardiae* (Vigran) Allen, McGregor and Owens, pl. 5, fig. 16.
1970 ?*Apiculiretusispora* sp. McGregor *et al.*, pl. 2, fig. 11.
1973 *Apiculiretusispora gaspiensis* McGregor, p. 28, pl. 3, figs 1–4.
1974 Large spores of *Chaleuria cirrosa* Andrews *et al.*, pp. 394, 398, pl. 56, figs 1, 2; pl. 57, fig. 4.
1976 *Apiculiretusispora gaspiensis* McGregor; McGregor and Camfield, p. 11, pl. 6, figs 11–12.

Dimensions. 53(65)80 μm; 16 specimens measured.

Remarks. McGregor (1973) described this form as two-layered. Nevertheless, specimens described here are single-layered, but the concentric equatorial folding sometimes gives the impression of a two-layered spore. This species is transferred to the genus *Dibolisporites* Richardson, 1965 because it is sculptured with short pila, bacula or biform elements up to 2 μm high and not spinae characteristic of the genus *Apiculiretusispora* (Streel) Streel, 1967.

Comparison. This species differs from *Apiculiretusispora arenorugosa* McGregor, 1973 and *A. plicata* (Allen) Streel, 1967 in possessing a thicker exine. *Geminospora svalbardiae* (Vigran) Allen, 1965 is distinctly two-layered.

Occurrence. JNDL-1, JNDL-3, JNDL-4 and WELL-3; Jauf (Subbat to Murayr members) and Jubah formations; *lindlarensis-sextantii* to *svalbardiae-eximius* zones. A1-69; Awaynat Wanin I and Awaynat Wanin II formations; *svalbardiae-eximius* to

FIG. 23. Each figured specimen is identified by borehole, sample, slide number and England Finder Co-ordinate location. All figured specimens are at magnification ×1000 except where mentioned otherwise. A–B, *Dibolisporites echinaceus* (Eisenack) Richardson, 1965, magnification ×750. A, A1-69, 1596 ft, 26989, R48/3. B, JNDL-1, 495.0 ft, PPM014, V35. C–E, *Dibolisporites eifeliensis* (Lanninger) McGregor, 1973. C, BAQA-1, 223.5 ft, 03CW109, F48/2. D, BAQA-1, 223.5 ft, 03CW109, J42. E, WELL-2, 15886.3 ft, 63102, G29/4. F–G, *Dibolisporites farraginis* McGregor and Camfield, 1982. F, A1-69, 1530 ft, 26984, O32/4. G, A1-69, 1416 ft, 26993, M44/2. H–J, *Dibolisporites* (*Apiculiretusispora*) *gaspiensis* (McGregor) comb. nov. H, JNDL-1, 172.7 ft, PPM007, N42. I, JNDL-1, 174.6 ft, 60848, L33/1. J, JNDL-1, 155.6 ft, 60838, T28/4. K, *Dibolisporites pilatus* Breuer *et al.*, 2007c. JNDL-1, 167.8 ft, 60843, L38/3.

rugulata-libyensis zones. MG-1; Awaynat Wanin I Formation; *incognita* Zone.

Previous records. From Emsian–Givetian of Canada (McGregor and Owens 1966; McGregor 1973; McGregor and Camfield 1976).

Dibolisporites pilatus Breuer et al., 2007c
Figure 23K

2007c *Dibolisporites pilatus* Breuer et al., p. 50,
pl. 6, figs 10–13.

Dimensions. 54(68)79 μm; 29 specimens measured.

Occurrence. JNDL-1; Jubah Formation; *svalbardiae-eximius* Zone.

Dibolisporites tuberculatus sp. nov.
Figure 24A–F

1988 *Dibolisporites eifeliensis* (Lanninger) McGregor;
Ravn and Benson, pl. 3, figs 16–23.

Derivation of name. From *tuberculatus* (Latin), meaning made up of tubercules; refers to the baculate distal surface.

Holotype. EFC O43/2 (Fig. 24B), slide 60849.

Paratype. EFC Q39 (Fig. 24F), slide 60845; JNDL-1 core hole, sample 172.7 ft.

Type locality and horizon. JNDL-1 core hole, sample 177.0 ft; Jubah Formation at Domat Al-Jandal, Saudi Arabia.

Diagnosis. A robust *Dibolisporites* sculptured with bacula, tubercules and biform elements. Sculptural elements consisting of a tapering conical stem, surmounted by an expanded tip.

Description. Amb is circular to sub-circular. Laesurae straight, simple or accompanied by labra, up to 3 μm in overall width, three-fifths to nine-tenths of the amb radius in length. Curvaturae are often visible or perceptible. Exine is 1.5–3 μm thick equatorially. Proximo-equatorial regions are sculptured with biform elements, tubercules, bacula 1.5–4.5 μm high, 1–3.5 μm

wide at base, 1–5 μm apart. Biform elements, sub-circular in plan view, consist of a tapering conical stem, surmounted by an expanded tip less than 2 μm in diameter. Contact areas are laevigate, scabrate or granulate.

Dimensions. 43(56)69 μm; 24 specimens measured.

Comparison. Ravn and Benson (1988) figured specimens as *D. eifeliensis* (Lanninger) McGregor, 1973. Although no description of the latter is present, they have sculptural elements that are too coarse for *D. eifeliensis* (Lanninger) McGregor, 1973 *sensu stricto.* On the contrary, they have the same sculpture as *D. tuberculatus* sp. nov. described here. *D. quebecensis* McGregor, 1973 has smaller tubercules and bacula. Sculptural elements of *D. eifeliensis* have either a smaller or no expansion at the tip.

Occurrence. JNDL-1, JNDL-3, JNDL-4 and S-462; Jauf (Hammamiyat and Murayr members) and Jubah formations; *lindlarensis-sextantii* to *rugulata-libyensis* Zone. A1-69; Awaynat Wanin II Formation; *incognita* to *triangulatus-catillus* zones. MG-1; Awaynat Wanin I Formation; *rugulata* Zone.

Previous records. From upper Eifelian of Parnaíba Basin, Brazil (Breuer and Grahn 2011); and ?Emsian–Eifelian of Georgia, USA (Ravn and Benson, 1988).

Dibolisporites turriculatus Balme, 1988
Figure 24G

1962 *Apiculatisporis* sp. Balme, p. 4, pl. 1, fig. 14.
1975 *Dibolisporites* sp. Grey, fig. 61 h.
1988 *Dibolisporites turriculatus* Balme, p. 128, pl. 5,
figs 10–14.
1992 *Dibolisporites* sp. cf. *D. turriculatus* Balme; Grey,
pl. 14, figs 5–6.

Dimensions. 63(82)114 μm; 20 specimens measured.

Remarks. Balme (1988) described some specimens showing laesurae set in a slightly thinner proximal concavotriangular area, but this character is not recognized in the population studied here.

Comparison. *Dibolisporites eifeliensis* (Lanninger) McGregor, 1973 is smaller. Although *D. tuberculatus* sp. nov. is smaller and has more massive sculptural elements, extreme forms characterized

FIG. 24. Each figured specimen is identified by borehole, sample, slide number and England Finder Co-ordinate location. All figured specimens are at magnification ×1000 except where mentioned otherwise. A–F, *Dibolisporites tuberculatus* sp. nov. A, JNDL-1, 172.7 ft, 60845, D50/3. B, Holotype, JNDL-1, 177.0 ft, 60849, O43/2. C, JNDL-1, 177.0 ft, 60850, P32. D, JNDL-1, 177.0 ft, 60850, J31/1. E, JNDL-1, 174.6 ft, 60847, C41/3. F, Paratype, JNDL-1, 172.7 ft, 60845, Q39. G, *Dibolisporites turriculatus* Balme, 1988. A1-69, 1277 ft, 62636, O53/3. H–I, *Dibolisporites uncatus* (Naumova) McGregor and Camfield, 1982. H, A1-69, 1322 ft, 27126, M40. I, A1-69, 1277 ft, 62637, V-W42. J–M, *Dibolisporites verecundus* sp. nov. J, BAQA-2, 50.8 ft, 66813, G35. K, BAQA-2, 133.0 ft, 03CW136, N30. L, Holotype, BAQA-2, 52.0 ft, 03CW128, L44/3. M, Paratype, BAQA-2, 64.5 ft, 66818, K33/3.

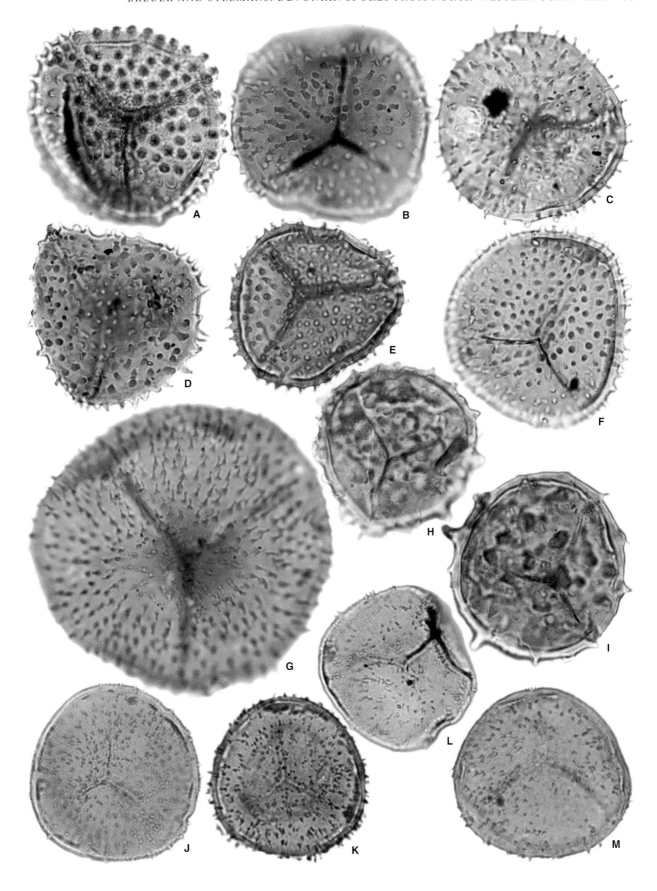

by thinner sculpture have the same ornamentation as *D. turriculatus*.

Occurrence. WELL-1; Jubah Formation; *lemurata-langii* to *triangulatus-catillus* zones. A1-69; Awaynat Wanin II Formation; *lemurata-langii* to *triangulatus-catillus* zones. MG-1; Awaynat Wanin II and Awaynat Wanin III formations; *lemurata-langii* to *langii-concinna* zones.

Previous records. From lower Eifelian – lower Frasnian of Australia (Balme 1962, 1988; Hashemi and Playford 2005).

Dibolisporites uncatus (Naumova) McGregor and Camfield, 1982
Figure 24H–I

 1953 *Acanthotriletes uncatus* Naumova, pp. 26, 28, pl. 1, figs 23–24; pl. 5, fig. 36.
 1960 *Verrucosisporites variabilis* McGregor, p. 30, pl. 11, fig. 15.
 1964 *Lophotriletes uncatus* (Naumova) Vigran, p. 13, pl. 1, figs 3–4.
 1964 Unnamed; Regali, text-fig. 2–6.
 1965 *Verrucosisporites* cf. *uncatus* (Naumova) Richardson, p. 572, pl. 89, fig. 13.
 1967 *Raistrickia* sp. Scott and Doher, fig. 3m.
 1971 *Dibolisporites variabilis* (McGregor) Smith, p. 83.
 1982 *Dibolisporites uncatus* (Naumova) McGregor and Camfield, p. 38, pl. 8, figs 5–6, 11; text-fig. 55.

Dimensions. 30(52)68 µm; 13 specimens measured.

Remarks. This taxon belongs to the *V. scurrus* Morphon (Table 1) defined by McGregor and Playford (1992). Specimens assigned to this species belong to a more or less intergrading series from those with predominantly conate and small verrucose sculpture (*D. farraginis* McGregor and Camfield, 1982) to those with large verrucate sculptural elements, and thus conform rather closely to the definition of *V. scurrus*.

Comparison. Extreme forms of *Verrucosisporites scurrus* (Naumova) McGregor and Camfield, 1982 intergrade with *D. uncatus*. *V. scurrus* has slightly larger, more closely spaced sculptural elements, a greater proportion of them flat-topped. *V. premnus* Richardson, 1965 has much larger, mostly flat-topped, non-biform verrucae and bacula. *D. farraginis* McGregor and Camfield, 1982 has smaller sculptural elements but intergrades with *D. uncatus* within the *D. farraginis* Morphon. *Convolutispora crassata?* (Naumova) McGregor and Camfield, 1982 has larger, commonly more closely crowded sculptural elements fused into ridges.

Occurrence. S-462 and WELL-8; Jubah Formation; *lemurata-langii* to *langii-concinna* zones. A1-69; Awaynat Wanin II Formation; *rugulata-libyensis* to *triangulatus-catillus* zones. MG-1;

Ouan-Kasa, Awaynat Wanin I, Awaynat Wanin II and Awaynat Wanin III formations; *annulatus-protea* to *langii-concinna* zones.

Previous records. From Givetian–lower Frasnian of Paraná Basin, Brazil (Loboziak *et al.* 1988); upper Givetian – lower Frasnian of France (Brice *et al.* 1979; Loboziak and Streel, 1980, 1988); upper Eifelian – lower Givetian of Canada (McGregor and Camfield, 1982); uppermost Eifelian of Germany (Loboziak *et al.* 1990); Frasnian of Spitsbergen, Norway (Vigran 1964); and Givetian of Scotland (Richardson 1965).

Dibolisporites verecundus sp. nov.
Figure 24J–M

Derivation of name. From *verecundus* (Latin) meaning inconspicuous, moderate; refers to the small scattered distal ornamentation.

Holotype. EFC L44/3 (Fig. 24L), slide 03CW128.

Paratype. EFC K33 (Fig. 24M), slide 66818; BAQA-2 core hole, sample 64.5 ft.

Type locality and horizon. BAQA-2 core hole, sample 52.0 ft; Jauf Formation at Baq'a, Saudi Arabia.

Diagnosis. A *Dibolisporites* sculptured with irregularly scattered small spinae and coni.

Description. Amb is circular to sub-circular. Laesurae are straight, simple, around four-fifths of the amb radius in length. Exine 1–2 µm thick equatorially. Proximo-equatorial and distal regions are sculptured with irregularly scattered spinae and coni, up to 2 µm high, 0.5–1 µm wide at base. Sculptural elements are sub-circular in plan view and 0.5–6 µm apart. Contact areas are laevigate to infragranulate.

Dimensions. 37(43)50 µm; 16 specimens measured.

Comparison. *Cymbosporites rarispinosus* Steemans, 1989 is patinate and sculptured with larger spinae.

Occurrence. BAQA-1 and BAQA-2; Jauf Formation (Sha'iba to Subbat members); *papillensis-baqaensis* to *ovalis* zone.

Dibolisporites sp. 1
Figure 25A–C

 2007c *Dibolisporites echinaceus* (Eisenack) Richardson; Breuer *et al.*, pl. 6, fig. 8.

Description. Amb is sub-circular to sub-triangular. Laesurae are straight, simple and two-thirds to three-quarters of the amb radius in length. Exine is 1–2 μm thick equatorially. Proximo-equatorial and distal regions are sculptured with a mixture of densely distributed parallel-sided elements, tubercules, bacula, spinae, 0.5–3 μm high, 0.5–1 μm wide at base. Sculptural elements are sub-circular to polygonal in plan view and generally *c.* 0.5 μm apart. Different types of ornament may occur on a single specimen. Contact areas are granulate.

Dimensions. 28(42)50 μm; nine specimens measured.

Remarks. On some specimens (Fig. 25A), ornamentation appears longer in interradial areas.

Comparison. Dibolisporites echinaceus (Eisenack) Richardson, 1965 has a larger amb (61–129 μm) and laevigate or infragranular contact areas.

Occurrence. BAQA-1 and BAQA-2; Jauf Formation (Sha'iba to Subbat members); *papillensis-baqaensis* to *milleri* zone.

Dibolisporites sp. 2
Figure 25D–E

Description. Amb is sub-circular to sub-circular. Laesurae are straight, simple, about three-quarters of the amb radius in length. Curvaturae often are visible or perceptible. Exine is 1–2 μm thick equatorially. Proximo-equatorial and distal regions are sculptured with scattered spinae 1.5–3.5 μm high, 1–2 μm wide at base. Sculptural elements are sub-circular in plan view and 1–3 μm apart. Contact areas are granulate.

Dimensions. 23(30)37 μm; four specimens measured.

Comparison. Dibolisporites eifeliensis (Lanninger) McGregor, 1973 is larger with no accuminate-tipped elements.

Occurrence. BAQA-2; Jauf Formation (Sha'iba Member); *papillensis-baqaensis* to *ovalis* zone.

Dibolisporites sp. 3
Figure 25F–G

Description. Amb is circular to sub-circular. Laesurae are straight, simple, extending to, or almost to, the equator. Exine 1–2 μm thick equatorially. Proximo-equatorial and distal regions are sculptured with irregularly scattered spinae and coni, 1.5–3.5 μm high, 1–3 μm wide at base. Sculptural elements are sub-circular in plan view and 0.5–10 μm apart. Contact areas are laevigate.

Dimensions. 32–44 μm; two specimens measured.

Comparison. Cymbosporites rarispinosus Steemans, 1989 is patinate and sculptured with spinae that flare more at the base and show a more flexuous appearance.

Occurrence. A1-69; Awaynat Wanin I Formation; *svalbardiae-eximius* to *rugulata-libyensis* zones.

Genus DICTYOTRILETES Naumova, 1939 ex Ishchenko, 1952

Type species. Dictyotriletes bireticulatus (Ibrahim) Potonié and Kremp, 1955.

Dictyotriletes biornatus Breuer *et al.,* 2007c var. *biornatus*
Figure 25H–J

2007c *Dictyotriletes biornatus* Breuer, p. 50, pl. 7, figs 1–9.

Description. Amb is sub-circular to triangular. Laesurae are rarely visible. Exine is 1–3 μm thick and thinner proximally, laevigate. Proximo-equatorial and distal are regions are reticulate. Muri of reticulum formed by orientated discrete rows of grana (1–2 μm wide and high) that are commonly slightly merged at the base. Polygonal lumina of reticulum are 4–9 μm in greatest diameter, about 30 to 40 in total number.

Dimensions. 47(58)72 μm; 55 specimens measured.

Comparison. Dictyotriletes biornatus Breuer *et al.,* 2007c var. *murinatus* var. nov. show a further merger of grana so as to form solid muri. Specimens of *Cymbosporites variabilis* var. *variabilis* sp. et var. nov. have no real lumina but rather pseudolumina with some grana occurring inside. All these forms are related and belong to the *D. biornatus* Morphon (Table 1; see discussion in *C. variabilis* sp. nov.).

Occurrence. BAQA-1, BAQA-2, JNDL-4 and WELL-7; Jauf Formation (Sha'iba to Subbat members); *ovalis-biornatus* to *lindlarensis-sextantii* zones.

Dictyotriletes biornatus Breuer *et al.,* 2007c var. *murinatus* nov. var.
Figure 25K–N

2007a *Dictyotriletes* sp. 1 Breuer *et al.,* text-figs 1–3E.
2007b Unnamed spore Breuer *et al.,* text-figs 1–5.

Derivation of name. From *murinatus* (Latin), meaning made up of muri; refers to the reticulate distal surface.

Holotype. EFC G37/3 (Fig. 25N), slide 03CW124.

Paratype. EFC J27/1 (Fig. 25M), slide 62275; BAQA-1 core hole, sample 395.2 ft.

Type locality and horizon. BAQA-1 core hole, sample 408.3 ft; Jauf Formation at Baq'a, Saudi Arabia.

Diagnosis. A *Dictyotriletes biornatus* with muri of reticulum constituted by discontinuous or irregular, low elongated elements, generally resulting of the merger of several discrete grana.

Description. Amb is sub-circular to sub-triangular. Exine is 1–3 μm thick equatorially. Exine is thinner proximally, laevigate. Proximo-equatorial and distal regions are reticulate. Muri of reticulum are comprised of discontinuous or irregular elongated elements, 0.5–2 μm high and wide. The elongated elements appear generally to be the result of the merger of several discrete elements. In addition, some discrete elements (grana) may be present in low numbers lined up on the reticulum. Sculptural elements vary in thickness and shape. Polygonal lumina of reticulum are 4–9 μm in greatest diameter, about 30 to 40 in number.

Dimensions. 40(53)67 μm; 27 specimens measured.

Remarks. Dictyotriletes biornatus var. *murinatus* represents the end-member of a lineage comprising several form-species (see discussion in *Cymbosporites variabilis* sp. nov.).

Comparison. Dictyotriletes biornatus Breuer *et al.*, 2007 var. *biornatus* is considered as sculptured with muri only formed by grana, which are slightly merged at base. *D. emsiensis* (Allen) McGregor, 1973 has more robust, continuous and regular muri.

Occurrence. BAQA-1, BAQA-2 and JNDL-4; Jauf Formation (Sha'iba to Subbat members); *ovalis-biornatus* to *lindlarensis-sextantii* zones.

Dictyotriletes emsiensis (Allen) McGregor, 1973
Figure 25O

1965 *Reticulatisporites emsiensis* Allen, p. 705, pl. 97, figs 9–11.

? 1967 *?Dictyotriletes* Richardson, pl. 4, fig. c.

non 1968 *Reticulatisporites emsiensis* Allen; Lanninger, p. 131, pl. 23, fig. 4.

1973 *Dictyotriletes emsiensis* (Allen) McGregor, p. 42, pl. 5, fig. 15.

1976 *Dictyotriletes* cf. *D. emsiensis* (Allen) McGregor; McGregor and Camfield, p. 21, pl. 3, figs 5–6.

Dimensions. 44(64)80 μm; 21 specimens measured.

Remarks. The *D. emsiensis* Morphon defined by Rubinstein *et al.* (2005) includes the species *D. emsiensis, D. granulatus* Steemans, 1989, some specimens incorrectly attributed to *D. subgranifer* McGregor, 1973, and other specimens in open nomenclature. These species share very similar morphological features. They are characterized by a proximo-equatorial reticulum, with robust muri that are commonly widened and bear papillae or spinae at the junctions.

Comparison. Dictyotriletes subgranifer McGregor, 1973 has narrower, lower muri that are serrated along the upper edge, and not widened or papillate at the junctions. In addition, the lumina tend to be smaller in diameter in *D. subgranifer*. Although *D. granulatus* Steemans, 1989 is closely comparable with *D. emsiensis*, it is smaller and has a granulate proximal face. *D. biornatus* Breuer *et al.*, 2007c var. *murinatus* var. nov. has lower muri.

Occurrence. BAQA-1, BAQA-2, JNDL-3, JNDL-4, WELL-2, WELL-3, WELL-4 and WELL-7; Jauf Formation (Sha'iba to Hammamiyat members); *papillensis-baqaensis* to *lindlarensis-sextantii* zones. A1-69; Ouan-Kasa Formation; *lindlarensis-sextantii* Zone. MG-1; Ouan-Kasa and Awaynat Wanin I formations; *annulatus-protea* to *svalbardiae-eximius* zones.

Previous records. Dictyotriletes emsiensis is eponymous for the Pragian polygonalis-emsiensis Assemblage Zone of the Old Red Sandstone Continent and adjacent regions (Richardson and McGregor 1986) and the early Pragian E Interval Zone of Western Europe (Streel *et al.* 1987). *D. emsiensis* has an almost worldwide distribution and has been reported from Pragian of Algeria (Boumendjel *et al.* 1988); upper Pragian – lower Eifelian of Argentina (Le Hérissé *et al.* 1997; Rubinstein and Steemans

FIG. 25. Each figured specimen is identified by borehole, sample, slide number and England Finder Co-ordinate location. All figured specimens are at magnification ×1000 except where mentioned otherwise. A–C, *Dibolisporites* sp. 1. A, BAQA-1, 406.0 ft, 03CW123, D23. B, BAQA-2, 50.8 ft, 03CW127, P25/3. C, BAQA-1, 406.0 ft, 03CW123, E50. D–E, *Dibolisporites* sp. 2. D, BAQA-2, 50.2 ft, 03CW126, G44/4. E, BAQA-2, 52.0 ft, 03CW128, F32/3. F–G, *Dibolisporites* sp. 3. F, A1-69, 1830 ft, 26962, L43. G, A1-69, 1596 ft, 26990, H48/4. H–J, *Dictyotriletes biornatus* Breuer *et al.*, 2007c var. *biornatus*. H, JNDL-4, 495.2 ft, 03CW275, L41/4. I, BAQA-1, 366.9 ft, 62259, D44/3. J, BAQA-1, 308.3 ft, 62244, D46. K–N, *Dictyotriletes biornatus* Breuer *et al.*, 2007c var. *murinatus* var. nov. K, BAQA-1, 345.5 ft, 62251, U32/1. L, BAQA-1, 345.5 ft, 62252, H46. M, Paratype, BAQA-1, 395.2 ft, 62275, J27/1. N, Holotype, BAQA-1, 408.3 ft, 03CW124, G37/3. O, *Dictyotriletes emsiensis* (Allen) McGregor, 1973. MG-1, 2180 m, 62973, H34/2. P–Q, *Dictyotriletes favosus* McGregor and Camfield, 1976. P, WELL-7, 13614.1 ft, 62372, S45. Q, JNDL-4, 364.6 ft, 03CW252, T37. R, *Dictyotriletes ?gorgoneus* Cramer, 1966a *in* McGregor (1973). BAQA-2, 64.5 ft, 66818, J42. S, *Dictyotriletes granulatus* Steemans, 1989. BAQA-2, 133.0 ft, 66825, L28. T–U, *Dictyotriletes hemeri* sp. nov. T, Paratype, S-462, 2460–2465 ft, 63293, W41. U, WELL-8, 16642.3 ft, 62407, E37.

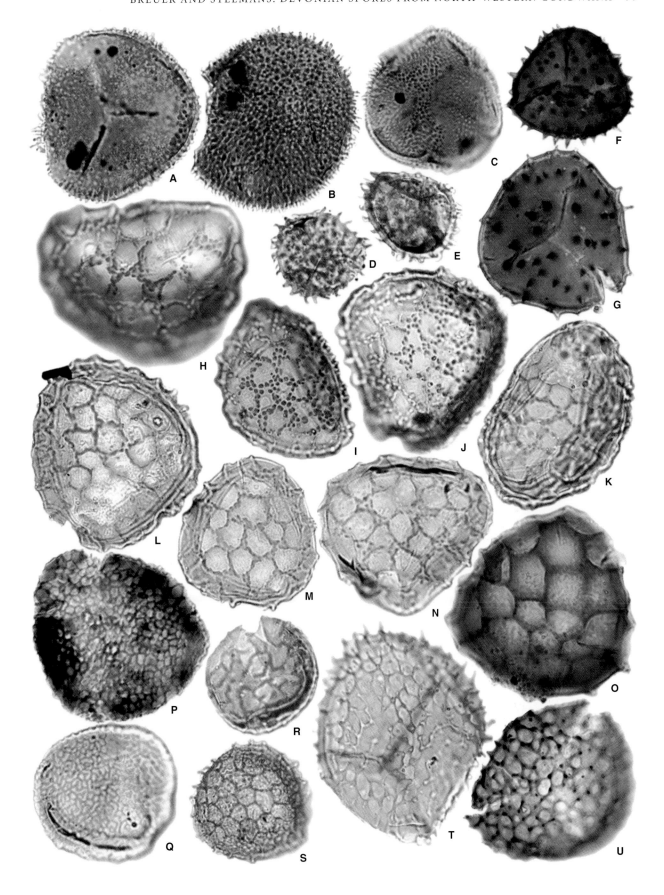

2007; Amenábar 2009); ?upper Lochkovian – upper Pragian of Belgium (Steemans 1989); Lochkovian–Eifelian of Bolivia (McGregor 1984); lower Lochkovian – lower Eifelian of Amazon, Parnaíba and Solimões basins, Brazil (Melo and Loboziak 2003; Grahn *et al.* 2005; Rubinstein *et al.* 2005; Steemans *et al.* 2008); Pragian–Emsian of Canada (McGregor 1973; McGregor and Camfield 1976); upper Pragian – lower Emsian of China (Gao Lianda 1981); Pragian–lower Emsian of Armorican Massif, France (Le Hérissé 1983); upper Lochkovian–Emsian of Germany (Steemans 1989); upper Pragian–Emsian of Morocco (Rahmani-Antari and Lachkar 2001); Emsian of Spitsbergen, Norway (Allen 1965); and Pragian of Saudi Arabia (Steemans, 1995).

Dictyotriletes favosus McGregor and Camfield, 1976
Figure 25P–Q

? 1973	*Dictyotriletes* sp. McGregor, p. 44, pl. 6, figs 1–2.	
1976	*Dictyotriletes favosus* McGregor and Camfield, p. 21, pl. 2, figs 5–6.	
1989	*Chelinospora favosa* (McGregor and Camfield) Steemans, p. 117, pl. 29, figs 1–3.	
non 2005	*Chelinospora favosa* (McGregor and Camfield) Steemans; Rubinstein *et al.*, pl. 2, fig. 4.	

Dimensions. 36(43)52 µm; 22 specimens measured.

Remarks. The Saudi Arabian population of *D. favosus* is smaller in overall size than that from Canada described by McGregor and Camfield (1976).

Comparison. Synonymy with *Dictyotriletes* sp. *in* McGregor (1973) is uncertain because it has slightly more robust muri with more pronounced elongations at their junctions.

Occurrence. BAQA-1, JNDL-3, JNDL-4, WELL-4 and WELL-7; Jauf Formation (Qasr to Hammamiyat members); *ovalis-biornatus* to *lindlarensis-sextantii* zones. A1-69; Ouan-Kasa Formation; *lindlarensis-sextantii* Zone.

Previous records. From uppermost Pragian – lowermost Emsian of Parnaíba Basin, Brazil (Grahn *et al.* 2005); upper Lochkovian – ?lowermost Pragian of Belgium (Steemans 1989); Pragian–Emsian of Canada (McGregor and Camfield 1976); Lochkovian of

France (Steemans 1989); upper Lochkovian–Emsian of Germany (Steemans 1989); and Pragian of Saudi Arabia (Steemans 1995).

Dictyotriletes ?gorgoneus Cramer, 1966a in McGregor (1973)
Figures 25R, 26A

1954	Unnamed; Radforth and McGregor, pl. 2, fig. 61.	
1965	*Reticulatisporites* sp. cf. *Dictyotriletes minor* Naumova; Allen, p. 706, pl. 97, figs 12–13.	
1966	*Reticulatisporites* sp. cf. *Dictyotriletes minor* Naumova *in* Allen; McGregor and Owens, pl. 3, fig. 18.	
? 1966a	*Dictyotriletes gorgoneus* Cramer; p. 265, pl. 3, figs 69, 72.	
? 1966	*Dictyotriletes minor* Naumova var. *nigritellus* Nadler, pl. 1, fig. 8.	
1967	*Dictyotriletes* sp. McGregor, pl. 1, fig. 4.	
1973	*Dictyotriletes ?gorgoneus* Cramer; McGregor, p. 43, pl. 5, figs 12, 17.	
1979	*Dictyotriletes ?gorgoneus* Cramer; Lessuise *et al.*, p. 337, pl. 5, figs 16–17.	

Dimensions. 26(33)37 µm; 16 specimens measured.

Remarks. In McGregor (1973), assignment to *D. gorgoneus* Cramer, 1966a is questioned because Cramer did not report a trilete mark on this species as in the Canadian specimens.

Comparison. *Reticulatisporites* sp. cf. *Dictyotriletes minor* Naumova, 1953 *in* Allen (1965) is considered here as similar to *D. ?gorgoneus* Cramer, 1966a *in* McGregor (1973). Allen (1965) mentioned the higher muri of *D. minor* Naumova, 1953 but expressed some doubt that this feature is of sufficient importance to distinguish his specimens from those assigned by Naumova to *D. minor*. The latter also has wider muri than *D. ?gorgoneus* Cramer, 1966a *in* McGregor (1973).

Occurrence. BAQA-1, BAQA-2 and WELL-4; Jauf Formation (Sha'iba to Subbat members); *ovalis* Zone.

Previous records. From upper Lochkovian – upper Emsian of Belgium (Lessuise *et al.* 1979; Steemans 1989); Emsian of Canada (McGregor and Owens 1966; McGregor 1973); lower Pragian–Emsian of Germany (Steemans, 1989); and Emsian–Givetian of Sipstbergen, Norway (Allen 1965).

FIG. 26. Each figured specimen is identified by borehole, sample, slide number and England Finder Co-ordinate location. All figured specimens are at magnification ×1000 except where mentioned otherwise. A, *Dictyotriletes ?gorgoneus* Cramer, 1966a *in* McGregor (1973). BAQA-2, 50.8 ft, 03CW127, W40/3. B–D, *Dictyotriletes hemeri* sp. nov. B, S-462, 2510–2515 ft, 63297, T41/4. C, S-462, 2460–2465 ft, 63293, J32. D, Holotype, S-462, 2460–2465 ft, 63293, L45/3. E–K, *Dictyotriletes marshallii* sp. nov. E, Paratype, JNDL-4, 355.4 ft, 68673, D26/3. F, MG-1, 2639 m, 62780, B48/4. G, JNDL-3, 413.2 ft, 68567, E36/4. H, JNDL-4, 341.2 ft, 68671, Q53. I, JNDL-4, 179.9 ft, 68634, F45. J, JNDL-4, 163.7 ft, 68626, S54. K, Holotype, JNDL-4, 402.4 ft, 68682, M36. L–M, *Dictyotriletes subgranifer* McGregor, 1973. L, BAQA-1, 366.9 ft, 03CW117, G28. M, BAQA-1, 376.4 ft, 03CW119, O26/3. N–O, *Dictyotriletes* sp. 1. N, BAQA-1, 366.9 ft, 62259, F39/4. O, BAQA-1, 366.9 ft, 62259, T41-42.

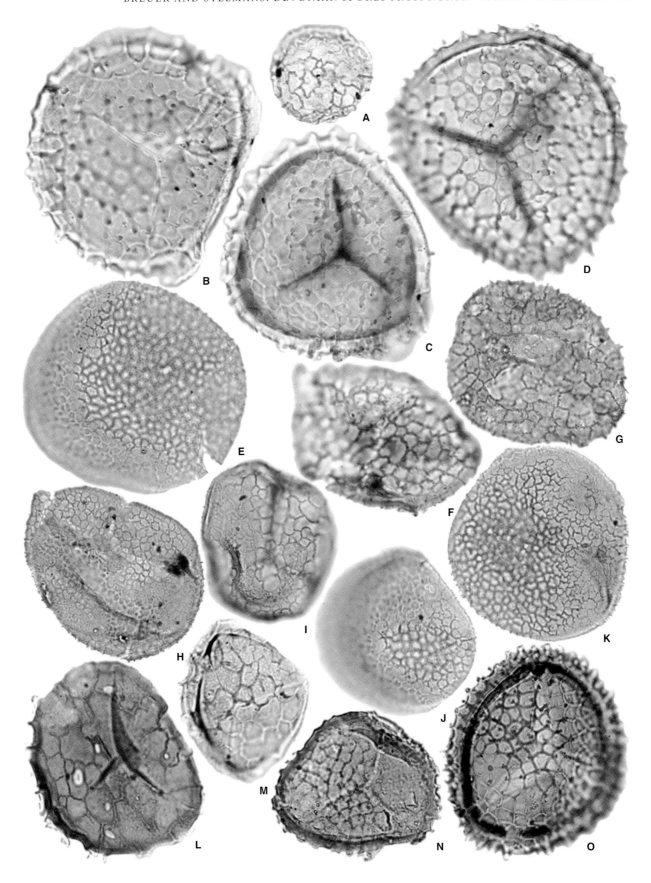

Dictyotriletes granulatus Steemans, 1989
Figure 25S

? 1968 Spore no 2565 Magloire, pl. 5, fig. 11.
? 1975 *Dictyotriletes* sp. L Rauscher and Robardet, pl. 11, figs 13–14.
  1979 *Dictyotriletes* sp. A D'Erceville, p. 94, pl. 3, fig. 6.
  1989 *Dictyotriletes granulatus* Steemans, p. 133, pl. 36, figs 6–10.
  2005 Dictyotriletes emsiensis Morphon (*pars*); Rubinstein *et al.*

Dimensions. 26–32 μm; two specimens measured.

Remarks. True specimens of *D. granulatus* Steemans, 1989 are included in the *D. emsiensis* Morphon defined by Rubinstein *et al.* (2005).

Comparison. Dictyotriletes sp. A *in* Richardson and Lister (1969) seems very similar but exhibits less elevated muri. *D. subgranifer* McGregor, 1973 has also granulate contact areas, but the grana are usually smaller. In addition, *D. subgranifer* has serrated muri along the upper edge. *D. emsiensis* (Allen) McGregor, 1973 is larger and does not show a granulate proximal face.

Occurrence. BAQA-2; Jauf Formation (Sha'iba Member); *papillensis-baqaensis* Zone.

Previous records. From upper Lochkovian of Solimões Basin, Brazil (Rubinstein *et al.* 2005); and lower Lochkovian of Armorican Massif, France (Steemans 1989).

Dictyotriletes hemeri sp. nov.
Figures 25T–U, 26B–D

? 1966 *Dictyotriletes* sp. McGregor and Owens, pl. 4, figs 1–2.
  1969 Indeterminate spore Cramer, pl. 2, fig. 28.
? 1973 *Dictyotriletes* sp. McGregor, p. 44, pl. 6 figs 1–2.
? 1992 *Dictyotriletes australis*? de Jersey; McGregor and Playford, pl. 7, figs 15–16.

Derivation of name. In honour of the American palynologist employed by Saudi Aramco, Darwin O. Hemer, who first studied the S-462 borehole.

Holotype. EFC L45/3 (Fig. 26D), slide 63293.

Paratype. EFC W41 (Fig. 25T), slide 63293; S-462 borehole, sample 2460–2465 ft.

Type locality and horizon. S-462 borehole, sample 2460–2465 ft; Jubah Formation in S-462, Saudi Arabia.

Diagnosis. A *Dictyotriletes* with lumina sub-circular to polygonal in plan view. Muri low and wider at junctions. Bacula with flared bases at the muri junctions. The tops of bacula flat or slightly concave, with bifurcate shape.

Description. Amb is sub-circular. Laesurae are straight and simple, three-fifths to four-fifths of the amb radius in length. Exine is 1–4 μm thick equatorially, sexine locally detached on some specimens. Contact areas are laevigate or infragranular. Proximo-equatorial and distal regions are reticulate. Lumina of reticulum are sub-circular to polygonal in plan view, 2–8 μm (rarely up to 12 μm) in greatest diameter. Muri are low, less than 1.5 μm wide at base, wider at junctions. Bacula occur with flared bases at the muri junctions, 0.5–3.5 μm high, 0.5–2.0 wide at base. The tops of bacula are flat or slightly concave, with bifurcate shape.

Dimensions. 46(59)80 μm; 24 specimens measured.

Comparison. Cramer (1969) illustrated, though not described, an indeterminate spore similar to *D. hemeri*. *Dictyotriletes* sp. figured in McGregor and Owens (1966) and McGregor (1973) resemble the specimens described here, but they have rounded and smaller protrusions at the muri junctions. McGregor and Playford (1992) figure a specimen resembling *D. hemeri* (except for the proximal dark triangular area) without giving a description. *D. australis* de Jersey, 1966 is smaller and devoid of discrete ornamentation at muri junctions. Some extreme specimens of *D. hemeri* may have a foveolate appearance and somewhat resemble *Brochotriletes bellatulus* Steemans, 1989, but they are distinguished from it by the tops of sculptural elements commonly being bifurcate.

Occurrence. WELL-1, S-462 and WELL-8; Jubah Formation; *rugulata-libyensis* to *triangulatus-catillus* zones, some specimens from S-462 may be caved into older strata.

Previous record. From Eifelian–Givetian of Spain (Cramer 1969).

Dictyotriletes marshallii sp. nov.
Figure 26E–K

Derivation of name. In honour of the British palynologist, Dr. John E. A. Marshall, for his outstanding contribution to the understanding of the Devonian spore assemblages and vegetational history.

Holotype. EFC M36 (Fig. 26K), slide 68682.

Paratype. D26/3 (Fig. 26E), slide 68673; JNDL-4 core hole, sample 355.4 ft.

Type locality and horizon. JNDL-4 core hole, sample 402.4 ft; Jauf Formation at Domat Al-Jandal, Saudi Arabia.

Diagnosis. A *Dictyotriletes* with small irregular lumina in plan view. Muri narrow, low and serrated. Proximal surface infragranular.

Description. Amb is sub-circular. Laesurae are straight, simple and seven-tenths to nine-tenths of the amb radius in length. Exine is 1–2 μm thick equatorially. Contact areas is infragranulate. Grana are less than 0.5 μm wide. Distal surface is reticulate. Muri are 0.5–1 μm high and wide at base, serrated along the upper edge. Sculptural elements (spines or coni), which constitute the serration, are less than 0.5 μm wide, often *c.* 0.5 μm high (rarely as high as 1.5 μm). Lumina are irregular in plan view, 1.5–6 μm in diameter, various in size and shape on a same specimen, *c.* 30–70 situated at equator.

Dimensions. 39(51)64 μm; 13 specimens measured.

Comparison. *Dictyotriletes subgranifer* McGregor, 1973 has larger lumina and their number is also fewer.

Occurrence. JNDL-3 and JNDL-4; Jauf Formation (Subbat and Hammamiyat members); *lindlarensis-sextantii* Zone. MG-1; Ouan-Kasa Formation; *svalbardiae-eximius* Zone but occurrences are probably reworked.

Dictyotriletes subgranifer McGregor, 1973
Figure 26L–M

1973 *Dictyotriletes subgranifer* McGregor, p. 43 (*cum syn.*), pl. 5, figs 16, 18–20.

Dimensions. 34(54)80 μm; 17 specimens measured.

Comparison. *Dictyotriletes emsiensis* (Allen) McGregor, 1973 has more robust muri that are commonly widened at the junctions, and not serrated along the upper edge.

Occurrence. BAQA-1, BAQA-2, JNDL-3 and JNDL-4; Jauf Formation; *papillensis-baqaensis* to *lindlarensis-sextantii* zones. A1-69; Ouan-Kasa, Awaynat Wanin I and Awaynat Wanin II formations; *lindlarensis-sextantii* Zone, some isolated occurrence of specimens in the youngest part of the section are due to reworking. MG-1; Ouan-Kasa, Awaynat Wanin, and Awaynat Wanin II formations; *annulatus-protea* Zone, the isolated occurrence of specimens in the youngest part of the section are probably due to reworking.

Previous records. *Dictyotriletes subgranifer* is eponymous for the upper Pragian – lower Emsian Su Interval Zone of Western Europe (Streel *et al.* 1987). *D. subgranifer* has an almost worldwide distribution extending from Pragian to upper Emsian. It has been reported from many parts of the world; e.g. Belgium (Steemans 1989), Brazil (Grahn *et al.* 2005; Mendlowicz Mauller *et al.* 2007), Canada (McGregor and Owens 1966; McGregor 1973; McGregor and Camfield 1976), China

(Gao Lianda 1981), Armorican Massif, France (Le Hérissé 1983), Germany (Steemans 1989), Luxembourg (Steemans *et al.* 2000a), Morocco (Rahmani-Antari and Lachkar 2001), Poland (Turnau 1986; Turnau *et al.* 2005), Saudi Arabia (Al-Ghazi 2007), Scotland (Wellman 2006) and USA (Ravn and Benson 1988).

Dictyotriletes sp. 1
Figure 26N–O

Description. Trilete patinate spores with sub-circular to sub-triangular amb. Laesurae straight and simple, extending to the inner edge of the patina. Exine 1–4 μm equatorially thick. Contact areas laevigate. Patina reticulate. Lumina of reticulum, polygonal in plan view, 2–6 μm in greatest diameter. Muri low, less than 1 μm wide at base, Bacula with bifurcate tips at the muri junctions, 1–3 μm high, 0.5–2 μm wide at base.

Dimensions. 36(43)55 μm; four specimens measured.

Comparison. The reticulum of described specimens looks like very much that of *D. hemeri* sp. nov., but ornamentation at muri junctions of the latter is less clearly bifurcate. In addition, it is not patinate. We have thus separated these specimens from those of *D. hemeri*.

Occurrence. BAQA-1; Jauf Formation (Subbat Member); *ovalis* Zone.

Genus ELENISPORIS Arkhangelskaya, 1985

Type species. *Elenisporis biformis* (Arkhangelskaya) Jansonius and Hills, 1987.

Comparison. This genus differs from *Emphanisporites* McGregor, 1961 in having an equatorial cingulum, the less clearly delimited muri on the proximal face, which are not radially oriented on the proximal surface, but rather abut against the laesurae.

Elenisporis gondwanensis sp. nov.
Figure 27A–C

Derivation of name. From *gondwanensis* (Latin), meaning from Gondwana; refers to its palaeogeographical occurrence.

Holotype. EFC K36/2 (Fig. 27C), slide 62845.

Paratype. EFC R39 (Fig. 27B), slide 62821, MG-1 borehole, sample 2405 m.

Type locality and horizon. MG-1 borehole, sample 2285 m; Awaynat Wanin II Formation at Mechiguig, Tunisia.

Diagnosis. An *Elenisporis* with laesurae characterized by triradiate sinuous fold-like labra. Proximal surface supporting radial sculpture of somewhat tortuous muri and distal surface sculptured with various and variable elements such as coni, spines, bacula or rounded verrucae.

Description. Amb is sub-circular. Laesurae are simple and straight but characterized by triradiate sinuous fold-like labra from the outer layer, *c.* 2–7 μm high in total width, about three-quarters of the amb radius in length. Curvaturae join the laesurae and are suggested by a narrow crassitude, 2–4 μm thick along the curvaturae, the outer edge of which can be denticulate. Proximal surface supports sculpture of somewhat tortuous muri, of uneven width but approximately uniform to their very end, 1–2.5 μm (rarely up to 4 μm) wide and more or less parallel to each other from the crassitude to laesurae. As they are closely spaced, they are often barely distinguishable. Distal surface is sculptured with various elements such as coni, spines, bacula or rounded verrucae, 1–6 μm high and up to 1–2 μm wide at their base, commonly 1–3 μm apart. Exine is 2–7 μm thick. Some specimens show two clearly differentiated wall layers; layers are tightly appressed distally, but local detachments sometimes occur equatorially.

Dimensions. 65(81)92 μm; 17 specimens measured.

Comparison. *Elenisporis biformis* (Arkhangelskaya) Jansonius and Hills, 1987 shows wider muri. It is finely granulate and is sometimes sculptured distally with small, short elements. *Elenisporis* sp. 1 has also wider rolls and is more poorly sculptured distally.

Occurrence. S-462; Jubah Formation; *rugulata-libyensis* to *lemurata-langii* zones. MG-1; Awaynat Wanin I and Awaynat Wanin II formations; *rugulata-libyensis* to *triangulatus-catillus* zones.

Elenisporis sp. 1
Figure 27D–E

Description. Amb is sub-circular to sub-triangular. Laesurae are straight to sinuous, characterized by triradiate fold-like labra *c.* 4–9 μm high in total width, extending to the crassitude. Curvaturae join the laesurae and are suggested by a narrow crassitude, 2–4 μm thick along the curvaturae. Proximal surface supports sculpture of somewhat tortuous muri, of uneven width but approximately uniform to their very end, 2–6 μm wide and more or less paralleled from the crassitude to laesurae. Distal surface is sculptured with small, sometimes barely noticeable, widely spaced spines, 0.5–3 μm high and up to 1 μm wide at their base. Exine is 4–8 μm thick.

Dimensions. 71(92)110 μm; six specimens measured.

Comparison. *Elenisporis biformis* (Arkhangelskaya) Jansonius and Hills, 1987 is very similar to *Elenisporis* sp. 1 but has narrow, straight laesurae, and its cingulum is thicker. *E. gondwanensis* sp. nov. has thinner muri and is sculptured distally with larger elements.

Occurrence. S-462; Jubah Formation; *triangulatus-catillus* to *langii-concinna* zones. MG-1; Awaynat Wanin I and Awaynat Wanin II formations; *rugulata-libyensis* to *lemurata-langii* zones.

Genus EMPHANISPORITES McGregor, 1961

Type species. *Emphanisporites rotatus* McGregor emend. McGregor, 1973.

Remarks. Wall ultrastructure of some *Emphanisporites* species are discussed in details in Taylor *et al.* (2011). This study likely suggests that this genus comprises spores produced by various plant groups, which adopted the emphanoid condition (possessing proximal radially disposed muri) by convergence.

Emphanisporites annulatus McGregor, 1961
Figure 27F

1956 Unnamed Radforth and McGregor, pl. 1, fig. 6.
1961 *Emphanisporites annulatus* McGregor, p. 3, pl. 1, figs 5–6.
1962 *Radiaspora* sp. Balme, p. 6, pl. 1, fig. 13.
1963 *Emphanisporites erraticus* (Eisenack) McGregor; Chaloner, p. 103, fig. 1.
1967 *Emphanisporites* cf. *erraticus* McGregor; Daemon *et al.*, p. 106, pl. 1, fig. 10.

Dimensions. 41(48)60 μm; 15 specimens measured.

Occurrence. JNDL-1, JNDL-3, S-462, WELL-1 and WELL-8; Jauf (Murayr Member) and Jubah formations; *annulatus-protea* to *lemurata-langii* zones. A1-69; Ouan-Kasa, Awaynat Wanin I and Awaynat Wanin II formations; *annulatus-protea* to *lemurata-langii* zones. MG-1; Ouan-Kasa, Awaynat Wanin I and Awaynat Wanin II formations; *annulatus-protea* to *lemurata-langii* zones.

Previous records. *Emphanisporites annulatus* is eponymous for the Emsian *annulatus-sextantii* Assemblage Zone of the Old Red

FIG. 27. Each figured specimen is identified by borehole, sample, slide number and England Finder Co-ordinate location. All figured specimens are at magnification ×1000 except where mentioned otherwise. A–C, *Elenisporis gondwanensis* sp. nov. A, MG-1, 2270 m, 62849, L47. B, Paratype, MG-1, 2405 m, 62821, R39. C, Holotype, MG-1, 2285 m, 62845, K36/2. D–E, *Elenisporis* sp. 1. D, MG-1, 2456 m, 62739, O48. E, MG-1, 2270 m, 62848, X34. F, *Emphanisporites annulatus* McGregor, 1961. JNDL-1, 174.6 ft, 60847, G31. G–H, *Emphanisporites* cf. *E. biradiatus* Steemans, 1989. G, JNDL-4, 448.6 ft, 03CW267, G37/1. H, JNDL-1, 155.6 ft, 60838, M51.

Sandstone Continent and adjacent regions (Richardson and McGregor 1986) and the early Emsian AB Oppel Zone of Western Europe (Streel *et al.* 1987). *E. annulatus* has a worldwide distribution extending from Emsian to the lower Frasnian.

Emphanisporites cf. *E. biradiatus* Steemans, 1989
Figure 27G–H

cf. 1989 *Emphanisporites biradiatus* Steemans, p. 136, pl. 37, figs 11–13.

Dimensions. 37–47 µm; two specimens measured.

Comparison. Emphanisporites biradiatus Steemans, 1989 is granulate distally and possesses six to eight pairs of muri per each interradial area, whereas the present specimens are entirely laevigate and show about four pairs of muri per interradial area. *E. rotatus* McGregor emend. McGregor, 1973 occasionally presents some pairs of muri.

Occurrence. JNDL-1 and JNDL-4. Jauf (Subbat Member) and Jubah formations; *asymmetricus* and *svalbardiae-eximius* zones.

Emphanisporites decoratus Allen, 1965
Figure 28A

1965 *Emphanisporites decoratus* Allen, p. 708, pl. 97, figs 15–18.
1981 *Emphanisporites* sp. K Steemans, p. 53, pl. 2, fig. 3.
2006 *Emphanisporites* cf. *decoratus* Allen; Wellman, p. 186, pl. 15, figs g–i.

Dimensions. 31(44)59 µm; nine specimens measured.

Comparison. The characteristics of *Emphanisporites* cf. *decoratus* Allen, 1965 *in* Wellman (2006) can be accommodated in the variability shown by the population studied here. *E. neglectus* Vigran, 1964 is occasionally sculptured, but then, only with very fine granules. *E. novellus* McGregor and Camfield, 1976 and *E. micrornatus* Richardson and Lister, 1969 var. *micrornatus* Steemans and Gerrienne, 1984 differ in finer granulate ornamentation. *E. micrornatus* var. *sinuosus* Steemans and Gerrienne, 1984 exhibits sinuous radial muri.

Occurrence. BAQA-1, JNDL-1, JNDL-4, WELL-2, WELL-4 and WELL-7; Jauf Formation (Subbat to Murayr members); *ovalisbiornatus* to *annulatus-protea* zones. MG-1; Ouan-Kasa Formation; *annulatus-protea* Zone.

Previous records. From upper Emsian – lower Eifelian of Argentina (Amenábar 2009); upper Lochkovian – upper Pragian of Belgium (Steemans 1989); Pragian–Emsian of Canada (McGregor and Camfield 1976); upper Pragian of Armorican Massif, France (Le Hérissé 1983); Emsian of Germany (Schultz 1968); Pragian of Spitsbergen, Norway (Allen 1965); and Pragian–?lowermost Emsian of Scotland (Wellman 2006).

Emphanisporites cf. *E. edwardsiae* Wellman, 2006
Figure 28B–C

cf. 2006 *Emphanisporites edwardsiae* Wellman, p. 188, pl. 16, figs a–f.

Dimensions. 57(65)71 µm; four specimens measured.

Comparison. Emphanisporites edwardsiae Wellman, 2006 does not have a thin triangular area inside the apical darker area as showed by the present specimens. *Retusotriletes phillipsii* Clendening *et al.*, 1980 possesses scattered grana on the proximal face and muri are straight to sinuous. *Scylaspora rugulata* (Riegel) Breuer *et al.*, 2007c is sculptured with fine, more or less radially oriented rugulate or subreticulate muri.

Occurrence. BAQA-1; Jauf Formation (Subbat Member); *milleri* to *lindlarensis-sextantii* zones.

Emphanisporites erraticus (Eisenack) McGregor, 1961
Figure 28D

1944 *Triletes erraticus* Eisenack, p. 114, pl. 2, fig. 9.
1954 Unnamed Radforth and McGregor, pl. 2, fig. 59.
1956 Unnamed Radforth and McGregor, pl. 3, fig. 1.
1961 *Emphanisporites erraticus* (Eisenack) McGregor, p. 4, pl. 1, figs 7–11.
1967 *Emphanisporites* cf. *erraticus* (Eisenack) McGregor; Richardson, pl. 4, fig. a.

Dimensions. 41(48)55 µm; nine specimens measured.

FIG. 28. Each figured specimen is identified by borehole, sample, slide number and England Finder Co-ordinate location. All figured specimens are at magnification ×1000 except where mentioned otherwise. A, *Emphanisporites decoratus* Allen, 1965. WELL-7, 13689.7 ft, 62316, Q50/1. B–C, *Emphanisporites* cf. *E. edwardsiae* Wellman, 2006. B, BAQA-1, 285.5 ft, 03CW111, D-E23. C, BAQA-1, 161.0 ft, 66773, W47. D, *Emphanisporites erraticus* McGregor, 1961. JNDL-3, 327.0 ft, 68555, S40. E–H, *Emphanisporites laticostatus* sp. nov. E, MG-1, 2161.8 m, 62528, M33. F, Holotype, MG-1, 2161.8 m, 62528, S38. G, MG-1, 2181.2 m, 62524, S49. H, Paratype, MG-1, 2161.8 m, 62528, V36/4. I, *Emphanisporites mcgregorii* Cramer, 1966a. MG-1, 2181.2 ft, 62525, V42/3. J–M, *Emphanisporites plicatus* sp. nov. J, Paratype, JNDL-1, 155.6 ft, 60838, M52. K, JNDL-1, 495.0 ft, 60854, F43/3. L, JNDL-3, 249.0 ft, 68543, Q24/2. M, JNDL-4, 163.3 ft, 68625, L37/4.

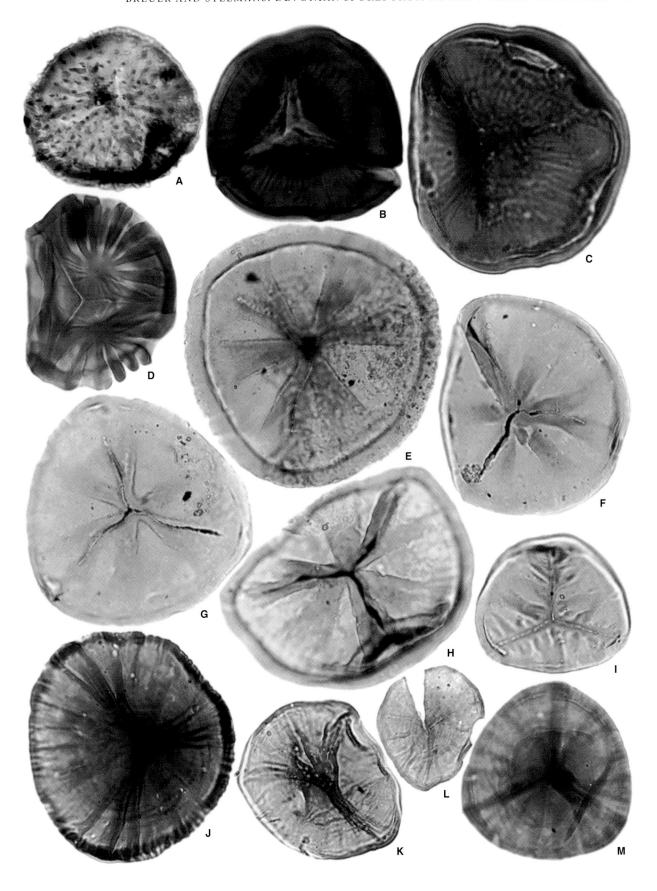

Comparison. *Emphanisporites schultzii* McGregor, 1973 possesses also a proximal rosette pattern but does not have a distal annulus.

Occurrence. JNDL-3 and JNDL-4; Jauf Formation (Hammamiyat Member); *lindlarensis-sextantii* Zone.

Previous records. From Emsian of Canada (McGregor and Owens 1966; McGregor 1973), Germany (Lanninger 1968; Schultz 1968) and Poland (Turnau 1986; Turnau *et al.* 2005); lower or middle Emsian of Libya (Paris *et al.* 1985); and Eifelian–Givetian of Spain (Cramer 1966*a*).

Emphanisporites laticostatus sp. nov.
Figure 28E–H

2011 *Emphanisporites* sp. 2 Breuer and Grahn, pl. 2, fig. q.

Derivation of name. From *lati-* and *costatus* (Latin), meaning sculptured with wide costae or muri; refers to the proximal muri.

Holotype. EFC S38 (Fig. 28F), slide 62528.

Paratype. EFC V36/4 (Fig. 28H), slide 62528; MG-1 borehole, sample 2161.8 m.

Type locality and horizon. MG-1 borehole, sample 2161.8 m; Awaynat Wanin III Formation at Mechiguig, Tunisia.

Diagnosis. A large, robust *Emphanisporites* with few and wide muri.

Description. Amb is circular to sub-triangular. Laesurae are straight or slightly undulating, simple, three-fifths to four-fifths of the amb radius in length. Laesurae are commonly bordered by muri. Exine is 2–5 μm thick equatorially. Proximal radially arranged muri are 3–12 μm in greatest width, low, tapering towards proximal pole. Muri extend from the proximal pole and fade inside the contact area before reaching the equator. Each contact area presents usually three muri. Distal surface is laevigate.

Dimensions. 61(67)75 μm; 13 specimens measured.

Comparison. *Emphanisporites laticostatus* is clearly distinguishable from all other species of *Emphanisporites* by the possession of few robust muri.

Occurrence. S-462; Jubah Formation; *langii-concinna* Zone. MG-1; Awaynat Wanin III Formation; *langii-concinna* Zone.

Previous record. From middle Givetian of Parnaíba Basin, Brazil (Breuer and Grahn 2011).

Emphanisporites mcgregorii Cramer, 1966*a*
Figure 28I

1961 *Emphanisporites* sp. McGregor, pl. 1, fig. 12.
1966*a* *Emphanisporites mcgregorii* Cramer, p. 263, pl. 3, fig. 59.
1968 *Emphanisporites spinaeformis* Schultz, p. 27, pl. 3, figs 10–10a.
1968 *Emphanisporites* sp. 1 Jardiné and Yapaudjian, pl. 1, fig. 3.
non 1968 *Emphanisporites macgregori* Schultz, p. 28, pl. 3, figs 12–12a.
non 1968 *Emphanisporites macgregori* Schultz; Lanninger, p. 136, pl. 23, fig. 15.

Dimensions. 36(46)55 μm; 12 specimens measured.

Comparison. Proximal muri of *E. mcgregorii* are aligned approximately parallel to one another, extending from the equator to the margin of laesurae forming a herringbone pattern, whereas muri of *E. rotatus* McGregor emend. McGregor, 1973 are radially distributed.

Occurrence. BAQA-1, BAQA-2, JNDL-1, JNDL-3, JNDL-4, S-462, WELL-1, WELL-4 and WELL-8; Jauf and Jubah formations; *papillensis-baqaensis* to *triangulatus-catillus* zones. A1-69; Awaynat Wanin I and Awaynat Wanin II formations; *svalbardiae-eximius* to *triangulatus-catillus* zones. MG-1; Ouan-Kasa, Awaynat Wanin I, Awaynat Wanin II and Awaynat Wanin III formations; *annulatus-protea* to *langii-concinna* zones.

Previous records. From upper Emsian – middle Givetian of Algeria (Moreau-Benoit *et al.* 1993); upper Lochkovian–Emsian of Belgium (Steemans 1989); upper Pragian – lower Emsian of Paraná Basin, Brazil (Mendlowicz Mauller *et al.* 2007); upper Pragian–Eifelian of Germany (Lanninger 1968; Riegel 1968; Steemans 1989); Pragian–lower Eifelian of Libya (Moreau-Benoit 1989); Emsian of Poland (Turnau 1986); and Pragian–Givetian of Spain (Cramer 1966*a*, 1969).

Emphanisporites plicatus sp. nov.
Figures 28J–M, 29A

Derivation of name. From *plicatus* (Latin), meaning folded; refers to the distal surface.

Holotype. EFC D43 (Fig. 29A), slide 60847.

Paratype. EFC M52 (Fig. 28J), slide 60838; JNDL-1 core hole, sample 155.6 ft.

Type locality and horizon. JNDL-1 core hole, sample 174.6 ft; Jubah Formation at Domat Al-Jandal, Saudi Arabia.

Diagnosis. An *Emphanisporites* with a distal surface concentrically folded.

Description. Amb is circular to sub-triangular. Laesurae are straight, simple or accompanied with low, narrow labra, up to 1.5 μm wide individually, extending to or almost to the equator. Exine is 1–3.5 μm equatorially thick. Proximal radially arranged muri, 0.5–2.5 μm wide at equator, are low and taper towards proximal pole. Each contact area contains between 8–16 muri. Distal surface laevigate and bears fine, concentric folds.

Dimensions. 35(47)63 μm; 19 specimens measured.

Remarks. Emphanisporites rotatus McGregor emend. McGregor, 1973 is similar. The presence of the distal concentric folds on *E. plicatus* occurs only on these specimens and is regarded as the principal characteristic feature of this new species.

Occurrence. JNDL-1, JNDL-3 and JNDL-4; Jauf (Hammamiyat and Murayr members) and Jubah formations; *lindlarensis-sextantii* to *svalbardiae-eximius* zones.

Emphanisporites rotatus McGregor emend. McGregor, 1973
Figures 6I–J, 29B

1973 *Emphanisporites rotatus* McGregor emend. McGregor, p. 46 (*cum syn.*), pl. 6, figs 9–13.

Dimensions. 34(44)55 μm; 32 specimens measured.

Remarks. Rare monolete specimens have been found (pl. 2, figs 9–10).

Occurrence. Emphanisporites rotatus is found in all sections from *papillensis-baqaensis* to *langii-concinna* zones.

Previous record. Emphanisporites rotatus has been widely reported from upper Silurian through the entire Devonian from many parts of the world.

Emphanisporites schultzii McGregor, 1973
Figure 29C

1966 *Emphanisporites* sp. McGregor and Owens, pl. 4, fig. 10.
non 1966a *Emphanisporites mcgregorii* Cramer, p. 263, pl. 3, fig. 59.
1967 *Emphanisporites* sp. McGregor, pl. 1, fig. 2.
1968 *Emphanisporites macgregori* Schultz, p. 28, pl. 3, figs 12–12a.
1968 *Emphanisporites pseudoerraticus* Schultz, p. 29, pl. 3, figs 15–15a.

1970 *Emphanisporites* sp. McGregor, pl. 31, fig. 2.
1973 *Emphanisporites schultzii* McGregor, p. 48, pl. 6, fig. 14.

Dimensions. 36(52)75 μm; 13 specimens measured.

Comparison. According to Schultz (1968), *E. schultzii* differs from *E. pseudoerraticus* Schultz, 1968 notably in that the centre of the rosette pattern of each contact area is not strongly displaced towards the proximal pole. *E. pseudoerraticus* is, however, considered here as a junior synonym of *E. schultzii* because the centre of the rosette pattern seems relatively variable in its position from the current material. *E. erraticus* (Eisenack) McGregor, 1961 has a distal annulus.

Occurrence. BAQA-1, JNDL-3, JNDL-4, WELL-3, WELL-4, WELL-7 and WELL-8; Jauf (Subbat to Hammamiyat members) and Jubah formations; *ovalis-biornatus* to *lindlarensis-sextantii* zones. MG-1; Awaynat Wanin I Formation; *rugulata* Zone but the specimen is probably reworked.

Previous records. From middle Pragian–Emsian of Belgium (Steemans 1989); Emsian of Canada (McGregor and Owens 1966; McGregor 1973; McGregor and Camfield 1976); upper Pragian–Emsian of Germany (Lanninger 1968; Schultz 1968; Steemans 1989); upper Pragian – lower Emsian of Luxembourg (Steemans *et al.* 2000a); and Emsian of Poland (Turnau 1986).

Emphanisporites sp. 1
Figure 29D–E

? 1981 *Emphanisporites rotatus* McGregor var. B; Streel *et al.*, pl. 1, figs 15–16.

Description. Amb is sub-circular. Laesurae are straight, simple and extend almost to the equator. Exine is laevigate or infragranular, 2–4 μm thick equatorially. Proximal radially arranged muri are 4–13 μm wide at equator, up to 5 μm high, tapering towards proximal pole. Muri extend from the proximal pole and over the equator, resulting in strongly undulating equator in plan view. Each contact area contains between 3–5 muri. Distal face laevigate.

Dimensions. 48(62)78 μm; four specimens measured.

Comparison. Emphanisporites rotatus McGregor, 1961 var. B in Streel *et al.* (1981), which shows muri extending over the equator, may be similar. *E. robustus* McGregor, 1961 has muri that extend from the proximal pole to, but not beyond, the equator.

Occurrence. A1-69; Awaynat Wanin II Formation; *lemurata-langii* Zone. MG-1; Awaynat Wanin II and Awaynat Wanin III formations; *lemurata-langii* to *langii-concinna* zones.

Genus GEMINOSPORA Balme, 1962

Type species. *Geminospora lemurata* Balme emend. Playford, 1983.

Geminospora convoluta sp. nov.
Figure 29F–J

Derivation of name. From *convolutus* (Latin), meaning convoluted; refers to the sculptured distal surface.

Holotype. EFC V29/2 (Fig. 29J), slide 60849.

Paratype. EFC M37/2 (Fig. 29F), slide PPM006; JNDL-1 core hole, sample 167.8 ft.

Type locality and horizon. JNDL-1 core hole, sample 177.0 ft; Jubah Formation at Domat Al-Jandal, Saudi Arabia.

Diagnosis. A *Geminospora* densely sculptured with cones, spines and biform processes round to polygonal in plan view, closely spaced and joined to form convoluted ridges of varied length.

Description. Amb is sub-circular to sub-triangular. Laesurae are straight, simple or bordered by labra up to 3 μm in total thickness, extending to the outer margin of the inner body. The inner body diameter equals three-quarters to nine-tenths of the total amb diameter. Nexine is 1–3 μm thick, laevigate. Sexine is 1.5–3 μm thick, densely sculptured proximo-equatorially and distally with cones, spines and biform processes (flask-shaped or mammate). Sculptural elements on the distal hemisphere are round to polygonal in plan view, 1–3 μm wide at their base, 1–2 μm high, closely spaced and joined to form ridges of varied length. Proximal surface is laevigate or with scattered grana. Specimens are very often folded.

Dimensions. 52(69)84 μm; nine specimens measured.

Comparison. *Acinosporites lindlarensis* Riegel, 1968 is morphologically similar, but its ornamentation is larger and the sexine is often partially separated from the nexine. *G. libyensis* Moreau-Benoit, 1980*b* is sculptured with shorter elements and has a nexine closely appressed to sexine.

Occurrence. JNDL-1; Jubah Formation; *annulatus-protea* to *svalbardiae-eximius* zones. A1-69; Awaynat Wanin II Formation; *svalbardiae-eximius* Zone.

Geminospora lemurata Balme emend. Playford, 1983
Figures 6K–L, 29K–L

1962		*Geminospora lemurata* Balme, p. 5, pl. 1, figs 5–10.
non	1965	*Geminospora svalbardiae* (Vigran) Allen, p. 696, pl. 94, figs 12–16.
1965		*Geminospora tuberculata* (Kedo) Allen, p. 696, pl. 94, figs 10–11.
1965		*Rhabdosporites parvulus* Richardson, p. 588 (*pars*), pl. 93, figs 5–6.
1982		*Geminospora micromanifesta* var. *minor* (Naumova) McGregor and Camfield, p. 40, pl. 8, figs 14–15, 19–22.
1983		*Geminospora lemurata* Balme emend. Playford, p. 316, text-figs 1–9.
non	2007c	*Geminospora lemurata* Balme emend. Playford; Breuer *et al.*, pl. 8, figs 7–9.

Dimensions. 40(53)72 μm; 47 specimens measured.

Remarks. Fairly uncommon monolete specimens (Figs 6K–L) and rare tetralete specimens have been found. Rare occurrence of asymmetrically trilete, dilete and monolete specimens was also reported by Playford (1983, text-fig. 4A–C). These forms accounted for less than 0.05 per cent of the *G. lemurata* population studied by Playford.

Comparison. Specimens of *G. micromanifesta* var. *minor* (Naumova) McGregor and Camfield, 1982 possess a spectrum of morphological variations comparable to specimens of *Geminospora lemurata* Balme, 1962 emended by Playford (1983). *G. lemurata* 'early form' *in* Marshall (1996) differ from typical *G. lemurata* in being both smaller (mean 55 μm), only very rarely showing any appreciable separation between the nexine and sexine, the two layers being appressed although still distinguishable and possessing a thicker sexine (2–7 μm) where measured at the equatorial margin. *Rhabdosporites langii* (Eisenack) Richardson, 1960 has a generally thinner sexine and a less rigid appearance. In addition, it is larger. *G. svalbardiae* (Vigran) Allen, 1965 is not considered here as synonymous with *G. lemurata* because it can be distinguished from the typical *lemurata* form. *G. svalbardiae* has a generally thinner sexine resulting in more numerous folds.

Occurrence. S-462, WELL-1 and WELL-8; Jubah Formation; *lemurata-langii* to *langii-concinna* zones, specimens from S-462 may be caved in older strata. A1-69; Awaynat Wanin II Formation; *lemurata-langii* to *langii-concinna* zones. MG-1; Awaynat

FIG. 29. Each figured specimen is identified by borehole, sample, slide number and England Finder Co-ordinate location. All figured specimens are at magnification ×1000 except where mentioned otherwise. A, *Emphanisporites plicatus* sp. nov. Holotype, JNDL-1, 174.6 ft, 60847, D43. B, *Emphanisporites rotatus* McGregor emend. McGregor, 1973. BAQA-1, 222.5 ft, 03CW108, L32/2. C, *Emphanisporites schultzii* McGregor, 1973. JNDL-1, 156.0 ft, 60839, H47/1. D–E, *Emphanisporites* sp. 1. D, MG-1, 2181.2 m, 62524, E45. E, MG-1, 2314 m, 62799, W39/2. F–J, *Geminospora convoluta* sp. nov. F, Paratype, JNDL-1, 167.8 ft, PPM006, M37/3. G, JNDL-1, 495.0 ft, PPM014, K37/1. H, JNDL-1, 495.0 ft, 60855, K48. I, JNDL-1, 177.0 ft, 60849, D42/2. J, Holotype, JNDL-1, 177.0 ft, 60849, V29/2. K–L, *Geminospora lemurata* Balme emend. Playford, 1983. K, S-462, 2060–2065 ft, 63270, N28/4. L, S-462, 1470–1475 ft, 63212, R32/3.

Wanin II and Awaynat Wanin III formations; *lemurata-langii* to *langii-concinna* zones.

Previous records. Geminospora lemurata is eponymous for the Givetian *lemurata-magnificus* Assemblage Zone of the Old Red Sandstone Continent and adjacent regions (Richardson and McGregor 1986) and the early Givetian Lem Interval Zone of Western Europe (Streel *et al.* 1987). *G. lemurata* has a worldwide distribution extending through the Givetian and Frasnian, possibly reaching the basal Famennian.

Geminospora libyensis Moreau-Benoit, 1980*b*
Figure 30A

1976   *Geminospora libyensis* n. sp. Massa and Moreau-Benoit, pl. 6, fig. 4.
1980*b*   *Geminospora libyensis* Moreau-Benoit, p. 44, pl. 13, fig. 6.

Dimensions. 66(94)115 μm; four specimens measured.

Comparison. Acinosporites lindlarensis Riegel, 1968 has the same sculpture, but the sexine is never separated as well as in specimens of *Geminospora libyensis.*

Occurrence. JNDL-1; Jubah Formation; *svalbardiae-eximius* Zone. A1-69; Awaynat Wanin I and Awaynat Wanin II formations; *svalbardiae-eximius* to *rugulata-libyensis* zones. MG-1; Awaynat Wanin I Formation; *rugulata-libyensis* Zone.

Previous records. From middle Givetian of Algeria (Moreau-Benoit *et al.* 1993); and upper Emsian – lower Givetian of Libya (Moreau-Benoit, 1989).

Geminospora punctata Owens, 1971
Figure 30B–C

1965   Unidentified spore types Kerr *et al.*, pl. 4, figs 15–16.
1966   *Geminospora* sp. McGregor and Owens, pl. 15, figs 7–10.
1971   *Geminospora punctata* Owens, p. 61, pl. 19, figs 1–9.

Dimensions. 43(57)80 μm; 16 specimens measured.

Comparison. Geminospora lemurata Balme emend. Playford, 1983 is closely comparable in general construction but has a discrete, positive ornamentation.

Occurrence. S-462 and WELL-8; Jubah Formation; *lemurata-langii* to *langii-concinna* zones. A1-69; Awaynat Wanin II Formation; *lemurata-langii* to *langii-concinna* zones. MG-1; Awaynat Wanin II and Awaynat Wanin III formations; *lemurata-langii* to *langii-concinna* zones.

Previous records. From Eifelian–upper Tournaisian of Brazil (Loboziak *et al.* 1988, 1992*b*; Melo and Loboziak 2003); Frasnian of Canada (McGregor and Owens 1966; Owens 1971) and Iran (Ghavidel-Syooki 2003); uppermost Eifelian–Givetian of Germany (Loboziak *et al.* 1990); and Givetian of Poland (Turnau and Racki 1999).

Geminospora svalbardiae (Vigran) Allen, 1965
Figure 30D–F

1964   *Lycospora svalbardiae* Vigran, p. 23, pl. 3, figs 4–5, pl. 4, figs 1–2.
1965   *Geminospora svalbardiae* (Vigran) Allen, p. 696, pl. 94, figs 12–16.
non 1974   *Geminospora svalbardiae* (Vigran) Allen; Becker *et al.*, pl. 16, figs 16–19.
1988   *Geminospora lemurata* Balme emend. Playford; Boumendjel *et al.*, pl. 1, figs 18–19.
? 1996   *Geminospora lemurata* Balme emend. Playford 'early form'; Marshall, p. 171, pl. 2, figs 2–5.
2007*c*   *Geminospora lemurata* Balme emend. Playford; Breuer *et al.*, pl. 8, figs 7–9.

Dimensions. 49(68)87 μm; 44 specimens measured.

Comparison. Geminospora lemurata Balme emend. Playford, 1983 is often smaller and has generally a thicker sexine resulting in a more rigid spore. Specimens of *G. lemurata* 'early form' *in* Marshall (1996) seems to be similar to those of *G. svalbardiae*, but they have an equatorially thicker sexine (2–7 μm).

Occurrence. Saudi Arabia: JNDL-1; Jubah Formation; *svalbardiae-eximius* Zone. A1-69; Awaynat Wanin I and Awaynat Wanin II formations; *svalbardiae-eximius* to *lemurata* zones. MG-1; Ouan-Kasa and Awaynat Wanin I formations; *svalbardiae-eximius* to *rugulata-libyensis* zones.

FIG. 30. Each figured specimen is identified by borehole, sample, slide number and England Finder Co-ordinate location. All figured specimens are at magnification ×1000 except where mentioned otherwise. A, *Geminospora libyensis* Moreau-Benoit, 1980*b*, magnification ×750. MG-1, 2483 m, 62802, S46/4. B–C, *Geminospora punctata* Owens, 1971. B, A1-69, 1322 ft, 27126, P51. C, A1-69, 1322 ft, 27125, O34/1. D–F, *Geminospora svalbardiae* (Vigran) Allen, 1965. D, JNDL-1, 172.7 ft, PPM007, W31/4. E, JNDL-1, 172.7 ft, PPM007, L37/1. F, JNDL-1, 167.8 ft, PPM006, H33. G–H, *Grandispora cassidea* (Owens) Massa and Moreau-Benoit, 1976, magnification ×500. G, MG-1, 2161.8 m, 62529, J49/1. H, MG-1, 2465 m, 62852, R43/2. I–J, *Grandispora douglastownensis* McGregor, 1973, magnification ×500. I, A1-69, 1962 ft, 27277, U54/3. J, JNDL-1, 174.6 ft, PPM008, T32.

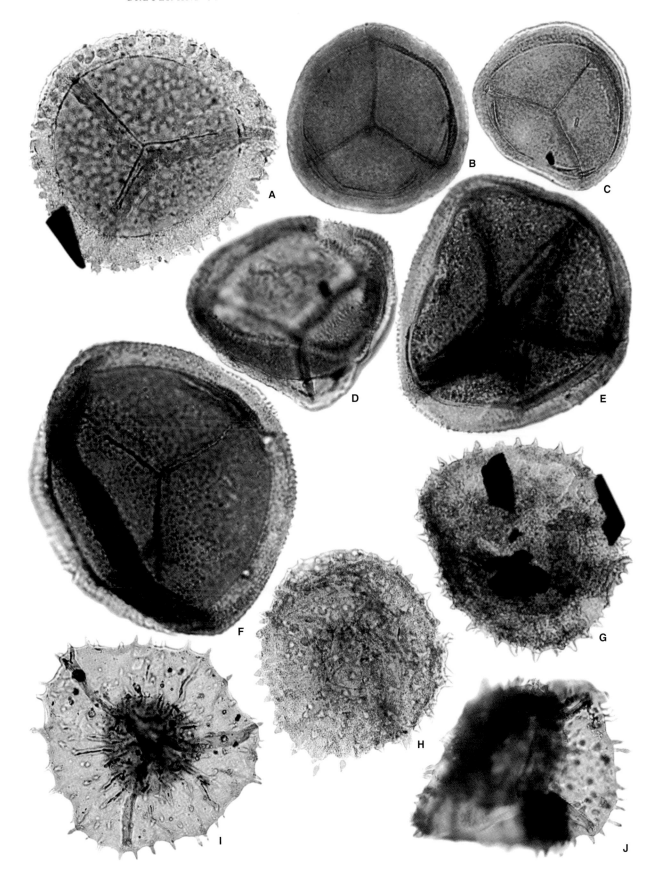

Previous records. From Emsian–lower Eifelian of Algeria (Moreau-Benoit *et al.* 1993); upper Eifelian – middle Givetian of Parnaíba Basin, Brazil (Breuer and Grahn 2011); Emsian–upper Frasnian of Libya (Moreau-Benoit 1989); Emsian–Frasnian of Spitsbergen, Norway (Vigran 1964; Allen 1965); and upper Givetian of Scotland (Marshall and Allen 1982).

Genus GRANDISPORA Hoffmeister *et al.* emend. Neves and Owens, 1966

Type species. *Grandispora spinosa* Hoffmeister *et al.*, 1955.

Remarks. Large, camerate spores with apparently similar exinal sculpture are common in Emsian to Givetian rocks, and considerable weight is ascribed to sculpture as a criterion for circumscribing species. However, most authors do not record the range of variation in shape and size of the sculptural elements. Consequently, it is difficult to make meaningful comparisons between many of the described species of this complex on the basis of ornamentation. In addition, relatively minor differences in sculpture form and distribution are apt to become untenable because of intergradation of many specimens from several populations of spores (McGregor 1973).

Grandispora cassidea (Owens) Massa and Moreau-Benoit, 1976
Figures 30G–H, 48J–O

1966 *Spinozonotriletes* sp. McGregor and Owens, pl. 18, fig. 8.
1971 *Spinozonotriletes cassideus* Owens, pl. 17, figs 3–5; text-fig. 11.
1976 *Grandispora cassidea* (Owens) Massa and Moreau-Benoit, table-fig. 5.
1980*b* *Grandispora cassidea* (Owens) Moreau-Benoit, p. 31, pl. 10, fig. 6; pl. 15, fig. 5.
1989 *Spinozonotriletes* cf. *cassideus* Owens; Moreau-Benoit, p. 13.

Dimensions. 95(123)163 μm; seven specimens measured.

Comparison. *Grandispora incognita* (Kedo) McGregor and Camfield, 1976 is comparable in general architecture, but the sculptural elements are more slender.

Ocurrence. MG-1; Ouan-Kasa, Awaynat Wanin I and Awaynat Wanin III formations; *svalbardiae-eximius* to *langii-concinna* zones.

Previous records. From Givetian–Frasnian of Paraná and Parnaíba basins, Brazil (Loboziak *et al.* 1988; Breuer and Grahn 2011); Frasnian of Canada (McGregor and Owens 1966;

Owens 1971); middle Givetian–lower Frasnian of Libya (Moreau-Benoit 1989); and Givetian of Saudi Arabia (PB, pers. obs.).

Grandispora douglastownensis McGregor, 1973
Figures 30I–J, 48P–X

? 1968 *Calyptosporites pilaspinosus* Lanninger, p. 153, pl. 26, fig. 1.
1973 *Grandispora douglastownense* McGregor, p. 62, pl. 8, figs 8–9, 12–14.
? 1988 *Grandispora* sp. B; Ravn and Benson, p. 191, pl. 6, figs 1–2.

Dimensions. 88(126)156 μm; 21 specimens measured.

Comparison. *Grandispora diamphida* Allen, 1965 is smaller, and the inner body is larger relative to the total diameter of the spore in the specimens illustrated. *G. naumovae* (Kedo) McGregor, 1973 has longer spines. *G. libyensis* Moreau-Benoit, 1980*b* is slightly larger with a strongly thickened sexine. Its general amb shape tends to be more triangular. *G. protea* (Naumova) Moreau-Benoit, 1980*b* has smaller, generally bulbous and widely distributed sculptural elements, but some specimens show intergradation with the population of *G. douglastownensis* described here. The two species are included in the *G. protea* Morphon defined here (Table 1). The possibility that *G. douglastownensis* and *G. ?macrotuberculata* (Arkangelskaya) McGregor, 1973, the synonymy of which is questioned herein with *G. protea* (Naumova) Moreau-Benoit, 1980*b*, represent end-members of the same complex is also supported by analysis of their gross structure and ultrastructure (Wellman and Gensel 2004).

Occurrence. JNDL-1 and WELL-1; Jubah Formation; *svalbardiae-eximius* to *triangulatus-catillus* zones. A1-69; Ouan-Kasa, Awaynat Wanin I and Awaynat Wanin II formations; *annulatus-protea* to *svalbardiae-eximius* zones. MG-1; Awaynat Wanin I Formation; *rugulata-libyensis* Zone.

Previous records. From upper Emsian – lower Eifelian of Algeria (Moreau-Benoit *et al.* 1993); Frasnian of Bolivia (Perez-Leyton 1990); upper Emsian – lower Givetian of Amazon and Parnaíba basins, Brazil (Loboziak *et al.* 1992*b*; Melo and Loboziak 2003); Emsian–Givetian of Canada: (McGregor 1973; McGregor and Camfield 1976, 1982); Givetian of France (Loboziak and Streel 1980); Eifelian of Germany (Loboziak *et al.* 1990); upper Emsian – lower Givetian (Ghavidel-Syooki 2003); and Emsian–middle Givetian of Libya (Moreau-Benoit 1989).

Grandispora fibrilabrata Balme, 1988
Figures 31A–C, 48Y–AA

1988 *Grandispora? fibrilabrata* Balme, p. 140, pl. 9, figs 6–8.

Dimensions. 87(95)105 µm; 12 specimens measured.

Remarks. The population of this species described by Balme (1988) is larger (144–255 µm). Although Balme (1988) was not certain of the allocation of this species to *Grandispora* Hoffmeister *et al.* emend. Neves and Owens, 1966, the specimens described here present all typical characters for this genus.

Occurrence. S-462; Jubah Formation; *triangulatus-catillus* to *langii-concinna* zones. A1-69; Awaynat Wanin II Formation; *triangulatus-catillus* Zone.

Previous record. From lower Frasnian of Carnarvon Basin, Australia (Balme 1988).

Grandispora gabesensis Loboziak and Streel, 1989
Figures 30D, 49A–C

1989 *Grandispora gabesensis* Loboziak and Streel, p. 181, pl. 6, figs 2–4; pl. 9, figs 17–20.

Dimensions. 65(96)133 µm; 13 specimens measured.

Occurrence. WELL-1; Jubah Formation; *lemurata-langii* to *triangulatus-catillus* zones. A1-69; Awaynat Wanin I and Awaynat Wanin II formations; *svalbardiae-eximius* to *lemurata-langii* zones. MG-1; Ouan-Kasa and Awaynat Wanin I formations; *svalbardiae-eximius* to *rugulata-libyensis* zones.

Previous records. From upper Emsian–Frasnian of Algeria (Moreau-Benoit *et al.* 1993); Givetian of Paraná Basin, Brazil (Loboziak *et al.* 1988); and Emsian of Morocco (Rahmani-Antari and Lachkar 2001).

Grandispora incognita (Kedo) McGregor and Camfield, 1976
Figures 31E–G, 49D–L

1955 *Archaeozonotriletes incognitus* Kedo, p. 33, pl. 4, fig. 9.
1976 *Grandispora incognita* (Kedo) McGregor and Camfield, p. 23, pl. 6, figs 9, 10.
1976 *Grandispora tomentosa* Taugourdeau-Lantz; McGregor and Camfield, p. 24, pl. 6, figs 4, 5, 8.
1976 *Grandispora* cf. *G. tomentosa* Taugourdeau-Lantz; McGregor and Camfield, p. 24, pl. 7, figs 2, 3.
1992 *Grandispora tomentosa* Taugourdeau-Lantz; McGregor and Playford, pl. 15, fig. 10.

Dimensions. 84(127)227 µm; 15 specimens measured.

Comparison. McGregor and Camfield (1976) illustrated both *Grandispora tomentosa* Taugourdeau-Lantz, 1967 and *Grandis-*

pora cf. *G. tomentosa* Taugourdeau-Lantz, 1967, which are similar to specimens figured as *G. incognita*. *G. cassidea* (Owens) Massa and Moreau-Benoit, 1976 has broad-based conate or spines with a bulbous appearance. *G. naumovae* (Kedo) McGregor, 1973 has longer spinae comparatively to the amb, but some specimens intergrade with *G. incognita*. The *G. incognita* Morphon is thus defined and include specimens characterized by slender spinae (Table 1).

Occurrence. S-462 and WELL-8; Jubah Formation; *lemurata-langii* to *triangulatus-catillus* zones. A1-69; Awaynat Wanin I and Awaynat Wanin II formations; *rugulata-libyensis* to *triangulatus-catillus* zones. MG-1; Awaynat Wanin I, Awaynat Wanin II and Awaynat Wanin III formations; *rugulata-libyensis* to *langii-concinna* zones.

Previous records. From Givetian–Frasnian of Paraná Basin, Brazil (Loboziak *et al.* 1988); Eifelian–Givetian of Canada (McGregor and Camfield 1976); and Givetian of Iran (Ghavidel-Syooki 2003).

Grandispora inculta Allen, 1965
Figures 31H, 49M

1965 *Grandispora inculta* Allen, p. 734, pl. 103, figs 7–9.

Dimensions. 57–64 µm; two specimens measured.

Occurrence. S-462, WELL-1 and WELL-8; Jubah Formation; *triangulatus-catillus* Zone. A1-69; Awaynat Wanin II Formation; *lemurata-langii* Zone.

Previous records. From Emsian–Frasnian of Algeria (Boumendjel *et al.* 1988; Moreau-Benoit *et al.* 1993); Frasnian of Bolivia (Perez-Leyton 1990); Givetian–lower Frasnian of Paraná Basin, Brazil (Loboziak *et al.* 1988); upper Eifelian – lower Givetian of Canada (McGregor and Camfield 1982); upper Givetian – upper Frasnian of France (Brice *et al.* 1979; Loboziak and Steel 1980, 1988; Loboziak *et al.* 1983); Emsian–middle Givetian of Libya (Paris *et al.* 1985; Streel *et al.* 1988; Moreau-Benoit 1989); Eifelian–Givetian of Morocco (Rahmani-Antari and Lachkar 2001); Givetian of Spitsbergen, Norway (Allen 1965); upper Eifelian–Givetian of Poland (Turnau 1996; Turnau and Racki 1999); upper Eifelian of Russian Platform (Avkhimovitch *et al.* 1993) and Scotland (Marshall 1988).

Grandispora libyensis Moreau-Benoit, 1980*b*
Figures 32A–B, 49N–Y, 50A–I

1967 *Spinozonotriletes echinatus* Moreau-Benoit, p. 230, pl. 3, fig. 51; pl. 4, figs 52–53.
1967 *Spinozonotriletes mamillatus* Moreau-Benoit, p. 231, pl. 4, figs 54–55.
1967 *Grandispora* sp. Daemon *et al.*, p. 115, pl. 3, figs 37–39.

1969 *Hymenozonotriletes* sp. no 2388 Lanzoni and Magloire, pl. 7, fig. 19, pl. 8, fig. 1.

1974 *Spinozonotriletes echinatus* Moreau-Benoit; Moreau-Benoit, p. 203, pl. 15, fig. 6.

1974 *Spinozonotriletes mamillatus* Moreau-Benoit; Moreau-Benoit, p. 203, pl. 15, fig. 7.

? 1974 *Spinozonotriletes* cf. *echinatus* Moreau-Benoit; Bär and Riegel, p. 44, pl. 1, fig. 14.

1976 *Grandispora echinata* (Moreau-Benoit) Massa and Moreau-Benoit, pl. 4, fig. 1; table-fig. 5.

1980b *Grandispora libyensis* Moreau-Benoit, p. 33, pl. 11, figs 2–3.

1989 *Spinozonotriletes libyensis* (Moreau-Benoit) Coquel and Moreau-Benoit, p. 96, pl. 3, fig. 5; pl. 4, fig. 3.

Dimensions. 133(166)194 μm; 22 specimens measured.

Remarks. It appears that specimens of *G. libyensis* show a continuous morphological variation in ornamentation, intergrading from morphotypes with rather slender spines to ones characterized by bulbous biform elements. Although the two end-members exist, all intermediate forms are present. The morphotype characterized by the most massive sculptural elements seems to appear later than the morphotype with more slender ornaments, but in the youngest samples, the two end-members co-occur (Breuer *et al.* 2007a).

Comparison. Grandispora douglastownensis McGregor, 1973 is slightly smaller and not as distinctly thickened equatorially. Moreover, the ornamentation is less bulbous at the base. *G. velata* (Richardson) McGregor, 1973 is more widely sculptured with commonly smaller pointed spinae and coni. In addition, the sexine is not equatorially thickened as in *G. libyensis*.

Occurrence. S-462; Jubah Formation; *triangulatus-catillus* to *langii-concinna* zones. A1-69; Awaynat Wanin II Formation; *lemurata-langii* to *triangulatus-catillus* zones. MG-1; Awaynat Wanin I, Awaynat Wanin II and Awaynat Wanin III formations; *rugulata-libyensis* to *langii-concinna* zones.

Previous records. From Emsian – lower Visean of Algeria (Boumendjel *et al.* 1988; Coquel and Moreau-Benoit 1989; Moreau-Benoit *et al.* 1993); upper Givetian – lower Frasnian of Argentina (Ottone 1996); upper Eifelian–Frasnian of Amazon and Paraná basins, Brazil (Loboziak *et al.* 1988; Melo and Loboziak 2003); late Emsian–?Tournaisian of Libya (Paris *et al.* 1985; Coquel and Moreau-Benoit 1986, 1989; Streel *et al.* 1988;

Moreau-Benoit 1989); and upper Famennian–Tournaisian of Morocco (Rahmani-Antari and Lachkar 2001).

Grandispora maura sp. nov.
Figures 32C–E, 50J–L

Derivation of name. From *maurus* (Latin), meaning from Maghreb; refers to its geographical occurrence.

Holotype. EFC K37/3 (Figs 32C, 50K), slide 62942.

Paratype. EFC Q51/4 (Fig. 32E), slide 63025; MG-1 borehole, sample 172.7 ft.

Type locality and horizon. MG-1 borehole, sample 2247 m; Awaynat Wanin II Formation at Mechiguig, Tunisia.

Diagnosis. A three-layered *Grandispora* sculptured with closely spaced small spines, parallel-sided or tapered upwards, rounded- or acute-tipped, or biform elements with a more bulbous base supporting a delicate minute tip.

Description. Amb is sub-circular. Laesurae are straight, accompanied by labra, 1.5–4 μm in total width and extending to equator of inner body. The inner body diameter three-quarters to nine-tenths of the total amb diameter. Diameter of the middle layer about nine-tenths of the total amb diameter. The middle layer is rarely appressed to the inner body. Inner body is laevigate and middle layer commonly infragranular to granular. Outer layer is laevigate, but sculptured distally and equatorially with closely spaced small spines, parallel-sided or tapered upwards, rounded- or acute-tipped, or biform elements with a more bulbous base supporting a delicate minute tip. Elements are up to 3 μm high, up to 1.5 μm wide at base and 1–2 μm apart. Ornamentation can be coalescent equatorially. Specimens are often folded distally.

Dimensions. 65(76)90 μm; eight specimens measured.

Comparison. Grandispora maura is clearly separable from all other species of *Grandispora* by the possession of three detached layers.

Occurrence. A1-69; Awaynat Wanin II Formation; *incognita* to *lemurata-langii* zones. MG-1; Awaynat Wanin I, Awaynat Wanin II and Awaynat Wanin III formations; *incognita* to *langii-concinna* zones.

FIG. 31. Each figured specimen is identified by borehole, sample, slide number and England Finder Co-ordinate location. All figured specimens are at magnification ×1000 except where mentioned otherwise. A–C, *Grandispora fibrilabrata* Balme, 1988, magnification ×750. A, S-462, 2010–2015 ft, 63266, S34. B, S-462, 1570–1575 ft, 63215, N45. C, S-462, 1860–1865 ft, 63259, P32/4. D, *Grandispora gabesensis* Loboziak and Streel, 1989, magnification ×750. A1-69, 1962 ft, 27278, H51. E–G, *Grandispora incognita* (Kedo) McGregor and Camfield, 1976, magnification ×500. E, A1-69, 1530 ft, 26984, S39. F, A1-69, 1540 ft, 26987, H37/1. G, A1-69, 1540 ft, 26988, N40/3. H, *Grandispora inculta* Allen, 1965. S-462, 2060–2065 ft, 63270. N30/3.

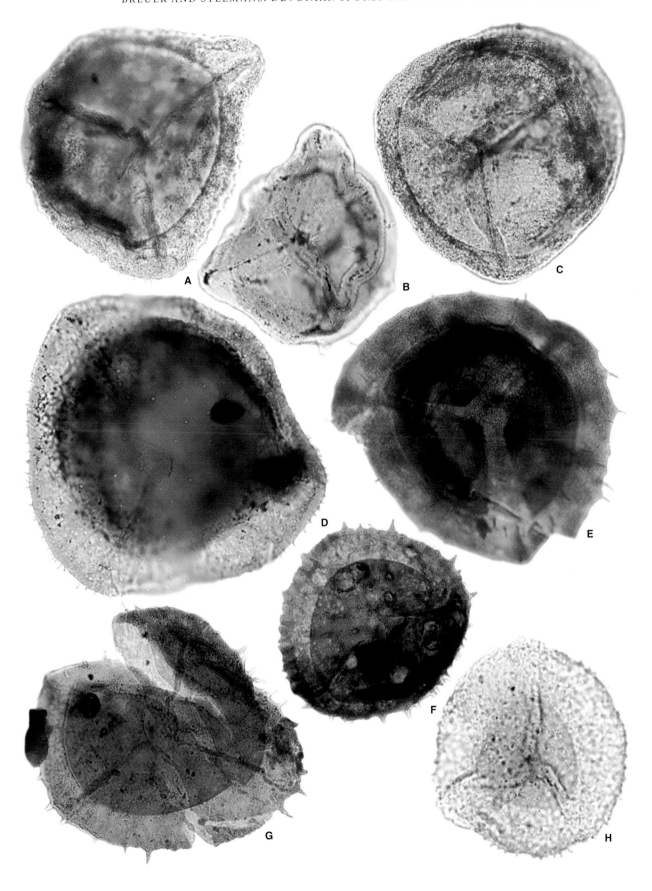

Previous record. From Frasnian of Saudi Arabia (PB, pers. obs.).

Grandispora naumovae (Kedo) McGregor, 1973
Figures 32F, 33A and 50M–U

1925 Spore type H Lang, p. 257, pl. 1, figs 18–19.
1955 *Archaeozonotriletes naumovii* Kedo, p. 33, pl. 4, fig. 8.
1965 ?*Spinozonotriletes* cf. *naumovii* (Kedo) Richardson, p. 583, pl. 92, figs 3–5; text-fig. 7.
1966 ?*Spinozonotriletes* cf. *naumovii* (Kedo) Richardson; McGregor and Owens, pl. 7, fig. 5.
1966 *Spinozonotriletes* sp. cf. *S. naumovii* (Kedo) Richardson; de Jersey, p. 18, pl. 9, figs 1–4, 6.
1966 *Spinozonotriletes tuberculatus* Neves and Owens, p. 356, pl. 3, figs 4–5.
? 1967 *Spinozonotriletes naumovii* (Kedo); Hemer and Nygreen, pl. 2, fig. 9.
1968 *Spinozonotriletes* cf. *naumovii* (Kedo) Richardson; Lanninger, p. 150, pl. 25, fig. 10.
? 1969 *Spinozonotriletes* cf. *tuberculatus* Neves and Owens; Peppers and Damberger, p. 16, pl. 5, fig. 1.
? 1969 *Spinozonotriletes* cf. *naumovii* (Kedo) Richardson; Peppers and Damberger, p. 16, pl. 5, fig. 2.
1970 *Spinozonotriletes naumovii* (Kedo) Richardson; McGregor *et al.*, pl. 2, fig. 13.
1973 *Grandispora* ?*naumovii* (Kedo) McGregor, p. 61, pl. 9, figs 1–3.
1986 *Grandispora naumovii* (Kedo) McGregor; Richardson and McGregor, pl. 13, fig. 1.

Dimensions. 87(110)135 µm; 10 specimens measured.

Remarks. The range of variation in this species includes those with mostly rather small, delicate spines and those with large, rigid spines. Mostly commonly, but not invariably, the larger spores bear the most robust spines.

Comparison. *Grandispora incognita* (Kedo) McGregor and Camfield, 1976 has generally smaller spinae, which are also shorter comparatively to amb, but the two species intergrade in the *G. incognita* Morphon (Table 1).

Occurrence. S-462; Jubah Formation; *langii-concinna* Zone. A1-69; Awaynat Wanin I and Awaynat Wanin II formations; *svalbardiae-eximius* to *langii-concinna* zones.

Previous records. From lower Frasnian of Australia (Balme 1988); Emsian–Eifelian of Canada (McGregor and Owens 1966; McGregor 1973); Emsian of Germany (Lanninger 1968); Givetian of Poland (Turnau and Racki 1999) and Scotland (Richardson 1965; Marshall and Allen 1982).

Grandispora permulta (Daemon) Loboziak et al., 1999
Figures 33B–C, 50V–X and 51A–F

1967 *Calyptosporites* sp. A Daemon *et al.*, p. 114, pl. 3, figs 31–34.
1967 *Calyptosporites* sp. B Daemon *et al.*, p. 114, pl. 3, figs 35–36.
1974 *Calyptosporites* sp. B Bär and Riegel, pl. 1, fig. 13.
1974 *Contagisporites permultus* Daemon, p. 574, pl. 3, figs 4–5.
1980b *Grandispora velata* (Eisenack) McGregor (*pars*); Moreau-Benoit, pl. 12, fig. 3.
1985 *Grandispora macrotuberculata* (Arkhangelskaya) McGregor; Massa and Moreau-Benoit, pl. 1, fig. 6.
1985 *Grandispora* sp. A Paris *et al.*, pl. 24, figs 8–9.
1985 *Grandispora* sp. B Paris *et al.*, pl. 24, fig. 10.
1987 *Grandispora* sp. A Schrank, pl. 1, fig. 11.
1989 *Grandispora riegelii* Loboziak and Steel, p. 190, pl. 5, figs 1–5, pl. 9, figs 10–13.
1992 *Grandispora riegelii* Loboziak and Steel; Loboziak *et al.*, pl. 1, fig. 16.
1995a *Grandispora riegelii* Loboziak and Steel; Loboziak and Steel, pl. 1, fig. 1.
1995b *Grandispora riegelii* Loboziak and Steel; Loboziak and Steel, pl. 1, fig. 1.
1999 *Grandispora permulta* (Daemon) Loboziak *et al.*, p. 99, pl. 1, figs 1–6.

Dimensions. 68(99)125 µm; 30 specimens measured.

Comparison. *Grandispora inculta* Allen, 1965 is smaller and bears mainly coni. *G. gabesensis* Loboziak and Steel, 1989 has spinae and capilli in addition to coni and biform elements. Its ornamentation is also larger than in specimens of *G. permulta* (Daemon) Loboziak *et al.*, 1999.

Occurrence. S-462, WELL-1 and WELL-8; Jubah Formation; *rugulata-libyensis* to *langii-concinna* zones. A1-69; Awaynat Wanin I and Awaynat Wanin II formations; *rugulata-libyensis* to *triangulatus-catillus* zones. MG-1; Ouan-Kasa, Awaynat Wanin I, Awaynat Wanin II and Awaynat Wanin III formations; *annulatus-protea* to *langii-concinna* zones.

FIG. 32. Each figured specimen is identified by borehole, sample, slide number and England Finder Co-ordinate location. All figured specimens are at magnification ×1000 except where mentioned otherwise. A–B, *Grandispora libyensis* Moreau-Benoit, 1980*b*, magnification ×500. A, A1-69, 1416 ft, 26993, K31/3. B, A1-69, 1416 ft, 26993, G31/3. C–E, *Grandispora maura* sp. nov. C, Holotype, MG-1, 2247 m, 62942, K37/3. D, MG-1, 2181.2 m, 62525, N34/4. E, Paratype, MG-1, 2292 m, 63025, Q51/4. F, *Grandispora naumovae* (Kedo) McGregor, 1973, magnification ×500. A1-69, 1530 ft, 26984, J34/2.

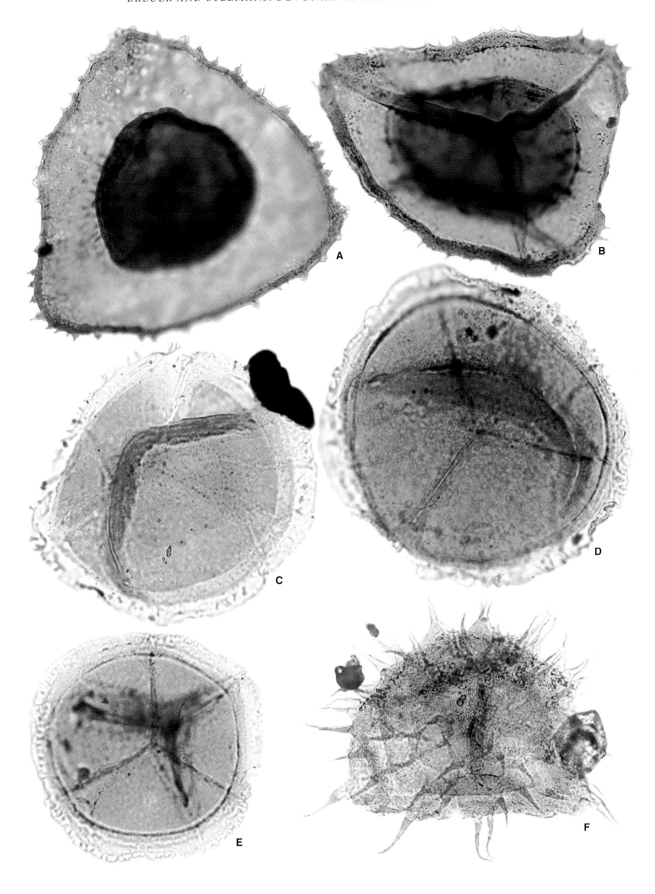

Previous records. From middle Givetian of Algeria (Moreau-Benoit *et al.* 1993); upper Givetian – lower Frasnian of Argentina (Ottone 1996); upper Eifelian – lower Frasnian of Bolivia (Perez-Leyton 1990); Eifelian–Frasnian of Amazon and Paraná basins, Brazil (Loboziak *et al.* 1988; Melo and Loboziak 2003); and lower Eifelian – lower Givetian of Libya (Paris *et al.* 1985; Streel *et al.* 1988).

Grandispora protea (Naumova) Moreau-Benoit, 1980*b*
Figures 33D–E, 51G–R

1953   *Hymenozonotriletes proteus* Naumova, p. 40, pl. 4. fig. 5.

1955   *Hymenozonotriletes proteus* var. *eximius* Kedo, p. 31, pl. 4, fig. 3.

1965   *Calyptosporites proteus* (Naumova) Allen, p. 735, pl. 103, figs 10–11.

1968   *Calyptosporites proteus* (Naumova) Allen, Lanninger, p. 153, pl. 25, fig. 14.

1968   *Calyptosporites proteus* (Naumova) Allen; Riegel, p. 91, pl. 20, figs 2–4.

? 1968   ?*Hymenozonotriletes* sp.; Jardiné and Yapaudjian, pl. 2, fig. 9.

? 1973   *Grandispora* ?*macrotuberculata* (Arkhangelskaya) McGregor, p. 59, pl. 8, figs 1–5.

1975   *Calyptosporites proteus* (Naumova) Allen; Tiwari and Schaarschmidt, p. 41, pl. 23, fig. 1; text-fig. 30.

1976   *Grandispora protea* (Naumova) Massa and Moreau-Benoit, pl. 4, fig. 1; table-fig. 5.

1980*b*   *Grandispora protea* (Naumova), Moreau-Benoit p. 37, pl. 11, fig. 6.

Dimensions. 73(122)152 μm; 22 specimens measured.

Comparison. Forms that may be similar are ?*Hymenozonotriletes* sp. *in* Jardiné and Yapaudjian (1968) and *G.* ?*macrotuberculata* (Arkhangelskaya) McGregor, 1973. *G. megaformis* (Richardson) McGregor, 1973 is larger and has wider ornaments. *G. velata* (Richardson) McGregor, 1973 is very similar in size, but has an ornamentation of much smaller spinae and coni, which have acute rather than rounded apices. *G. megaformis* has the same ornamentation but is larger in diameter. *Calyptosporites microspinosus* (Richardson) Richardson, 1962 is considerably larger and has small bifurcate spinae. *G. douglastownensis* McGregor, 1973 has longer spinae and is more densely sculptured, but some specimens intergrade with *G. protea*; the two species are thus included here in the *G. protea* Morphon (Table 1).

Occurrence. JNDL-1; Jauf (Murayr Member) and Jubah formations; *annulatus-protea* to *rugulata-libyensis* zones. A1-69; Ouan-Kasa, Awaynat Wanin I and Awaynat Wanin II formations; *annulatus-protea* to *rugulata-libyensis* zones. MG-1; Ouan-Kasa, Awaynat Wanin I and Awaynat Wanin II formations; *annulatus-protea* to *rugulata-libyensis* zones.

Previous records. *Grandispora protea* is eponymous for the upper Emsian – lower Eifelian AP Oppel Zone of Western Europe (Streel *et al.* 1987). *G. protea* has an almost worldwide distribution extending from upper Emsian into the Frasnian. It is notably reported in Algeria (Moreau-Benoit *et al.* 1993), Argentina (Ottone 1996), Belgium (Gerrienne *et al.* 2004), Bolivia (Perez-Leyton 1990), Brazil (Loboziak *et al.* 1988; Melo and Loboziak 2003), Canada (McGregor and Uyeno 1972; McGregor and Camfield 1976, 1982), Germany (Riegel 1968, 1973; Tiwari and Schaarschmidt 1975; Loboziak *et al.* 1990), Libya (Moreau-Benoit 1989), Spitsbergen, Norway (Allen 1965), Poland (Turnau 1986, 1996), Russian Platform (Avkhimovitch *et al.* 1993) and Scotland (Marshall and Allen 1982; Marshall 1988; Marshall and Fletcher 2002).

Grandispora rarispinosa Moreau-Benoit, 1980*b*
Figures 33F–H, 51S–X

1976   *Grandispora rarispinosa* Massa and Moreau-Benoit, pl. 5, fig. 7.

1980*b*   *Grandispora rarispinosa* Moreau-Benoit, p. 38, pl. 12, fig. 1.

Dimensions. 66(86)108 μm; seven specimens measured.

Comparison. *Grandispora protea* (Naumova) Moreau-Benoit, 1980*b* is larger and does not have the spongy appearing sexine.

Occurrence. S-462; Jubah Formation; *langii-concinna* Zone. MG-1; Awaynat Wanin I, Awaynat Wanin II and Awaynat Wanin III formations; *incognita* to *langii-concinna* zones.

Previous records. From lower–middle Givetian of Algeria (Boumendjel *et al.* 1988; Moreau-Benoit *et al.* 1993); and lower Eifelian – upper Frasnian of Libya (Moreau-Benoit 1989).

FIG. 33. Each figured specimen is identified by borehole, sample, slide number and England Finder Co-ordinate location. All figured specimens are at magnification ×1000 except where mentioned otherwise. A, *Grandispora naumovae* (Kedo) McGregor, 1973, magnification ×500. S-462, 1660–1665 ft, 63219, L38/4. B–C, *Grandispora permulta* (Daemon) Loboziak *et al.*, 1999, magnification ×750. B, A1-69, 1277 ft, 62637, V49/1. C, A1-69, 1296 ft, 62643, G34/1. D–E, *Grandispora protea* (Naumova) Moreau-Benoit, 1980*b*, magnification ×500. D, A1-69, 1962 ft, 27278, T36/3. E, JNDL-1, 155.6 ft, PPM003, D32. F–H, *Grandispora rarispinosa* Moreau-Benoit, 1980*b*, magnification ×500. F, MG-1, 2182.4 m, 62527, Q32. G, MG-1, 2285 m, 62845, L26/3. H, MG-1, 2413 m, 62776, R35/2. I–L, *Grandispora (Calyptosporites) stolidota* (Balme) comb. nov., magnification ×500. I, MG-1, 2375 m, 62772, O25. J, A1-69, 1322 ft, 27126, Q39/3. K, MG-1, 2476 m, 63015, F44. L, MG-1, 2476 m, 63015, P38.

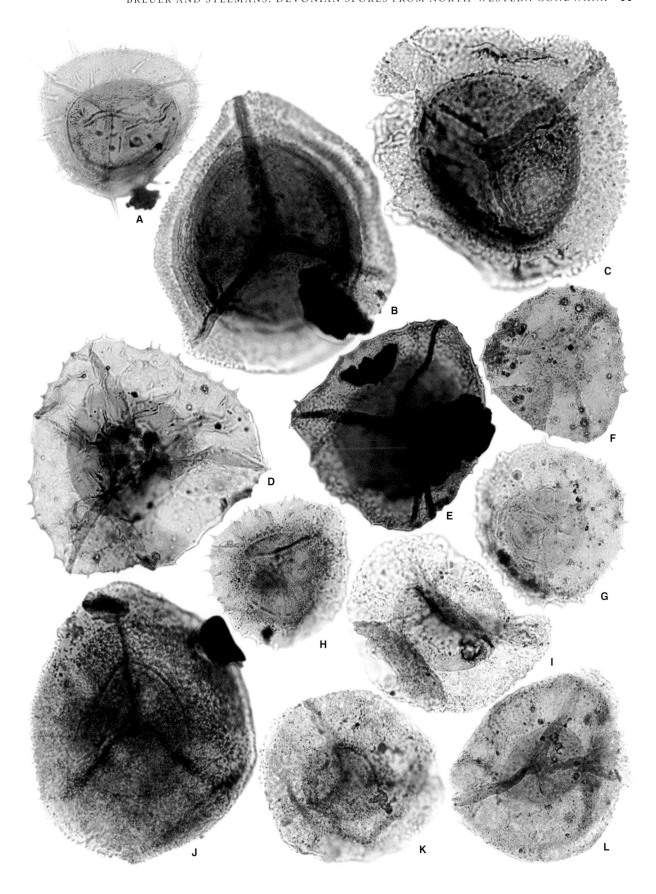

Grandispora (Calyptosporites) stolidota (Balme) comb. nov.
Figures 33I–L, 51Y–AA

1988 *Calyptosporites stolidotus* Balme, p. 141, pl. 10,
figs 8–10.

Dimensions. 103(125)161 μm; 14 specimens measured.

Remarks. Since *Calyptosporites* Richardson, 1962 is considered as a junior synonym of *Grandispora* Hoffmeister *et al.* emend. Neves and Owens, 1966, *C. stolidotus* Balme, 1988 is here transferred.

Comparison. According to Balme (1988), *G. uyenoi* McGregor and Camfield, 1982 and *Rhabdosporites* sp. *in* McGregor and Camfield, 1982 were the same morphology as *G. stolidota*. *G. uyenoi* differs in possessing curvaturate contact faces, and *Rhabdosporites* sp. *in* McGregor and Camfield, 1982 has a more uniform sculpture without biform elements.

Occurrence. S-462 and WELL-1; Jubah Formation; *rugulata-libyensis* to *triangulatus-catillus* zones but some specimens from S-462 may be caved in older strata. A1-69; Awaynat Wanin II Formation; *rugulata-libyensis* to *triangulatus-catillus* zones. MG-1; Awaynat Wanin I, Awaynat Wanin II and Awaynat Wanin III formations; *rugulata-libyensis* to *langii-concinna* zones.

Previous records. From middle Givetian – lower Frasnian of Australia (Balme 1988; Grey 1991).

Grandispora velata (Richardson) McGregor, 1973
Figures 34A–B, 51AB–AD and 52A–C

1944 *Triletes velatus* Eisenack p. 108 (*pars*), pl. 1,
figs 1–3.
1960 *Cosmosporites velatus* (Eisenack); Richardson, p. 52,
pl. 14, fig. 4.
1962 *Calyptosporites velatus* (Eisenack); Richardson,
p. 192.
1965 *Calyptosporites velatus* (Eisenack); Richardson,
p. 587, pl. 93, fig. 4.
1973 *Grandispora velata* (Richardson) McGregor,
p. 61, pl. 8, figs 10–11.

Dimensions. 106(131)170 μm; 18 specimens measured.

Comparison. *Grandispora libyensis* Moreau-Benoit, 1980*b* is densely sculptured with commonly larger spinae which can be bulbous. In addition, the sexine is thickened equatorially.

Occurrence. JNDL-1 and S-462; Jubah Formation; *svalbardiae-eximius* to *lemurata-langii* zones. A1-69; Awaynat Wanin I and Awaynat Wanin II formations; *svalbardiae-eximius* to *lemurata-langii* zones. MG-1; Awaynat Wanin I Formation; *rugulata-libyensis* Zone.

Previous records. *Grandispora velata* is eponymous for the lower Eifelian *velata-langii* Assemblage Zone of the Old Red Sandstone Continent and adjacent regions (Richardson and McGregor 1986). *R. langii* has a worldwide distribution extending into the Frasnian. *G. velata* is widely reported from upper Emsian through Frasnian from many parts of the world; e.g. Argentina (Amenábar 2009), Australia (Hashemi and Playford 2005), Russian Platform (Avkhimovitch *et al.* 1993), Belgium (Lessuise *et al.* 1979), Bolivia (Perez-Leyton 1990), Brazil (Loboziak *et al.* 1988), Canada (McGregor and Owens 1966; McGregor and Uyeno 1972; McGregor 1973; McGregor and Camfield 1976, 1982), China (Gao Lianda 1981), France (Brice *et al.* 1979; Loboziak and Streel 1980, 1988), Germany (Riegel 1968, 1973; Tiwari and Schaarschmidt 1975), Greenland (Friend *et al.* 1983; Marshall and Hemsley 2003), Iran (Ghavidel-Syooki 2003), Libya (Streel *et al.* 1988), Poland (Turnau 1996), Portugal (Lake *et al.* 1988), Spain (Cramer 1969), Scotland (Richardson 1965; Marshall and Allen 1982; Marshall 1988; Marshall and Fletcher 2002) and Georgia, USA (Ravn and Benson 1988).

Genus GRANULATISPORITES Ibrahim emend. Potonié and Kremp, 1954

Type species. *Granulatisporites granulatus* Ibrahim, 1933.

Granulatisporites concavus sp. nov.
Figure 34C–G

? 1976 *Granulatisporites muninensis* Allen; Massa and Moreau-Benoit, pl. 3, fig. 5.
? 1991 *Granulatisporites muninensis* Allen; Grigani *et al.*, pl. 10, figs 3–4.
? 2003 *Granulatisporites granulatus* Ibrahim; Melo and Loboziak, pl. 5, fig. 13.

Derivation of name. From *concavus* (Latin); refers to the concave shape of the spore body.

FIG. 34. Each figured specimen is identified by borehole, sample, slide number and England Finder Co-ordinate location. All figured specimens are at magnification ×1000 except where mentioned otherwise. A–B, *Grandispora velata* (Richardson) McGregor, 1973, magnification ×500. A, A1-69, 1530 ft, 26984, H40/1. B, A1-69, 1416 ft, 26992, U47/2. C–G, *Granulatisporites concavus* sp. nov. C, A1-69, 2039–2041 ft, 27279, M-N43. D, Paratype, JNDL-1, 162.3 ft, PPM005, C29/4. E, JNDL-1, 156.0 ft, 60840, N27-28. F, JNDL-1, 495.0 ft, 60855, L36/3. G, Holotype, A1-69, 1483 ft, 26995, R33/3. H–J, *Hystricosporites brevispinus* sp. nov., magnification ×750. H, Holotype, MG-1, 2295 m, 63007, P37. I, Paratype, MG-1, 2375 m, 62772, H34/1. J, MG-1, 2222.7 m, 62680, P37; focus on the distal face. K–L, *Hystricosporites* sp. 1, magnification ×750. K, MG-1, 2160.6 m, 62746, N49. L, MG-1, 2160.6 m, 62727, E49.

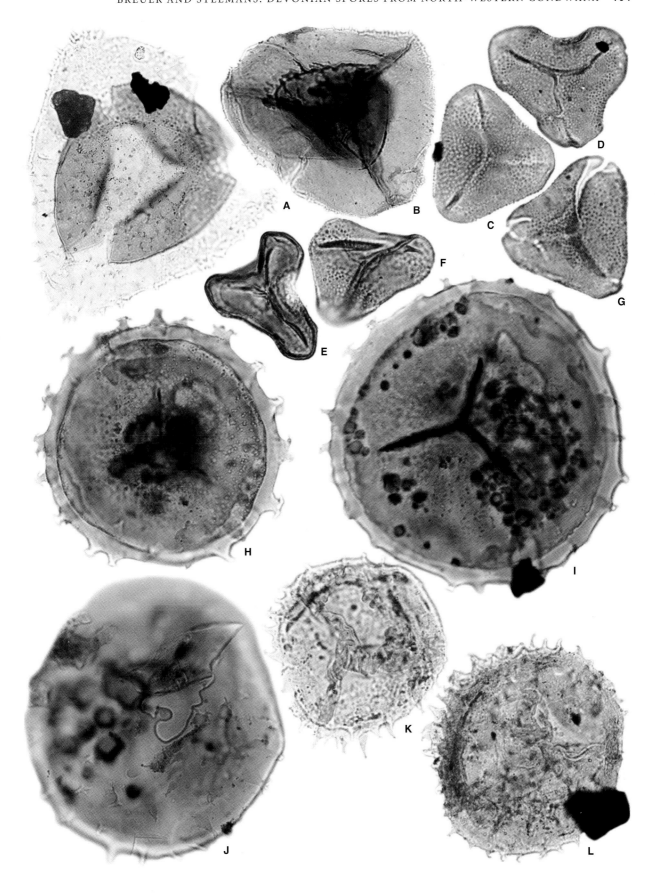

Holotype. EFC R33/3 (Fig. 34G), slide 26995.

Paratype. EFC F41 (Fig. 34D), slide 26987; A1-69 borehole, sample 1540 ft.

Type locality and horizon. A1-69 borehole, sample 1483 ft; Awaynat Wanin II Formation in A1-69, Libya.

Diagnosis. A triangular *Granulatisporites* with the margins concave or straight. Distal and equatorial regions sculptured with densely distributed grana.

Description. Amb is triangular. The corners are rounded, while the margins are concave or straight. Exine 0.5–1 µm thick. Laesurae are simple, straight, three-quarters to nine-tenths of the amb radius in length. Curvaturae are visible. Distal and equatorial sculpture are composed of densely distributed grana less than 0.5 µm high, less than 1 µm wide at their base, and 0.5–1 µm apart.

Dimensions. 31(36)40 µm; six specimens measured.

Comparison. *Granulatisporites muninensis* Allen, 1965 has a triangular amb with straight to slightly convex margins. *G. muninensis* illustrated in Massa and Moreau-Benoit (1976) and Grignani *et al.* (1991) may correspond to species described herein; however, no description was given. As the original paper where *G. granulatus* Ibrahim, 1933 was defined is not available, it cannot be compared. The specimen of *G. granulatus* illustrated in Melo and Loboziak (2003) may be similar to *G. concavus*, but no description was given.

Occurrence. Saudi Arabia. JNDL-1; Jauf (Murayr Member) and Jubah formations; *annulatus-protea* to *svalbardiae-eximius* zones. A1-69; Ouan-Kasa, Awaynat Wanin I and Awaynat Wanin II formations; *lindlarensis-sextantii* to *lemurata* Zone.

Genus HYSTRICOSPORITES McGregor, 1960

Type species. Hystricosporites delectabilis McGregor, 1960.

Hystricosporites brevispinus sp. nov.
Figures 34H–J, 52D–F

Derivation of name. From *brevi-* and *spinus* (Latin); refers to the sculptural elements of short spines.

Holotype. EFC P37 (Figs 34H, 52D), slide 63007.

Paratype. EFC H34/1 (Figs 34I, 52E), slide 62772; MG-1 borehole, sample 2375 m.

Type locality and horizon. MG-1 borehole, sample 2295 m; Awaynat Wanin II Formation at Mechiguig, Tunisia.

Diagnosis. A thick-walled *Hystricosporites* sculptured with short multifurcate grapnel-tipped process terminations.

Description. Amb is circular to sub-circular. Laesurae are straight, approximately one-half to two-thirds of the amb radius in length, accompanied by smooth, labra, 2–4.5 µm in total width. Curvaturae are indistinct. Exine is 3–6 µm thick equatorially and laevigate. Proximo-equatorial and distal surfaces are sculptured with multifurcate grapnel-tipped ornament, 2–9 µm long, 2–10 µm wide at their base. Sculptural elements are 15–25 around the equatorial margin; their bases are robust, grapnel-tipped or often divided upwards into two or three parts, which are themselves grapnel-tipped. The first divided branches are 1–3 µm long and their grapnel-shaped tips are smaller. On the less well-preserved specimens, the grapnel-shaped tips are commonly broken.

Dimensions. 77(91)108 µm; four specimens measured.

Comparison. This thick-walled species does not resemble any species of *Hystricosporites* McGregor, 1960. The unique feature, which defines this genus, is the very short divided sculptural elements.

Occurrence. MG-1; Awaynat Wanin II Formation; *incognita* to *triangulatus-catillus* zones.

Hystricosporites sp. 1
Figures 34K–L, 52G–I

Description. Amb is circular to sub-circular. Laesurae are straight to sinuous, generally accompanied by elevated labra, 7–25 µm high. Exine is 2.5–5 µm thick equatorially and laevigate. Proximo-equatorial and distal surfaces are ornamented with grapnel-tipped spines, commonly 4–8 µm (rarely as much as 10 µm) long, 1.5–4 µm wide at their base, 3–6 µm apart.

Dimensions. 63(79)95 µm; four specimens measured.

Remarks. This form may correspond rather to the definition of the genus *Nikitinsporites* Chaloner, 1959, it is placed in the genus *Hystricosporites* McGregor, 1960 because the elevated labra do not form here an apical prominence as required for *Nikitinsporites*.

Occurrence. MG-1; Awaynat Wanin III Formation; *langii-concinna* Zone.

Genus IBEROESPORA Cramer and Díez, 1975

Type species. Iberoespora cantabrica Cramer and Díez, 1975.

Iberoespora cantabrica Cramer and Díez, 1975
Figure 35A

1968   Spore trilète à papilles proximales sp. 2 Jardiné
        and Yapaudjian, pl. 1, figs 8–9.
1975   *Iberoespora cantabrica* Cramer and Díez, p. 339,
        pl. 2, figs 24, 26–28, 30–31.
1980a   ?*Geminospora* sp. A Moreau-Benoit, p. 73, pl. 10,
        fig. 8.
1980a   ?*Geminospora* sp. B Moreau-Benoit, p. 73, pl. 10,
        fig. 9.
1981   *Iberoespora glabella* Cramer and Díez; Steemans,
        pl. 1, figs 8–9.

Dimensions. 26(33)41 µm; seven specimens measured.

Comparison. Specimens identified here as *Iberoespora* cf. *I. guzmani* Cramer and Díez, 1975 have a cingulum well-divided by short radially oriented muri and do not seem to have proximal inspissations.

Occurrence. BAQA-2; Jauf Formation (Sha'iba Member); *papillensis-baqaensis* to *ovalis* zones. MG-1; Ouan-Kasa Formation; *svalbardiae-eximius* Zone but occurrences are probably reworked.

Previous records. From lower Lochkovian – upper Pragian of Belgium (Steemans 1989); upper Lochkovian of Solimões Basin, Brazil (Rubinstein *et al.* 2005); lower Lochkovian – lower Emsian of Armorican Massif, France (Le Hérissé 1983; Steemans 1989); lower Lochkovian of Germany (Steemans 1989); middle Přídolí of Libya (Rubinstein and Steemans 2002); and Lochkovian–lower Pragian of Spain (Cramer and Díez 1975; Rodriguez 1978*b*).

Iberoespora cf. *I. guzmani* Cramer and Díez, 1975
Figure 35B–E

cf. 1975   *Iberoespora guzmani* Cramer and Díez, p. 340,
            pl. 2, figs 23, 25, 32.

Description. Amb is circular to sub-triangular. Laesurae are rarely visible, straight, accompanied by narrow labra, up to 1 µm wide in overall width, extending to the inner margin of cingulum. Cingulum is 2–4 µm wide, divided in short radially oriented muri, 0.5–2 µm wide giving a crenulate appearance to the cingulum. A narrow, straight-edged and flat-bottomed, slightly sinuous furrow, generally 1–2 µm wide, separates the cingulum from the distal face of the spore body. Distal face is sculptured with convolute rugulae, generally 1–2.5 µm wide and less than 1 µm apart, resulting in a pseudoreticulum.

Dimensions. 32(39)45 µm; eight specimens measured.

Remarks. *Iberoespora guzmani* Cramer and Díez, 1975 clearly exhibits inspissations on the interradial areas, whereas the Saudi Arabian specimens seem to possess none. The proximal face, however, may have been torn because the laesurae are rarely observed.

Comparison. *Iberoespora noninspissatosa* Steemans, 1989 does not have inspissations and seems very similar to specimens described here, but it does not possess short radially oriented muri on the cingulum. The studied specimens belonging to *I. cantabrica* Cramer and Díez, 1975 are proximally sculptured with inspissations and do not have a crenulate cingulum.

Occurrence. JNDL-4, WELL-4 and WELL-7; Jauf Formation (Hammamiyat Member); *lindlarensis-sextantii* Zone.

Genus JHARIATRILETES Bharadwaj and Tiwari, 1970

Type species. *Jhariatriletes baculosus* Bharadwaj and Tiwari, 1970.

Comparison. *Bacutriletes* Potonié, 1956 lacks a well-defined contact areas; its inner structure is not known. *Biharisporites* Potonié, 1956 is spinose (spinae, coni rather than bacula). *Raistrickia* Schopf *et al.* emend. Potonié and Kremp, 1954 is single-layered and is usually used for baculate microspores.

Jhariatriletes (*Verruciretusispora*) *emsiensis* (Moreau-Benoit) comb. nov.
Figures 35F–H, 52J–L

1976   Mégaspore 1 Massa and Moreau-Benoit,
        table-fig. 5.
1979   *Verruciretusispora emsiensis* Moreau-Benoit, p. 43,
        pl. 6, figs 1–2.

Dimensions. 92(177)223 µm; eight specimens measured.

Comparison. Mégaspore 1 *in* Massa and Moreau-Benoit (1976) and *Verruciretusispora emsiensis* Moreau-Benoit, 1979, which are described from the same area, are similar to the specimens described here. The genus *Jhariatriletes* Bharadwaj and Tiwari, 1970 is more appropriate because its species are two-layered unlike *Verruciretusispora* Owens, 1971. In addition, this last genus is used for microspores. Although *Dibolisporites pilatus* Breuer *et al.*, 2007*c* is a microspore, it is also relatively thick walled and has the same kind of distal sculpture. It differs from *J. emsiensis* by its smaller size and single-layered homogenous exine.

Occurrence. A1-69; Awaynat Wanin I Formation; *svalbardiae-eximius* Zone.

Genus KNOXISPORITES Potonié and Kremp emend. Neves, 1961

Type species. *Knoxisporites hagenii* Potonié and Kremp, 1954.

?*Knoxisporites riondae* Cramer and Díez, 1975
Figure 35I–J

1968 Spore trilète à papilles soudées sp. 3 Jardiné and Yapaudjian, pl. 1, fig. 10.
1968 Spore trilète à papilles distinctes sp. 3 Jardiné and Yapaudjian, pl. 1, fig. 11.
1972 ?*Aneurospora* sp. Kemp, p. 115 (*pars*), pl. 55, fig. 10.
1975 ?*Knoxisporites riondae* Cramer and Díez, p. 341, pl. 1, figs 14, 16–17.
1983 *Knoxisporites*? *riondae* Cramer and Díez; Le Hérissé, p. 45, pl. 8, figs 10–12.
1983 *Knoxisporites*? cf. *riondae* Cramer and Díez; Le Hérissé, p. 45, pl. 8, figs 16–19.
? 1985 *Aneurospora* sp. B Paris *et al.*, pl. 18, fig. 7.

Dimensions. 28(31)35 μm; 15 specimens measured.

Comparison. *Knoxisporites*? *riondae* Cramer and Díez, 1975 and *Knoxisporites*? cf. *riondae* Cramer and Díez, 1975 described by Le Hérissé (1983) seem to represent extreme forms of the same species because they differ from each other only by the presence of additional distal verrucae. Paris *et al.* (1985) figured a specimen that resembles those of the population described here by having a distal annulus, but no description was given. *Synorisporites papillensis* McGregor, 1973 does not exhibit a distal annulus but sometimes sub-circular verrucae. As these two species intergrade, they are included in the *S. papillensis* Morphon defined here (Table 1).

Occurrence. BAQA-1, BAQA-2, JNDL-4, WELL-3 and WELL-6; Jauf Formation (Sha'iba to Subbat members); *papillensis-baqaensis* to *lindlarensis-sextantii* zones. MG-1; Ouan-Kasa Formation; *svalbardiae-eximius* Zone but occurrences are probably reworked.

Previous records. From the Horlick Formation of Antarctica (Kemp 1972), the age of which is considered as Pragian by Troth *et al.* (2011) based on correlation of chitinozoans and

spore assemblages from South America; upper Pragian of Belgium (Steemans 1989); upper Pragian – lower Emsian of Paraná and Parnaíba basins, Brazil (Grahn *et al.* 2005; Mendlowicz Mauller *et al.* 2007); upper Pragian – lower Emsian of Armorican Massif, France (Le Hérissé 1983); and Ludlow–lowermost Lochkovian of Spain (Cramer and Díez 1975; Rodriguez 1978*b*).

Genus LEIOZOSTEROSPORA Wellman, 2006

Type species. *Leiozosterospora andersonii* Wellman, 2006.

Leiozosterospora cf. *L. andersonii* Wellman, 2006
Figure 35K–N

cf. 2006 *Leiozosterospora andersonii* Wellman, p. 194, pl. 18, figs g–i.

Description. Amb is sub-circular. Laesurae are straight, accompanied by labra, up to 1 μm wide, and associated with a narrow strip, 0.5–1 μm wide, of thinner exine on either side, extending to, or almost to, the edge of the zona. Curvaturae are sometimes visible. The inner body is sub-circular, and its diameter usually equals about one-half to three-quarters of the total amb diameter.

Dimensions. 43(67)77 μm; 16 specimens measured.

Comparison. According to C. H. Wellman (pers. comm. 2007) *L. andersonii* Wellman, 2006 is subtly different from the specimens described here. Indeed, the latter does not possess curvaturae and laesurae extend to the equator of the zona where they join a narrow limbus. *Auroraspora minuta* Richardson, 1965 is pseudosaccate with an inner body, only slightly smaller than the sexine. Its sexine is laevigate, infrapunctate or occasionally infragranular.

Occurrence. BAQA-2, JNDL-4 and WELL-5; Jauf Formation (Sha'iba to Hammamiyat members); *ovalis-biornatus* to *lindlarensis-sextantii* zones.

Genus LOPHOTRILETES (Naumova) Potonié and Kremp, 1954

Type species. *Lophotriletes gibbosus* (Ibrahim) Potonié and Kremp, 1954.

FIG. 35. Each figured specimen is identified by borehole, sample, slide number and England Finder Co-ordinate location. All figured specimens are at magnification ×1000 except where mentioned otherwise. A, *Iberoespora cantabrica* Cramer and Díez, 1975. MG-1, 2631.2 m, 62252, V50.5. B–E, *Iberoespora* cf. *I. guzmani* Cramer and Díez, 1975. B, WELL-7, 13614.1 ft, 62372, K25. C, WELL-7, 13689.7 ft, 62319, O47/3. D, JNDL-4, 120.0 ft, 68612, K47. E, WELL-4, 16224.7 ft, 62090, O39. F–H, *Jhariatriletes (Verruciretusispora) emsiensis* (Moreau-Benoit) comb. nov., magnification ×500. F, A1-69, 1962 ft, 27278, J42. G, A1-69, 1962 ft, 27277, T45/4. H, A1-69, 1962 ft, 27277, Q40. I–J, ?*Knoxisporites riondae* Cramer and Díez, 1975. I, BAQA-1, 366.9 ft, 03CW117, S42-43. J, BAQA-1, 366.9 ft, 03CW117, K27/1. K–N, *Leiozosterospora* cf. *L. andersonii* Wellman, 2006. K, JNDL-4, 120.0 ft, 68612, Q59. L, BAQA-2, 64.5 ft, 66818, T48. M, JNDL-4, 221.8 ft, 68646, G33; the inner body diameter is abnormally small compared to the total amb diameter. N, BAQA-2, 64.5 ft, 03CW132, N49.

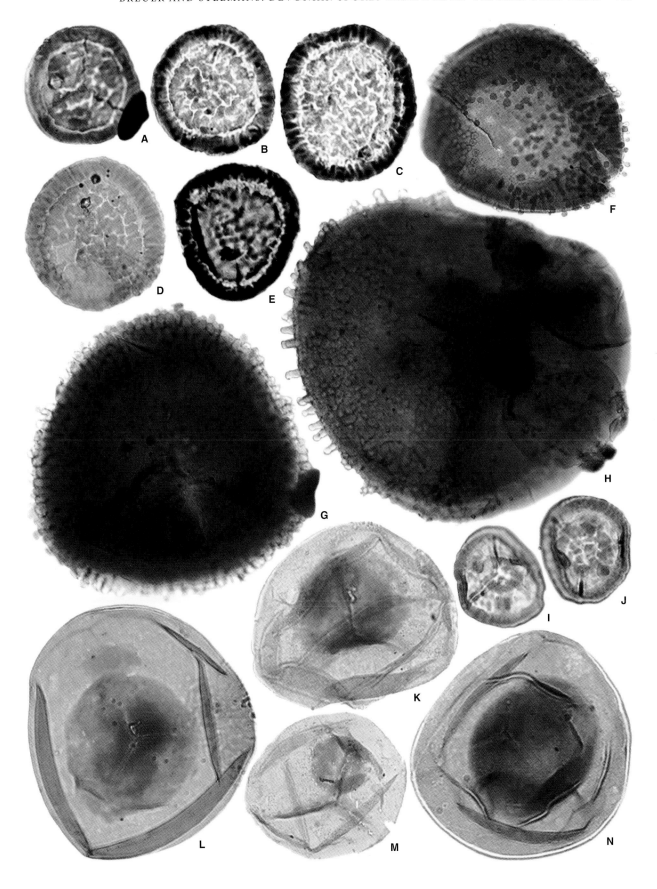

Lophotriletes devonicus (Naumova ex Chibrikova) McGregor
and Camfield, 1982
Figure 36A–B

1982 *Lophotriletes devonicus* (Naumova ex Chibrikova)
McGregor and Camfield, p. 54 (*cum syn.*), pl. 15,
figs 5–11; text-fig. 86.

Dimensions. 38(43)48 μm; four specimens measured.

Remarks. The distal sculptural elements of this species commonly resemble small verrucae in plan view. In lateral compression, however, they prove to be commonly broad-based, blunt coni in profile (McGregor and Camfield 1982).

Occurrence. A1-69; Ouan-Kasa, Awaynat Wanin I and Awaynat Wanin II formations; *lindlarensis-sextantii* to *lemurata-langii* zones.

Previous records. From upper Emsian – lower Eifelian of Algeria (Moreau-Benoit *et al.* 1993); ?Emsian–Eifelian of Bolivia (McGregor 1984); upper Eifelian of the Parnaíba Basin, Brazil (Breuer and Grahn 2011); upper Eifelian – lower Givetian of Canada (McGregor and Uyeno 1972; McGregor and Camfield 1982); Emsian–lower Eifelian of Libya (Moreau-Benoit 1989); Emsian–Eifelian of Morocco (Rahmani-Antari and Lachkar 2001); and ?upper Eifelian–Givetian of Saudi Arabia (PB, pers. obs.).

Genus LOPHOZONOTRILETES Naumova, 1953

Type species. *Lophozonotriletes lebedianensis* Naumova, 1953.

Lophozonotriletes media Taugourdeau-Lantz, 1967
Figure 36C–D

1965 Unidentified spores types Kerr *et al.*, pl. 4,
figs 9, 11.
1967 *Lophozonotriletes media* Taugourdeau-Lantz, p. 52,
pl. 2, fig. 6.
1971 *Geminospora verrucosa* Owens, p. 63, pl. 19,
figs 10–12.

1989 *Spinozonotriletes verrucosus* (Owens) Coquel and
Moreau-Benoit, p. 93.

Dimensions. 40(65)92 μm; 18 specimens measured.

Comparison. The specimens figured by Owens (1971) of *Geminospora verrucosa* Owens, 1971 do not seem to have an inner body as alleged in the diagnosis. These specimens seem to correspond rather to *L. media*. Note that many species of *Lophozonotriletes* Naumova, 1953 were described in the literature and notably from Libya (Massa and Moreau-Benoit 1976; Moreau-Benoit 1979, 1980b). It is impossible to compare all because their diagnoses are inadequate. In addition, some of these species could be grouped together or be synonymous with *L. media* since the latter is a highly variable form. *Archaeozonotriletes variabilis* Naumova emend. Allen, 1965 is also finely punctate without protuberances. Some extreme variants of *L. media*, which show a much reduced ornamentation, could intergrade with *A. variabilis* in the *A. variabilis* Morphon (Table 1). As the latter includes variable and intergrading morphotypes, the concept of *L. media* may appear challenging to constrain and larger here than in the literature. *Cyrtospora tumida* sp. nov. may be infrapunctate and commonly has larger protuberances. It may also intergrade with *L. media*.

Occurrence. A1-69; Awaynat Wanin II Formation; *triangulatus-catillus* to *langii-concinna* zones. MG-1; Awaynat Wanin II and Awaynat Wanin III formations; *undulatus* to *langii-concinna* zones.

Previous records. From lower Givetian – upper Frasnian of Brazil (Loboziak *et al.* 1988; Melo and Loboziak 2003; Breuer and Grahn 2011); upper Eifelian–Frasnian of Canada (Owens 1971; McGregor and Camfield 1982); Frasnian of France (Loboziak *et al.* 1983, 1988); middle Givetian of Greenland (Friend *et al.* 1983; Marshall and Hemsley 2003); middle Eifelian – upper Frasnian (Streel *et al.* 1988; Moreau-Benoit 1989); Givetian of Poland (Turnau and Racki 1999); upper Givetian – lower Frasnian of Portugal (Lake *et al.* 1988); Givetian–Frasnian of Saudi Arabia (PB, pers. obs.); and uppermost Givetian – lower Frasnian of Scotland (Marshall *et al.* 1996).

Genus LYCOSPORA Schopf *et al.* emend. Potonié and Kremp, 1954

Type species. *Lycospora micropapillata* (Wilson and Coe) Schopf *et al.*, 1944.

FIG. 36. Each figured specimen is identified by borehole, sample, slide number and England Finder Co-ordinate location. All figured specimens are at magnification ×1000 except where mentioned otherwise. A–B, *Lophotriletes devonicus* (Naumova ex Chibrikova) McGregor and Camfield, 1982. 1, A1-69, 1486 ft, 26977, R39. 2, A1-69, 2108–2111 ft, 26913, G44. C–D, *Lophozonotriletes media* Taugourdeau-Lantz, 1967. C, A1-69, 1174 ft, 62673, K-L32. D, A1-69, 971 ft, 62640, X-Y49. E–H, *Lycospora culpa* Allen, 1965. E, BAQA-2, 64.5 ft, 03CW132, R31. F, BAQA-2, 54.8 ft, 03CW129, R38/3. G, BAQA-2, 52.0 ft, 03CW128, N27/1. H, BAQA-2, 134.4 ft, 03CW137, H35/4. I–J, *Perotrilites caperatus* (McGregor) Steemans, 1989. I, JNDL-3, 389.0 ft, 68563, D48/3. J, JNDL-4, 120.0 ft, 68612, H50/3. K–M, *Raistrickia commutata* sp. nov. K, MG-1, 2194 m, 63013, G29/4. L, Holotype, MG-1, 2247 m, 62942, J30. M, MG-1, 2258 m, 62948, E42.

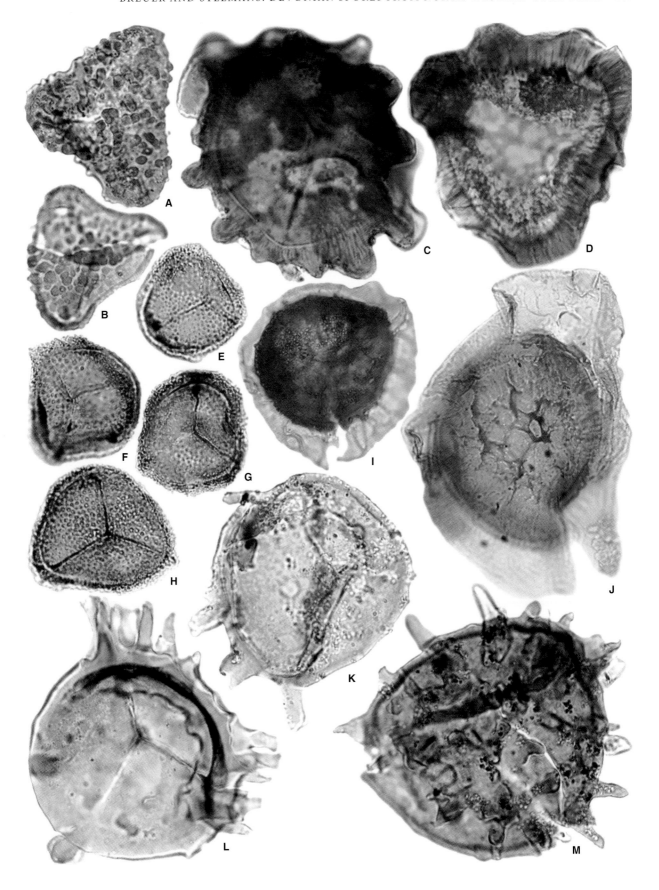

Lycospora culpa Allen, 1965
Figure 36E–H

1965 *Lycospora culpa* Allen, p. 713, pl. 98, figs 7–8.
? 1968 *Zonotriletes* sp. 5 Jardiné and Yapaudjian, pl. 2, figs 4–5.

Dimensions. 32(40)48 μm; 15 specimens measured.

Remarks. According to Somers (1972), *Lycospora culpa* Allen, 1965 should be excluded from this genus because of the zonate nature of the equatorial flange. This statement is rejected here as the definition of the genus matches the species described here.

Comparison. The figured specimens of *Zonotriletes* sp. 5 in Jardiné and Yapaudjian (1968) are very similar to the present specimens, but they are not described.

Occurrence. BAQA-1 and BAQA-2; Jauf Formation (Sha'iba to Subbat members); *papillensis-baqaensis* to *ovalis* zones. MG-1; Ouan-Kasa Formation; *svalbardiae-eximius* Zone but occurrences are probably reworked.

Previous record. From Pragian of Spistbergen, Norway (Allen 1965).

Genus PEROTRILITES Couper emend. Evans, 1970

Type species. Perotrilites granulatus Couper, 1953.

Perotrilites caperatus (McGregor) Steemans, 1989
Figure 36I–J

1989 *Perotrilites caperatus* (McGregor) Steemans, pp. 150 –151 (*cum syn.*), pl. 42, figs 12–14; pl. 43, figs 1–2.
2006 *Camptozonotriletes? caperatus* McGregor; Wellman, p. 190, pl. 18, figs a–c.

Dimensions. 37(65)95 μm; 16 specimens measured.

Remarks. The zonate, zonate-camerate and camerate conditions are difficult to differentiate in compressed specimens.

Comparison. Camptozonotriletes aliquantus Allen, 1965 is both zonate and camerate, like *P. caperatus*. According to McGregor (1973), the radially directed muri of *C. aliquantus* are not similar to the fold-like structures of *P. caperatus. C. aliquantus* also differs in having prominent distal anastomosing muri. *Zonotriletes venatus* sp. nov. is larger, only zonate and does not exhibit such a distal sculpture on the central body.

Occurrence. BAQA-1, JNDL-3 and JNDL-4; Jauf Formation (Subbat to Muray members); *milleri* to *annulatus-protea* zones. MG-1, Ouan-Kasa Formation, *lindlarensis-sextantii* to *svalbardiae-eximius* zones.

Previous records. From upper Pragian – lower Emsian of Argentina (Rubinstein and Steemans 2007); upper Lochkovian–Emsian of Belgium (Steemans 1989); upper Pragian–lower Emsian of Paraná and Parnaíba basins, Brazil (Grahn *et al.* 2005; Mendlowicz Mauller *et al.* 2007); Pragian–Emsian of Canada (McGregor and Owens 1966; McGregor 1973, 1977; McGregor and Camfield 1976); upper Lochkovian–Emsian of Germany (Steemans 1989); upper Pragian of Morocco (Rahmani-Antari and Lachkar 2001); Pragian–uppermost Emsian of Poland (Turnau 1986; Turnau *et al.* 2005); upper Lochkovian of Romania (Steemans 1989); and upper Pragian – ?lowermost Emsian of Scotland (Wellman 2006).

Genus RAISTRICKIA Schopf *et al.* emend. Potonié and Kremp, 1954

Type species. Raistrickia grovensis Schopf *et al.*, 1944.

Raistrikia commutata sp. nov.
Figures 36K–M, 37A–B

? 1968 *Raistrickia* sp. Jardiné and Yapaudjian, pl. 2, fig. 7.

Derivation of name. From *commutatus* (Latin), meaning variable; refers to the proximo-equatorial and distal sculpture.

Holotype. EFC J30 (Fig. 36L), slide 62942.

Paratype. EFC G34 (Fig. 37A), slide 62746; MG-1 borehole, sample 2160.6 ft.

FIG. 37. Each figured specimen is identified by borehole, sample, slide number and England Finder Co-ordinate location. All figured specimens are at magnification ×1000 except where mentioned otherwise. A–B, *Raistrickia commutata* sp. nov. A, Paratype, MG-1, 2160.6 m, 62746, G34. B, MG-1, 2258 m, 62948, T49. C–H, *Raistrickia jaufensis* sp. nov. C, Paratype, BAQA-1, 390.6 ft, 03CW120, N38. D, BAQA-1, 371.1 ft, 03CW118, H47. E, Holotype, BAQA-1, 366.9 ft, 03CW117, T35/2. F, BAQA-1, 222.5 ft, 03CW108, M31. G, BAQA-1, 371.1 ft, 03CW118, N26/3. H, BAQA-1, 285.5 ft, 03CW111, H31/4. I–L, *Retusotriletes atratus* sp. nov. I, JNDL-4, 272.0 ft, 68656, Q41/1. J, Paratype, WELL-4, 16273.0 ft, 62122, O34-35. K, JNDL-1, 153.8 ft, 60834, O38/2. L, Holotype, WELL-4, 16316.6 ft, 62157, T48. M–R, *Retusotriletes celatus* sp. nov. M, Holotype, BAQA-1, 175.9 ft, 66778, T38/3. N, JNDL-4, 448.6 ft, 68693, S25/4. O, WELL-3, 14195.3 ft, 66836, R53/1. P, WELL-3, 14195.2 ft, 60550, K35/1. Q, WELL-7, 13689.7 ft, 62317, J37/2. R, Paratype, WELL-7, 13689.7 ft, 62316, D-E57.

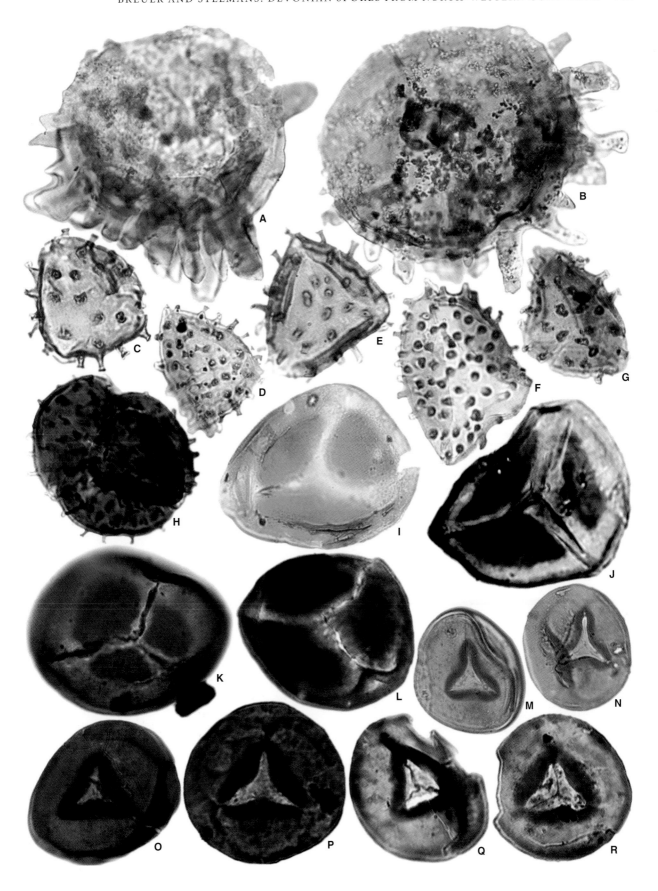

Type locality and horizon. MG-1 borehole, sample 2247 m; Awaynat Wanin II Formation at Mechiguig, Tunisia.

Diagnosis. A *Raistrickia* irregularly sculptured with a variable mixture of coni, spinae, bacula and verrucae. Sculptural elements mostly longer than wide, very variable in shape and in density.

Description. Amb is sub-circular to sub-triangular, or occasionally oval. Laesurae are straight, simple, one-half to nine-tenths of the amb radius in length. Exine is commonly 1.5–4 µm thick equatorially, homogeneous or punctate. Proximo-equatorial and distal regions are irregularly sculptured with a variable mixture of coni, spinae, bacula and verrucae with blunt, flat-topped, pointed or widened tips. Sculptural elements are 1–12 µm wide at base, 2–20 µm high, mostly longer than wide. They are very variable in shape and in density. Contact areas are laevigate or with sculpture like that of the distal face but reduced in size.

Dimensions. 55(68)96 µm; 13 specimens measured.

Comparison. *Raistrickia aratra* Allen, 1965 is more densely sculptured with less elongated elements. *Raistrickia* cf. *clavata* Hacquebard, 1957 *in* Richardson (1965) has few sculptural elements around the equator and these are not really elongated as in *R. commutata*. *Raistrickia* sp. *in* Jardiné and Yapaudjian (1968) could be assignable to *R. commutata*, but no diagnosis was given.

Occurrence. A1-69; Awaynat Wanin II Formation; *rugulata-libyensis* to *lemurata-langii* zones. MG-1; Awaynat Wanin II and Awaynat Wanin III formations; *triangulatus-catillus* to *langii-concinna* zones.

Previous record. Similar specimens have been reported in ?upper Eifelian–Givetian of Saudi Arabia (PB, pers. obs.).

Raistrikia jaufensis sp. nov.
Figure 37C–H

Derivation of name. From *jaufensis* (Latin); refers to the lithostratigraphic unit (formation) where this species occurs.

Holotype. EFC T35/2 (Fig. 37E), slide 03CW117.

Paratype. EFC N38 (Fig. 37C), slide 03CW120; BAQA-1 core hole, sample 390.6 ft.

Type locality and horizon. BAQA-1 core hole, sample 366.9 ft; Jauf Formation at Baq'a, Saudi Arabia.

Diagnosis. A sub-triangular to triangular *Raistrickia* usually irregularly sculptured with bacula. Base of elements generally flared. Tops of elements flat or slightly concave, with generally a bifurcate shape.

Description. Amb is sub-triangular to triangular. Laesurae are straight, simple, three-fifths to nine-tenths of the amb radius in length, but often indistinct because the proximal face seems to be usually torn. Exine is 1–2 µm thick equatorially. Equatorial and distal regions are usually irregularly sculptured with bacula, 1–3 µm wide at base, 2–5 µm high, 1–8 µm apart. The base of elements is generally flared. The tops of elements are flat or slightly concave, with generally a bifurcate shape. Contact areas are laevigate.

Dimensions. 27(36)48 µm; 26 specimens measured.

Remarks. Some of these specimens are preserved as tetrads.

Comparison. *Raistrickia* sp. *in* McGregor (1973) may correspond to the description of the specimens presented here, but it is sub-circular and only a few specimens were recorded from the Gaspé assemblages. *Cymbosporites dammamensis* Steemans, 1995 is patinate and more densely sculptured with generally smaller bacula.

Occurrence. BAQA-1, BAQA-2, JNDL-4 and WELL-4; Jauf Formation (Sha'iba to Subbat members); *papillensis-baqaensis* to *ovalis-biornatus* zones.

Genus RETUSOTRILETES Naumova emend. Streel, 1964

Type species. *Retusotriletes simplex* Naumova, 1953.

Retusotriletes atratus sp. nov.
Figure 37I–L

2005 *Retusotriletes* sp. A Hashemi and Playford, p. 338, pl. 2, fig. 12.

Derivation of name. From *atratus* (Latin), meaning dark; refers to the proximal interradial thickened zones.

Holotype. EFC T48 (Fig. 37L), slide 62157.

Paratype. EFC O34-35 (Fig. 37J), slide 62122; WELL-4 well, sample 16273.0 ft.

Type locality and horizon. WELL-4 well, sample 16316.6 ft; Jauf Formation at Kharma, Saudi Arabia.

Diagnosis. A *Retusotriletes* with a darker proximal zone, variable in size and shape, present in each contact area.

Description. Amb is sub-circular. Laesurae are straight, simple or accompanied by labra, up to 3 µm in overall width, three-fifths to nine-tenths of the amb radius in length, connected by curvaturae perfectae. Exine is laevigate or scabrate, 1–2 µm thick.

A darker proximal zone is present in each contact area. These thickened zones are variable in size and shape, and situated towards the proximal pole between the laesurae.

Dimensions. 45(56)77 μm; 14 specimens measured.

Comparison. *Retusotriletes crassus* Clayton *et al.*, 1980 is slightly different because it has simple laesurae which are also rarely seen. *R. aureoladus* Rodriguez, 1978*a* is smaller and has a paler ring around the proximal thickened zones. *Retusotriletes* sp. A *in* Hashemi and Playford (2005) is similar but slightly smaller.

Occurrence. BAQA-1, BAQA-2, JNDL-1, JNDL-3, JNDL-4, WELL-4, WELL-7 and WELL-8; Jauf (Sha'iba to Hammamiyat members) and Jubah formations; *ovalis-biornatus* to *triangulatus-catillus* zones.

Previous record. From Emsian–lower Givetian of Adavale Basin, Australia (Hashemi and Playford 2005).

Retusotriletes celatus sp. nov.
Figure 37M–R

Derivation of name. From *celatus* (Latin), meaning hidden; refers to the fact that this species may have a delicate folded outer layer and thus have a different genus and species name (see below).

Holotype. EFC S25/4 (Fig. 37N), slide 68693.

Paratype. EFC D-E57 (Fig. 37R), slide 62316; WELL-7 well, sample 13689.7 ft.

Type locality and horizon. JNDL-4 core hole, sample 448.6 ft; Jauf Formation at Domat A1–Jandal, Saudi Arabia.

Diagnosis. A *Retusotriletes* with a darker apical sub-triangular band, with sharp margins and straight, slightly concave or convex sides, that extends almost to, or to the end of, laesurae. Inner lighter sub-triangular band generally with slightly concave sides also present at the proximal pole.

Description. Amb is sub-circular. Laesurae are straight, simple or rarely accompanied by narrow labra, *c.* 1 μm in overall width, two-thirds to three-quarters of the amb radius in length, connected by curvaturae perfectae not always well developed. Exine is laevigate, commonly 1–2 μm thick. A darker apical sub-triangular band, with sharp margins and straight, slightly concave or convex sides, extends to, or almost to, the end of the laesurae. This thickened area is up 3–6 μm wide interradially. An inner lighter sub-triangular area (with a thinner exine), generally with slightly concave sides, is present proximally and surrounded by the darkened band.

Dimensions. 28(39)54 μm; 15 specimens measured.

Remarks. *Retusotriletes celatus* is comparable with the specimens of *Diaphanospora milleri* sp. nov. in which the outer layer is absent. The very delicate outer layer of the second species could have been torn off by sedimentary or taphonomic processes. The two form-species *R. celatus* and *D. milleri* represent thus a unique biological species with the different states of preservation between both. They are grouped in the *D. milleri* Morphon (Table 1). They sometimes co-occur and have comparable stratigraphical ranges.

Comparison. *Retusotriletes tenerimedium* Chibrikova, 1959 has a lighter apical area that is sub-triangular with convex sides. Moreover, the darker area is diffuse at its outer margin and does not reach the end of laesurae. *R. rotundus* (Steel) Steel emend. Lele and Steel, 1969 and *R. triangulatus* (Steel) Steel, 1967 are larger, and their thickened apical areas do not extend to the end of laesurae.

Occurrence. BAQA-1, JNDL-3, JNDL-4, WELL-2, WELL-3 and WELL-7; Jauf Formation (Subbat and Hammamiyat members); *milleri* to *lindlarensis-sextantii* zones. A1-69; Awaynat Wanin I Formation; *svalbardiae-eximius* Zone.

Retusotriletes goensis Lele and Streel, 1969
Figure 38A

1969   *Retusotriletes goensis* Lele and Streel, p. 93, pl. 1, figs 12–16.
non 1978*b*   *Retusotriletes goensis* Lele and Streel; Rodriguez, p. 420, pl. 2, fig. 25.

Dimensions. 47(68)104 μm; four specimens measured.

Comparison. *Retusotriletes rotundus* (Steel) Steel emend. Lele and Streel, 1969 and *R. tenerimedium* Chibrikova, 1959 have a sub-triangular apical area which is differentiated into two zones. *Retusotriletes* cf. *microgranulatus* (Vigran) Steel, 1967 is sculptured with micropila.

Occurrence. BAQA-1, JNDL-4, WELL-3 and WELL-8; Jauf (Subbat and Hammamiyat members) and Jubah formations; *asymmetricus* to *lemurata-langii* zones.

Previous records. From upper Emsian (Lessuise *et al.* 1979) and the Pepinster Formation of Belgium (Lele and Streel 1969), the age of which is considered as late Eifelian by Laloux *et al.* (1996); upper Lochkovian of Solimões Basin, Brazil (Rubinstein *et al.* 2005); and lower Eifelian – lower Givetian of Libya (Moreau-Benoit 1989).

Retusotriletes maculatus McGregor and Camfield, 1976
Figure 38B

1967 *Leiotriletes* sp. Mortimer, pl. 1b.
1968 Spore trilète à papilles proximales sp. 4 Jardiné and
    Yapaudjian, pl. 1, fig. 16.
1976 *Retusotriletes maculatus* McGregor and Camfield,
    p. 26, pl. 1, fig. 6.

Dimensions. 29(41)55 µm; 12 specimens measured.

Remarks. Mortimer (1967) and Jardiné and Yapaudjian (1968) figure similar spores. *Ambitisporites* sp. B *in* Richardson and Ioannides (1973) is smaller and has an equatorial crassitude. *R. ocellatus* McGregor (1973) has also proximal papillae but is much larger. *Ambitisporites eslae* (Cramer and Díez) Richardson *et al.*, 2001 is cingulate. *Scylaspora elegans* Richardson *et al.*, 2001 is proximally microrugulate.

Occurrence. BAQA-1, BAQA-2, JNDL-3, JNDL-4, WELL-2, WELL-3, WELL-4, and WELL-7; Jauf Formation; *papillensis-baqaensis* to *annulatus-protea* zones. A1-69; Awaynat Wanin I and Awaynat Wanin II formations; *svalbardiae-eximius* to *lemurata-langii* zones, but the rare and isolated occurrence of some specimens are probably due to reworking. MG-1; Ouan-Kasa and Awaynat Wanin I formations; *lindlarensis-sextantii* to *rugulata-libyensis* zones.

Previous records. From Lochkovian–upper Emsian of Bolivia (McGregor 1984; Perez-Leyton 1990); from lower Lochkovian–lower Emsian of Brazil (Rubinstein *et al.* 2005; Mendlowicz Mauller *et al.* 2007; Steemans *et al.* 2008); Lochkovian–Emsian of Canada (McGregor and Camfield 1976); Pragian–lower Emsian of Armorican Massif, France (Le Hérissé 1983); middle Přídolí–lowermost Eifelian of Libya (Moreau-Benoit 1989; Rubinstein and Steemans 2002); Lochkovian of Poland (Turnau *et al.* 2005); and Pragian from UK (Mortimer 1967).

Retusotriletes rotundus (Streel) Streel emend. Lele and
Streel, 1969
Figure 38C

1964 *Phyllothecotriletes rotundus* Streel, pl. 1, figs 1–2.

1967 *Retusotriletes rotundus* (Streel) Streel, p. 25, pl. 1,
    fig. 11; pl. 2, figs 16–17.
1969 *Retusotriletes rotundus* (Streel) Streel emend. Lele
    and Streel, p. 94, pl. 1 figs 18–20.

Dimensions. 57(82)108 µm; 12 specimens measured.

Comparison. *Retusotriletes tenerimedium* Chibrikova, 1959 also has a differentiated proximal face but the thinner sub-triangular apical area extends to about one-third of the amb radius. Moreover, this species is smaller and equatorially thicker. The dark apical area of *R. goensis* Lele and Streel, 1969 may be larger and is not differentiated into two zones. *R. triangulatus* (Streel) Streel, 1967 has a sub-triangular thickened zone at the proximal pole but with concave sides.

Occurrence. BAQA-1, BAQA-2, JNDL-3, JNDL-4, S-462, WELL-3, WELL-4, WELL-7 and WELL-8; Jauf and Jubah formations; *papillensis-baqaensis* to *triangulatus-catillus* zones. A1-69; Awaynat Wanin I and Awaynat Wanin II formations; *svalbardiae-eximius* to *lemurata-langii* zones. MG-1; Awaynat Wanin I Formation; *incognita* Zone.

Previous records. Widely dispersed and often common in Devonian (particularly Early–Middle Devonian) assemblages (Streel 1964, 1967; Riegel 1968; Cramer 1969; Lele and Steel 1969; McGregor 1973; McGregor and Camfield 1976, 1982; Marshall and Allen 1982; Le Hérissé 1983; Balme 1988; Boumendjel *et al.* 1988; Ravn and Benson 1988; Moreau-Benoit *et al.* 1993; Turnau 1996; Rahmani-Antari and Lachkar 2001; Marshall and Fletcher 2002; Hashemi and Playford 2005).

Retusotriletes tenerimedium Chibrikova, 1959
Figure 38D–G

1959 *Retusotriletes tenerimedium* Chibrikova, p. 52, pl. 5,
    figs 9–10.
1966 *Retusotriletes tenerimedium* Chibrikova; de Jersey,
    p. 7, pl. 2, fig. 4.
1967 *Retusotriletes triangulatus* (Streel) Streel;
    Richardson, pl. 2, fig.a.
1968 *Retusotriletes tenerimedium* Chibrikova; Schultz,
    p. 14, pl. 1, fig. 15.

FIG. 38. Each figured specimen is identified by borehole, sample, slide number and England Finder Co-ordinate location. All figured specimens are at magnification ×1000 except where mentioned otherwise. A, *Retusotriletes goensis* Lele and Streel, 1969. JNDL-4, 285.5 ft, 68660, K27/1. B, *Retusotriletes maculatus* McGregor and Camfield, 1976. JNDL-4, 87.2 ft, 03CW195, P34/3. C, *Retusotriletes rotundus* (Streel) Streel emend. Lele and Streel, 1969. BAQA-1, 345.5 ft, 03CW114, O36. D–G, *Retusotriletes tenerimedium* Chibrikova, 1959. D, BAQA-1, 285.5 ft, 03CW111, C40/2. E, BAQA-1, 366.9 ft, 62257, M28. F, WELL-7, 13738.5 ft, 62322, U43. G, WELL-7, 13738.5 ft, 62325, G41. H–I, *Retusotriletes triangulatus* (Streel) Streel, 1967. H, BAQA-1, 371.1 ft, 03CW118, O23. I, BAQA-2, 50.8 ft, 03CW127, M33/2. J–L, *Retusotriletes* sp. 1. J, MG-1, 2205 m, 62597, S44/3. K, A1-69, 1293 ft, 63066, K32. L, S-462, 1910–1915 ft, 63260, E33/3.

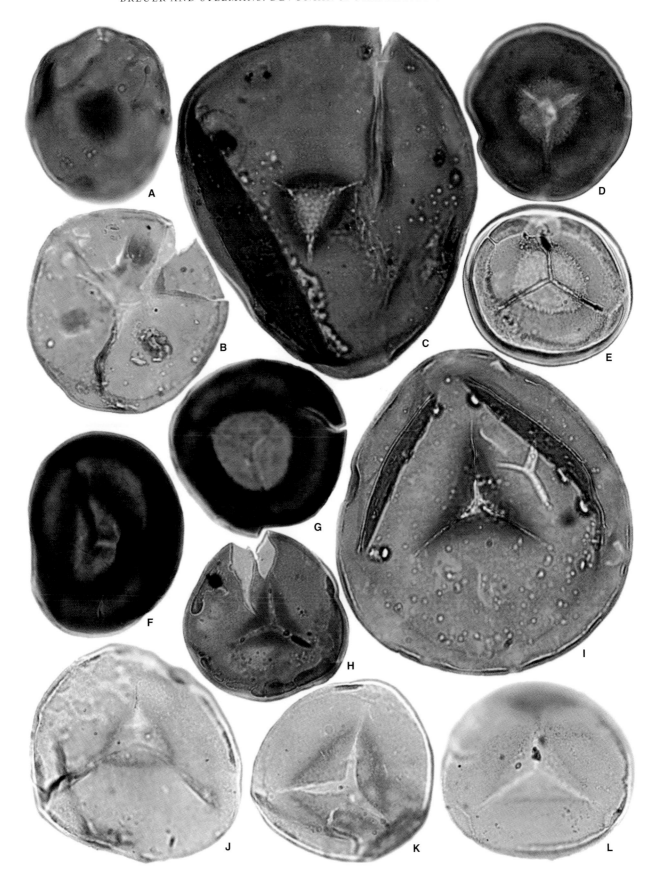

Dimensions. 36(53)86 µm; 30 specimens measured.

Comparison. Ambitisporites (Archaicusporites) asturicus (Rodriguez) comb. nov. has a less pronounced sub-triangular darker apical zone and elevated curvaturae and is commonly smaller.

Occurrence. BAQA-1, BAQA-2, WELL-1, WELL-2, WELL-3, WELL-5, WELL-6 and WELL-7; Jauf Formation; *ovalis-biornatus* to *annulatus-protea* zones. A1-69; Awaynat Wanin II Formation; *undulatus* Zone. MG-1; Ouan-Kasa Formation; *svalbardiae-eximius* Zone.

Previous records. From upper Emsian – lower Eifelian of the Russian Platform (Avkhimovitch *et al.* 1993); and Emsian of Germany (Schultz 1968).

Retusotriletes triangulatus (Streel) Streel, 1967
Figure 38H–I

1964 *Phyllothecotriletes triangulatus* Streel, p. 5, pl. 1, figs 3–5.
1967 *Retusotriletes triangulatus* (Streel) Streel, p. 24.
2006 *Retusotriletes* cf. *triangulatus* (Streel) Streel; Wellman, p. 175, pl. 10, figs i–k.

Dimensions. 40(66)87 µm; 22 specimens measured.

Comparison. Retusotriletes rotundus (Streel) Streel emend. Lele and Streel, 1969 has a sub-triangular apical area, with convex sides, which extends to one-fifth to one-third of the amb radius.

Occurrence. BAQA-1, BAQA-2, JNDL-4, S-462, WELL-1, WELL-2, WELL-3, WELL-4, WELL-5, WELL-6, WELL-7 and WELL-8; Jauf and Jubah formations; *papillensis-baqaensis* to *triangulatus-catillus* zones. A1-69; Ouan-Kasa, Awaynat Wanin I and Awaynat Wanin II formations; *lindlarensis-sextantii* to *triangulatus-catillus* zones. MG-1; Ouan-Kasa and Awaynat Wanin I formations; *lindlarensis-sextantii* to *rugulata-libyensis* zones.

Previous records. Widely dispersed and often common in Devonian (particularly Early–Middle Devonian) assemblages (Streel 1964, 1967; Riegel 1968; Lele and Streel 1969; Tiwari and Schaarschmidt 1975; Lu Lichang and Ouyang Shu 1976; Gao Lianda 1981; Streel *et al.* 1988; Turnau 1996; Turnau and Racki 1999; Turnau and Matyja 2001; Wellman 2006; Steemans *et al.* 2008).

Retusotriletes sp. 1
Figure 38J–L

Description. Amb is sub-circular. Laesurae are straight, simple, four-fifths to nine-tenths of the amb radius in length, connected near the equator by curvaturae perfectae. Exine is scabrate, 1–2 µm thick. An triangular apical area with more or less straight sides commonly occurs two-fifths to three-fifths of the amb radius along the laesurae. Outer margin of the apical area is delimited by a more or less diffuse darker zone, 1–4 µm wide. Interior to the triangular area at proximal pole has a wall thickness comparable to the remaining part of the spore body.

Dimensions. 51(62)81 µm; four specimens measured.

Comparison. Retusotriletes triangulatus (Streel) Streel, 1967 is characterized by a more pronounced dark sub-triangular apical area with concave sides. *R. rotundus* (Streel) Streel emend. Lele and Streel, 1969 has a proportionately less developed sub-triangular apical area with convex sides. *Calamospora atava* McGregor, 1973 figured in McGregor and Camfield, 1982 has a similar apical area, but does not have curvaturae and is commonly folded.

Occurrence. S-462; Jubah Formation; *triangulatus-catillus* to *langii-concinna* zones. A1-69; Awaynat Wanin II Formation; *triangulatus-catillus* Zone. MG-1; Awaynat Wanin II Formation; *langii-concinna* Zone.

Genus RHABDOSPORITES Richardson emend. Marshall and Allen, 1982

Type species. Rhabdosporites langii (Eisenack) Richardson, 1960.

Comparison. Camerate spores with coarser sculpture are included in other genera (e.g. *Grandispora* Hoffmeister *el al.* emend. Neves and Owens, 1966). In emending the generic concept of *Rhabdosporites*, Marshall and Allen (1982) were aware of the close similarity between small specimens of *Rhabdosporites* and *Geminospora* Balme (1962), which are mentioned by other authors (Lele and Streel 1969). *Geminospora* is typified by a thin-walled nexine either closely appressed to, or showing a variable degree of separation from, a sculptured sexine with a thickened distal surface. Further evidence for the similarity or inability to easily distinguish between *Rhabdosporites* and *Geminospora* comes from the study of *in situ* spores (Allen 1980), where spores assignable to both genera are recorded from closely related progymnosperms (Marshall and Allen 1982; Marshall 1996). Wellman (2009) concluded that there is

FIG. 39. Each figured specimen is identified by borehole, sample, slide number and England Finder Co-ordinate location. All figured specimens are at magnification ×1000 except where mentioned otherwise. A–B, *Rhabdosporites langii* (Eisenack) Richardson, 1960. A, MG-1, 2258 m, 62947, Q46/2. B, MG-1, 2180 m, 62973, N44. C. *Rhabdosporites minutus* Tiwari and Schaarschmidt, 1975. JNDL-4, 346.3 ft, 68672, J34/1. D, *Rhabdosporites streelii* Marshall, 1996, magnification ×750. A1-69, 1322 ft, 27125, L36-37. E, *Samarisporites angulatus* (Tiwari and Schaarschmidt) Loboziak and Streel, 1989, magnification ×750. A1-69, 1596 ft, 26990, K40/2. F, *Samarisporites eximius* (Allen) Loboziak and Streel, 1989, magnification ×750. JNDL-1, 162.3 ft, 60841, W28/3.

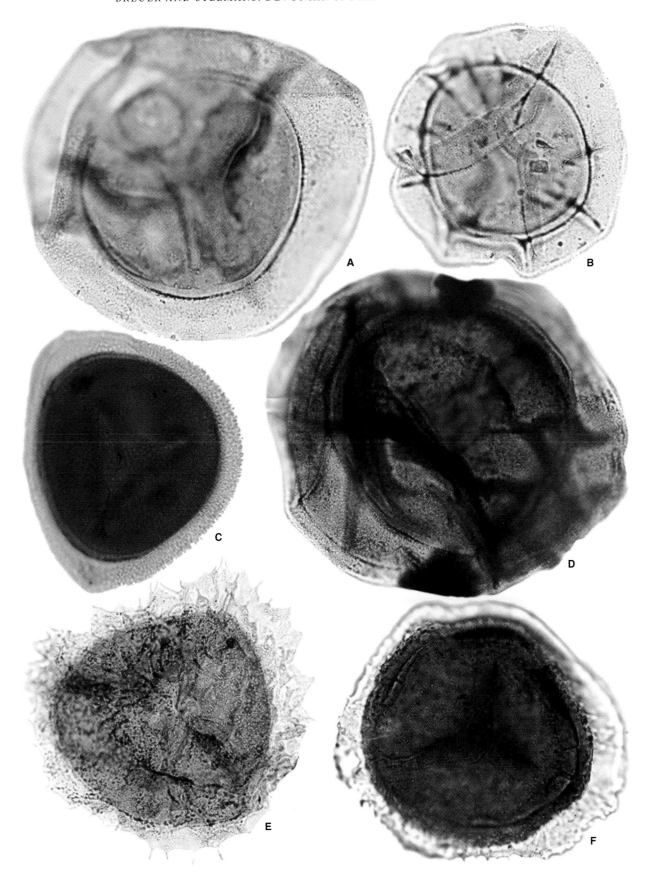

a smooth evolutionary transition between *Apiculiretusispora* (Streel) Streel, 1967, *Rhabdosporites*, *Geminospora* and *Contagisporites* Owens, 1971.

Rhabdosporites langii (Eisenack) Richardson, 1960
Figure 39A–B

1925 Type B Lang, p. 256, pl. 1, figs 3–6.
1944 *Triletes langi* Eisenack, p. 112, pl. 2, fig. 4.
1959 Spores of *Milleria* (*Protopteridium*) *thomsonii* (Dawson) Lang; Obrhel, p. 387, pl. 2, fig. 6.
1960 *Rhabdosporites langi* (Eisenack) Richardson, p. 54, pl. 14, figs 8, 9, text-figs 4, 6B.
1963 *Rhabdosporites firmus* Guennel, p. 256, fig. 12.
1964 *Calyptosporites plicatus* Vigran, p. 19, pl. 6, fig. 4.
1965 *Rhabdosporites parvulus* Richardson, p. 588 (*pars*), pl. 93, figs 5–6.
1967 Spores of *Tetraxylopteris schmidtii* Beck; Bonamo and Banks, p. 765, figs 34–36, 38, 40.
1969 ?*Rhabdosporites parvulus* Richardson; Lele and Streel, p. 103, pl. 3, fig. 66–68.
1971 *Rhabdosporites micropaxillus* Owens, p. 49, pl. 15, figs 3–7.
1971 Spores of *Milleria* (*Protopteridium*) *thomsonii* (Dawson) Lang; Leclercq and Bonamo, p. 98, pl. 36, figs 24–33.
1972 *Rhabdosporites* n. sp. McGregor and Uyeno, pl. 3, fig. 5.
1974 *Rhabdosporites* sp. Hamid, p. 202, pl. 10, fig. 1.
1975 *Rhabdosporites* sp. Tiwari and Schaarschmidt, p. 40, pl. 21, fig. 7.
1977 Spores of *Rellimia thomsonii* (Dawson) Leclercq and Bonamo; Bonamo, p. 1277, figs 6–7.

Dimensions. 62(93)128 µm; 69 specimens measured.

Remarks. *Rhabdosporites parvulus* Richardson, 1965 differs in being smaller and has in addition a nexine-to-sexine size ratio different from typical *R. langii*. Populations of *R. langii* in this study include continuous size variation including specimens as small as *R. parvulus*. As in Marshall and Allen (1982) and Marshall (1996), *R. parvulus* is regarded as a junior synonym of *R. langii*. However, the situation is more complex. According to Marshall (1996), the specimens attributed by Richardson (1965) to *R. parvulus* include both *Geminospora lemurata* Balme emend. Playford, 1983 and small specimens of *R. langii*.

Studies of *in situ* spores indicated that dispersed *R. langii* was produced by at least two different, but closely related, aneurophytalean progymnosperms. The spores of these two plant species cannot be distinguished on purely morphological grounds, although there are possibly some minor differences at the ultrastructural level (Wellman 2009).

Comparison. *Rhabdosporites minutus* Tiwari and Schaarschmidt, 1975 is commonly smaller and less folded. In addition, this species often has an asymmetrical appearance with an elongate sub-triangular or oval amb. Normally, the separation of *G. lemurata* and *R. langii* presents no difficulty as the latter is significantly larger and has a nexine that is significantly smaller than the sexine and thus well-developed cameration. In addition, its sexine is thinner and folded with the folds continuous across a significant proportion of the sexine diameter. Typically specimens of *G. lemurata* show only a small separation between nexine and sexine with the sexine being relatively rigid because of its greater thickness such that the nexine is usually centrally placed. *G. svalbardiae* (Vigran) Allen, 1965 is often folded as in *R. langii* but has a slightly thicker sexine and its ornamentation is coarser.

Occurrence. S-462, WELL-1 and WELL-8; Jubah Formation; *lemurata-langii* to *langii-concinna* zones. A1-69; Awaynat Wanin II Formation; *lemurata-langii* to *langii-concinna* zones. MG-1; Awaynat Wanin II and Awaynat Wanin III formations; *lemurata-langii* to *langii-concinna* zones.

Previous records. *Rhabdosporites langii* is eponymous for the early Eifelian *velata-langii* Assemblage Zone of the Old Red Sandstone Continent and adjacent regions (Richardson and McGregor 1986). *R. langii* has an almost worldwide distribution extending into the Frasnian.

Rhabdosporites minutus Tiwari and Schaarschmidt, 1975
Figure 39C

1975 *Rhabdosporites minutus* Tiwari and Schaarschmidt, p. 39, pl. 21, figs 4–6.

Dimensions. 42(60)84 µm; 39 specimens measured.

Comparison. *Rhabdosporites langii* (Eisenack) Richardson, 1960 is larger, commonly having smaller body in relation to the saccus, the latter being much folded. *R. scamnus* Allen, 1965 is larger and is typically folded distally. *Apiculiretusispora brandtii* Streel, 1964 has the same type of sculpture and the sexine is irregularly detached from nexine on some specimens, whereas *R. minutus* has a sexine completely detached equatorially from the spore body. The two species could belong to a same spore lineage as suggested by Wellman (2009). Therefore, the two species are included in the *A. brandtii* Morphon (Table 1).

Occurrence. BAQA-1, JNDL-1, JNDL-3, JNDL-4 and WELL-4; Jauf (Subbat to Murayr members) and Jubah formations; *asymmetricus* to *svalbardiae-eximius* zones. A1-69; Ouan-Kasa, Awaynat Wanin I and Awaynat Wanin II formations; *lindlarensis-sextantii* to *lemurata-langii* zones. MG-1; Ouan-Kasa, Awaynat Wanin I, Awaynat Wanin II and Awaynat Wanin III formations; *annulatus-protea* to *langii-concinna* zones.

Previous records. From lower Eifelian – lower Givetian of Germany (Tiwari and Schaarschmidt 1975); middle–upper

Emsian of Luxembourg (Steemans *et al.* 2000*a*); and Emsian of Saudi Arabia (Al-Ghazi 2007).

Rhabdosporites streelii Marshall, 1996
Figure 39D

1969 ?*Rhabdosporites Langi* (Eisenack) Richardson; Lele and Steel, p. 102, pl. 3, fig. 65.
1996 *Rhabdosporites streelii* Marshall, p. 177 (*cum syn.*), pl. 3, figs 1–5.

Dimensions. 111(133)184 μm; four specimens measured.

Remarks. As *R. streelii* occurs within sporangia (Bonamo and Banks 1967) containing also *R. langii* (Eisenack) Richardson, 1960, it could be regarded as a variant of *R. langii*. *R. streelii* is created by splitting of the nexine into ectonexine and endonexine. It represents a distinct evolutionary development within the sporangia of *Rhabdosporites* (Richardson) Marshall and Allen, 1982. As its stratigraphical range is short, it constitutes a potential stratigraphical index species (Marshall 1996).

Comparison. *Rhabdosporites streelii* is clearly separable from all other species of *Rhabdosporites* by the possession of three detached layers.

Occurrence. A1-69; Awaynat Wanin II Formation; *undulatus* Zone.

Previous records. From the Pepinster Formation of Belgium (Lele and Steel 1969), which is considered as upper Eifelian by Laloux *et al.* (1996); lower Givetian of Parnaíba Basin, Brazil (Breuer and Grahn 2011); middle Eifelian – middle Givetian of Libya (Moreau-Benoit 1989); and Givetian of Scotland (Richardson 1965; Marshall and Allen 1982).

Genus SAMARISPORITES Richardson, 1965

Type species. *Samarisporites orcadensis* (Richardson) Richardson, 1965.

Remarks. The nature of zonae-cingula is difficult to apprehend in compressed specimens under a transmitted light microscope. Indeed, different layers in multilayered spores may have differing optical properties, and this can be misleading when attempting to interpret wall thickness and spore structure. This problem is enhanced when coupled with optical effects resulting from compressional artefacts (Wellman 2001).

Comparison. The definition of *Cristatisporites* Potonié and Kremp, 1954 restricts the distal sculpture to being dominantly

mammoid, i.e. coni or spinae which are fused together at their bases into ridges carrying the spinose projections.

Samarisporites angulatus (Tiwari and Schaarschmidt) Loboziak and Streel, 1989
Figures 39E, 52M–O

1975 *Calyptosporites angulatus* Tiwari and Schaarschmidt, p. 44, pl. 26, figs 4–5, pl. 27, fig. 1.
1989 *Samarisporites angulatus* (Tiwari and Schaarschmidt) Loboziak and Streel, p. 191, pl. 5, figs 8–9.

Dimensions. 84(102)131 μm; 18 specimens measured.

Comparison. *Samarisporites eximius* (Allen) Loboziak and Streel, 1989 often has smaller spines and its equatorial flange is more or less uniform around the central body but some specimens intergrade with *S. angulatus*. Therefore, the two species are included in the *S. eximius* Morphon defined here (Table 1).

Occurrence. A1-69; Awaynat Wanin I and Awaynat Wanin II formations; *svalbardiae-eximius* to *rugulata-libyensis* zones. MG-1; Awaynat Wanin I Formation; *rugulata* Zone.

Previous records. From Eifelian of Paraná Basin, Brazil (Loboziak *et al.* 1988); upper Emsian–lower Givetian (Tiwari and Schaarschmidt 1975; Loboziak *et al.* 1990); and ?upper Eifelian–Givetian of Saudi Arabia (PB, pers. obs.).

Samarisporites eximius (Allen) Loboziak and Streel, 1989
Figures 39F, 40A and 52P–R

1965 *Perotrilites eximius* Allen, p. 731, pl. 102, figs 11–13.
1980*b* *Grandispora velata* (Eisenack) McGregor; Moreau-Benoit, p. 40, pl. 12, fig. 3.
1982 *Grandispora eximia* (Allen) McGregor and Camfield, p. 44, pl. 10, figs 2, 6–7; text- fig. 64.
1989 *Samarisporites eximius* (Allen) Loboziak and Streel, p. 192, pl. 4, figs 1–4.
non 1995 *Samarisporites eximius* (Allen) Loboziak and Streel; Rodriguez *et al.*, pl. 1, fig. 2.

Dimensions. 95(116)145 μm; 25 specimens measured.

Comparison. *Samarisporites praetervisus* (Naumova) Allen, 1965 possesses a pseudoreticulate pattern formed by rugulae comprised of spines similar to *S. eximius*. Extreme forms of *S. eximius* intergrade with *S. praetervisus* and *S. angulatus* (Tiwari and Schaarschmidt) Loboziak and Streel, 1989, which show larger spines and a more triangular flange. Therefore, the *S. eximius* Morphon is defined here (Table 1).

Occurrence. JNDL-1, S-462 and WELL-1; Jubah Formation; *svalbardiae-eximius* to *langii-concinna* zones. A1-69; Awaynat Wanin

I and Awaynat Wanin II formations; *svalbardiae-eximius* to *triangulatus-catillus* zones. MG-1; Ouan-Kasa, Awaynat Wanin I, Awaynat Wanin II and Awaynat Wanin III formations; *svalbardiae-eximius* to *langii-concinna* zones.

Previous records. From upper Eifelian–Famennian of Bolivia (Perez-Leyton 1990); Eifelian–Frasnian of Paraná Basin, Brazil (Loboziak *et al.* 1988); upper Eifelian – lower Givetian of Canada (McGregor and Camfield 1982); upper Emsian–Givetian of Germany (Riegel 1973; Loboziak *et al.* 1990); upper Emsian to lowermost Eifelian – lower Givetian of Libya (Moreau-Benoit 1989); and Emsian–Eifelian of Spistbergen, Norway (Allen 1965).

Samarisporites praetervisus (Naumova) Allen, 1965
Figures 40B–C, 52S–U

1953 *Hymenozonotriletes praetervisus* Naumova, p. 40, pl. 4, fig. 8.
1965 *Samarisporites praetervisus* (Naumova) Allen, p. 714, pl. 98, figs 9–10.
1980b *Grandispora praetervisa* (Naumova) Moreau-Benoit, p. 37, pl. 11, fig. 5.
1985 *?Calyptosporites* sp. A Paris *et al.*, pl. 20, fig. 8.
non 1995 *Samarisporites praetervisus* (Naumova) Allen; Rodriguez *et al.*, pl. 1, fig. 1.

Dimensions. 79(98)114 μm; 10 specimens measured.

Comparison. *Samarisporites eximius* (Allen) Loboziak and Streel, 1989 differs only by the possession of discrete elements on the distal surface of the central body however they can be locally fused in rugulae. Extreme forms of *S. praetervisus* (Naumova) Allen, 1965 probably intergrade with *S. eximius* in the *S. eximius* Morphon (Table 1). *S. orcadensis* Richardson, 1960 is considerably larger.

Occurrence. S-462; Jubah Formation; *lemurata-langii* to *triangulatus-catillus* zones. A1-69; Awaynat Wanin I and Awaynat Wanin II formations; *svalbardiae-eximius* to *lemurata-langii* zones. MG-1; Ouan-Kasa, Awaynat Wanin I and Awaynat Wanin II formations; *svalbardiae-eximius* to *triangulatus-catillus* zones.

Previous records. From lower Givetian of Algeria (Boumendjel *et al.* 1988); uppermost Eifelian–lower Givetian of Amazon Basin, Brazil (Melo and Loboziak 2003); Eifelian–middle Givetian of Libya (Paris *et al.* 1985; Streel *et al.* 1988; Moreau-Benoit 1989); and Givetian of Spistbergen, Norway (Allen 1965).

Samarisporites triangulatus Allen, 1965
Figures 40D–E, 52V–X

1964 *Cirratriradites* sp. Vigran, p. 24, pl. 3, fig. 6.
1965 *Samarisporites triangulatus* Allen, p. 716, pl. 99, figs 1–6.

Dimensions. 40(60)94 μm; 40 specimens measured.

Remarks. In the studied material, *S. triangulatus* is preserved in proximo-distal or in lateral compression when specimens have very high laesurae.

Comparison. The distinctive equatorial flange separates this species from other species assigned to *Samarisporites* Richardson, 1965.

Occurrence. S-462, WELL-1 and WELL-8; Jubah Formation; *triangulatus-catillus* to *langii-concinna* zones, specimens from S-462 may be caved in older strata. A1-69; Awaynat Wanin II Formation; *triangulatus-catillus* to *langii-concinna* zones. MG-1; Awaynat Wanin II and Awaynat Wanin III formations; *triangulatus-catillus* to *langii-concinna* zones.

Previous records. *Samarisporites triangulatus* is eponymous for the upper Givetian – lower Frasnian *optivus-triangulatus* Assemblage Zone of the Old Red Sandstone Continent and adjacent regions (Richardson and McGregor 1986), the middle Givetian – lower Frasnian TA and TCo Oppel zones of Western Europe (Streel *et al.* 1987). *S. triangulatus* has a worldwide distribution extending through the Givetian and Frasnian, locally reaching the Famennian. From middle Givetian of Algeria (Moreau-Benoit *et al.* 1993); upper Givetian – lower Frasnian of Argentina (Ottone 1996), Portugal (Lake *et al.* 1988) and Russian Platform (Avkhimovitch *et al.* 1993); middle Givetian – lower Frasnian of Australia (Balme 1988; Grey 1991; Hashemi and Playford 2005); middle Givetian–Famennian of Belgium (Becker *et al.* 1974; Streel and Loboziak 1987; Gerrienne *et al.* 2004); upper Givetian –Famennian of Bolivia (Perez-Leyton 1990); Givetian–upper Tournaisian of Brazil (Loboziak *et al.* 1988, 1992b; Melo and Loboziak 2003); upper Givetian – upper Frasnian of France (Brice *et al.* 1979; Loboziak and Streel 1980, 1988; Loboziak *et al.* 1983); Givetian of Germany (Loboziak *et al.* 1990) and Poland (Turnau 1986, 1996; Turnau and Racki 1999); middle Givetian of Greenland (Friend *et al.* 1983; Marshall and Hemsley 2003); Frasnian–?lower Famennian of Libya (Paris *et al.* 1985; Streel *et al.* 1988); Givetian–Frasnian of Spistbergen, Norway (Vigran 1964; Allen 1965); and uppermost Givetian–lower Frasnian of Scotland (Marshall *et al.* 1996).

FIG. 40. Each figured specimen is identified by borehole, sample, slide number and England Finder Co-ordinate location. All figured specimens are at magnification ×1000 except where mentioned otherwise. A, *Samarisporites eximius* (Allen) Loboziak and Streel, 1989, magnification ×750. JNDL-1, 167.8 ft, PPM006, O34/2. B–C, *Samarisporites praetervisus* (Naumova) Allen, 1965, magnification ×750. B, A1-69, 1334 ft, 27127, K48. C, MG-1, 2258 m, 62947, M39/2. D–E, *Samarisporites triangulatus* Allen, 1965. D, A1-69, 1277 ft, 62636, X31/1. E, A1-69, 1277 ft, 62637, Q37/3. F–G, *Samarisporites tunisiensis* sp. nov. F, MG-1, 2741.4 m, 62611, L28/1. G, Paratype, MG-1, 2741.4 m, 62612, W28.

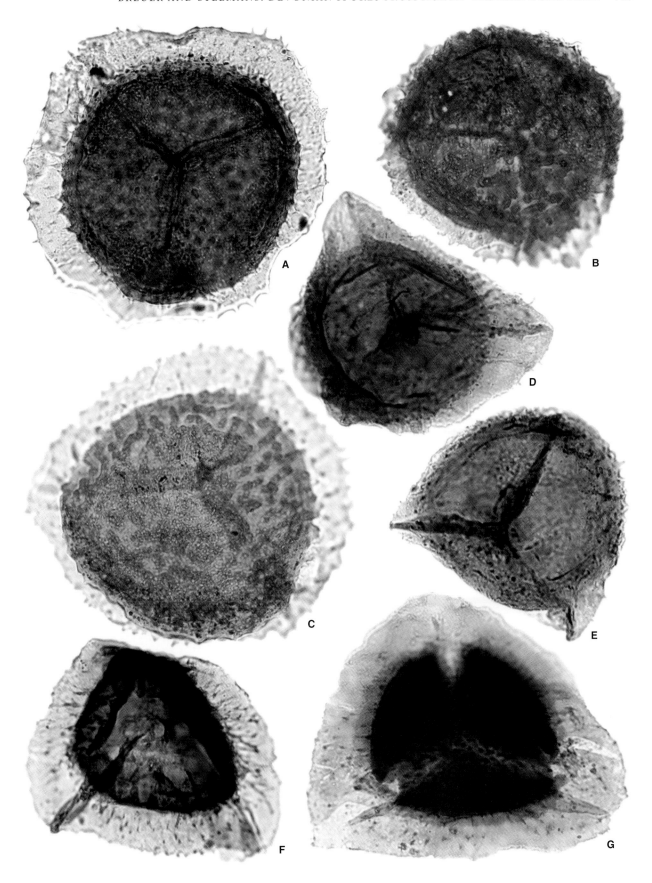

Samarisporites tunisiensis sp. nov.
Figures 40F–G, 41A and 52Y–AA

1992a *Samarisporites* sp. Loboziak *et al.*, pl. 1, fig. 1.

Derivation of name. From *tunisiensis* (Latin), meaning from Tunisia; refers to its geographical occurrence.

Holotype. EFC P39/2 (Figs 41A, 52Z), slide 62611.

Paratype. EFC W28 (Fig. 40G), slide 62612; MG-1 borehole, sample 2741.4 m.

Type locality and horizon. MG-1 borehole, sample 2741.4 m; Ouan-Kasa Formation at Mechiguig, Tunisia.

Diagnosis. A sub-triangular *Samarisporites* sometimes with a bizonate appearance. Distal surface sculptured with small coni or spines, more densely distributed on the distal hemisphere of the central body than on the zona.

Description. Amb is sub-triangular. Laesurae are straight or slightly sinuous, elevated, up to 7 μm high, often extending on to the equatorial flange, and frequently to the equatorial margin. The central body is sub-circular to sub-triangular, and its diameter equals three-fifths to two-thirds of the total amb diameter. The equatorial flange is wider opposite the laesurae (12–24 μm wide) than interradially (6–14 μm wide). Exine may sometimes be folded radially. Camerate structure can give a bizonate appearance. Camera width equals one-fifth to two-fifths of the zona width. Proximal surface laevigate, distal surface supporting coni or spines, 0.5–1.5 μm wide, 1–2.5 μm high and 1–3 μm apart. Ornament on the distal hemisphere of the central body is more densely distributed than those on the zona. Groups of sculptural elements on the central body can be fused at their bases making short ridges.

Dimensions. 73(82)88 μm; six specimens measured.

Comparison. *Samarisporites eximius* (Allen) Loboziak and Steel, 1989 possesses a thicker central body and its ornamentation is commonly larger. *S. angulatus* (Tiwari and Schaarschmidt) Loboziak and Steel, 1989 has considerably larger sculptural elements. *Perotrilites caperatus* (McGregor) Steemans, 1989 has smaller ornamentation, often barely visible.

Occurrence. MG-1; Ouan-Kasa Formation; *lindlarensis-sextantii* Zone.

Samarisporites sp. 1
Figure 41B

Description. Amb is sub-circular to sub-triangular. Laesurae are straight, simple, extending half or almost to the inner margin of the zona. Curvaturae are not visible. The central body diameter equals three-fifths to four-fifths of the total amb diameter. Maximum width of the equatorial flange is 12–22 μm. Zona is folded radially. Exine of the central body is punctate and 2.5–5 μm thick equatorially. Distal hemisphere of central body is sculptured with polygonal verrucae, commonly 2–5 μm wide (rarely up to 8 μm) and closely packed.

Dimensions. 90(98)107 μm; three specimens measured.

Occurrence. A1-69; Awaynat Wanin I Formation; *svalbardiae-eximius* Zone. MG-1; Ouan-Kasa Formation; *svalbardiae-eximius* Zone.

Samarisporites sp. 2
Figures 41C, 52AB–AD

Description. Amb is sub-circular to sub-triangular. Laesurae are straight, simple and elevated up to 8 μm high at the pole, decreasing in height to or almost to the outer margin of the zona. Curvaturae are not visible. The central body diameter equals more or less three-quarters of the total amb diameter. The equatorial flange is laevigate to infragranulate and can be slightly wider opposite the laesurae than interradially; its maximum width is up to 25 μm. Distal surface of zona is ornamented with spines, 1–2.5 μm wide at base, 2–5.5 μm high and 1.5–5 μm apart. Exine of the central body is 4–7 μm thick equatorially. Exine of body is entirely laevigate except the equatorial part, which is ornamented with the same spines as the zona.

Dimensions. 116(120)123 μm; three specimens measured.

Comparison. *Samarisporites eximius* (Allen) Loboziak and Steel, 1989 possesses a distally ornamented central body as *S. praetervisus* (Naumova) Allen, 1965, which has a smaller ornamentation in addition.

Occurrence. A1-69; Awaynat Wanin I Formation; *svalbardiae-eximius* to *rugulata-libyensis* zones.

Genus SCYLASPORA Burgess and Richardson, 1995

Type species. *Scylaspora scripta* Burgess and Richardson, 1995.

FIG. 41. Each figured specimen is identified by borehole, sample, slide number and England Finder Co-ordinate location. All figured specimens are at magnification ×1000 except where mentioned otherwise. A, *Samarisporites tunisiensis* sp. nov. Holotype, MG-1, 2741.4 m, 62611, P39/2. B, *Samarisporites* sp. 1, magnification ×750. A1-69, 1962 ft, 27277, K37/4. C, *Samarisporites* sp. 2, magnification ×750. A1-69, 1596 ft, 26989, G38/1. D, *Scylaspora costulosa* Breuer *et al.*, 2007c. JNDL-4, 419.3 ft, 03CW261, U37/3. E–F, *Scylaspora rugulata* (Riegel) Breuer *et al.*, 2007c. E, WELL-8, 16649.3 ft, 62416, Q47/2. F, A1-69, 1530 ft, 26985, E32/4. G–H, *Scylaspora* sp. 1. G, MG-1, 2631.2 m, 62551, O30. H, MG-1, 2631.2 m, 62553, J48.

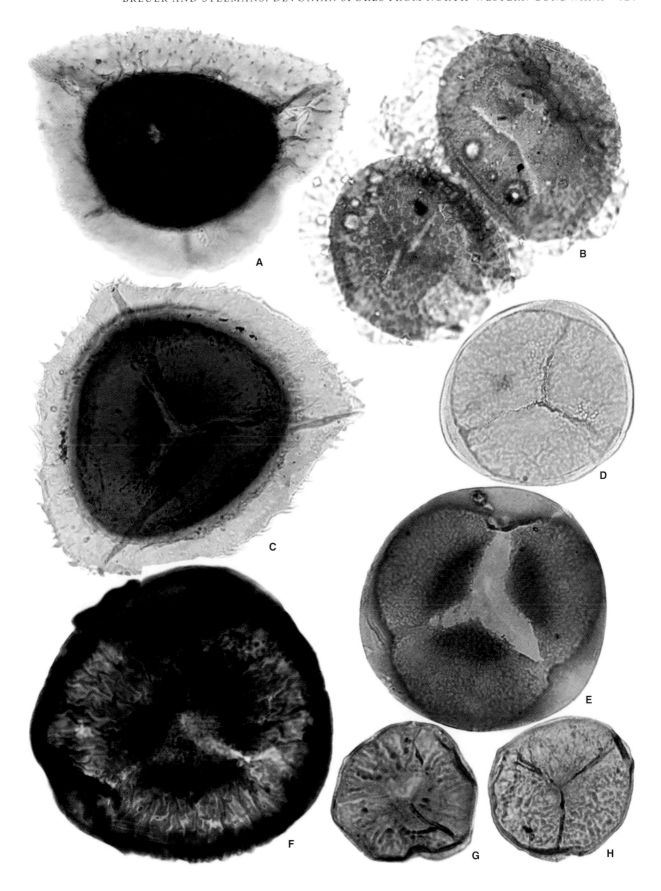

Comparison. Emphanisporites McGregor, 1961 has proximal sculpture of radially aligned muri.

Scylaspora costulosa Breuer et al., 2007c
Figure 41D

2005   *Retusotriletes rugulatus* Riegel; Hashemi and
        Playford, p. 337, pl. 2, fig. 1.
2007c *Scylaspora costulosa* Breuer *et al.*, p. 50, pl. 10,
        figs 1–5.

Dimensions. 45(65)91 μm; 24 specimens measured.

Comparison. The specimen of *Retusotriletes rugulatus* Riegel, 1973 illustrated in Hashemi and Playford (2005) has radially oriented sinuous muri and does not has a darker sub-triangular area at the proximal pole as described in Riegel (1973). *R. rugulatus in* Hashemi and Playford (2005) corresponds to the description of *S. costulosa* and it is considered as synonymous.

Occurrence. BAQA-1, BAQA-2, JNDL-1, JNDL-3, JNDL-4, WELL-2, WELL-3, WELL-4, WELL-6 and WELL-7; Jauf Formation; *papillensis-baqaensis* to *annulatus-protea* zones.

Previous records. From Emsian of Adavale Basin, Australia (Hashemi and Playford 2005); and upper Pragian – lower Emsian of Paraná Basin, Brazil (Mendlowicz Mauller *et al.*, 2007).

Scylaspora rugulata (Riegel) Breuer et al., 2007c
Figure 41E–F

1965   *Retusotriletes dubius* (Eisenack); Richardson p. 564
        (*pars*), pl. 88, fig. 6.
1966   *Retusotriletes* sp. McGregor and Owens (*pars*),
        pl. 9, fig. 2.
1967   *Retusotriletes* sp. 1 Hemer and Nygreen, pl. 2,
        fig. 6.
1971   *Retusotriletes dubius* (Eisenack) Richardson; Owens,
        p. 12 (*pars*), pl. 1, fig. 10.
1973   *Retusotriletes rugulatus* Riegel, p. 82, pl. 10,
        figs 2–5.
1974   *Stenozonotriletes extensus* Naumova; Hamid, pl. 10,
        fig. 3.
non 2005 *Retusotriletes rugulatus* Riegel; Hashemi and
        Playford, p. 337, pl. 2, fig. 1.
2007c *Scylaspora rugulata* (Riegel) Breuer *et al.*, p. 50.

Dimensions. 53(67)85 μm; 12 specimens measured.

Comparison. Emphanisporites sp. *in* Tiwari and Schaarschmidt (1975) appears similar to *S. rugulata* except for its prominent elevated labra. *Retusotriletes biarealis* McGregor, 1964 is thinner

and bears a faintly defined radial pattern in the contact areas. *S. costulosa* has less prominent and less angular rugulae.

Occurrence. S-462, WELL-1 and WELL-8; Jubah Formation; *rugulata-libyensis* to *triangulatus-catillus* zones. A1-69; Awaynat Wanin I and Awaynat Wanin II formations; *rugulata-libyensis* to *triangulatus-catillus* zones. MG-1; Awaynat Wanin I and Awaynat Wanin II formations; *rugulata-libyensis* to *triangulatus-catillus* zones.

Previous records. From middle Givetian of Belgium (Gerrienne *et al.* 2004); Eifelian–lower Givetian of Canada (McGregor and Owens 1966; Owens 1971; McGregor and Camfield 1982); upper Givetian – lower Frasnian of France (Brice *et al.* 1979; Loboziak and Streel 1980, 1988); Eifelian–Givetian of Germany (Riegel 1973; Loboziak *et al.* 1990); lower Eifelian – lower Givetian of Libya (Streel *et al.* 1988); lower Eifelian – upper Givetian of Poland (Turnau 1986, 1996; Turnau and Racki 1999; Turnau and Matyja 2001); and Eifelian – lower Frasnian of Scotland (Richardson 1965; Marshall 1988, 2000; Marshall *et al.* 1996; Marshall and Fletcher 2002).

Scylaspora sp. 1
Figure 41G–H

cf. 1995 *Scylaspora scripta* Burgess and Richardson, p. 17,
        pl. 7, figs 5–9.

Description. Amb is sub-circular to sub-triangular. Laesurae are straight, simple, up to 2 μm wide, extending almost to the equator. Exine is 1–2 μm thick equatorially. Contact areas are sculptured with low, rounded, sinuous to convolute and anastomosing muri, 0.5–1.5 μm wide and commonly less than 1 μm apart. Muri are radially aligned near equator, but randomly orientated towards the proximal pole. Distal surface is laevigate.

Dimensions. 40–42 μm; two specimens measured.

Comparison. Scylaspora scripta Burgess and Richardson, 1995 is equatorially and distally sculptured with scattered to regularly distributed grana and tiny verrucae. *S. downiei* Burgess and Richardson, 1991 is smaller with predominantly verrucate proximal sculpture. *S. costulosa* Breuer *et al.*, 2007c is larger.

Occurrence. MG-1; Ouan-Kasa Formation; *svalbardiae-eximius* Zone but occurrences are probably reworked.

Genus SQUAMISPORA Breuer et al., 2007c

Type species. Squamispora arabica Breuer *et al.*, 2007c.

Comparison. Diducites Van Veen, 1981 is two-layered with a very thin outer layer folding and wrinkling frequently resulting as a rugulate appearance. *Squamispora* cannot be compared to genera such *Verrucosisporites* Ibrahim, 1933 because the flakes do not constitute real positive sculptural elements.

Squamispora arabica Breuer et al., 2007c
Figure 42A–B

2007c *Squamispora arabica* Breuer *et al.*, p. 51, pl. 10, figs 6–10.

Dimensions. 39(51)63 μm; 43 specimens measured.

Remarks. *Squamispora arabica* may occur in clusters of several specimens. Some local detachment of the scaly layer can occur. Consequently, total detachment of this outer layer possibly could resemble specimens of *Leiotriletes* Naumova emend. Potonié and Kremp, 1954 in poorly preserved assemblages, but *S. arabica* specimens are also recorded in such assemblages from eastern Saudi Arabia (PB, pers. obs.).

Occurrence. JNDL-1; Jubah Formation; *svalbardiae-eximius* Zone.

Genus STELLATISPORA Burgess and Richardson, 1995

Type species. *Stellatispora inframurinata* (Richardson and Lister) Burgess and Richardson, 1995.

Stellatispora multicostata Breuer et al., 2007c
Figure 42C–D

2007c *Stellatispora multicostata* Breuer *et al.*, p. 51, pl. 11, figs 1–4.

Dimensions. 54(74)93 μm; 17 specimens measured.

Occurrence. BAQA-1, JNDL-3, JNDL-4, WELL-4 and WELL-7; Jauf Formation (Subbat and Hammamiyat members); *asymmetricus* to *lindlarensis-sextantii* zones.

Genus SYNORISPORITES Richardson and Lister, 1969

Type species. *Synorisporites downtonensis* Richardson and Lister, 1969.

Synorisporites cf. *S. lobatus* Rodriguez, 1978a
Figure 42E–F

cf. 1978a *Synorisporites lobatus* Rodriguez, p. 216, pl. 1, figs 13–14.

Description. Trilete spores with sub-circular to sub-triangular amb. Laesurae straight, accompanied by labra, up to 2.5 μm wide individually, extending to the inner margin of the cingulum. Cingulum 3–5 μm wide. Proximal surface laevigate and distal surface sculptured with a convoluted sub-circular thickening, about three-fifths of the amb diameter. Additional verrucae, 1.5–5 μm wide may be present on the distal surface.

Dimensions. 39(48.5)58 μm; three specimens measured.

Remarks. According to its diagnosis, *S. lobatus* Rodriguez, 1978a is smaller and its sub-circular thickening is developed on the proximal face. The location of the thickening may be misinterpreted as it was observed on compressed specimens.

Occurrence. MG-1; Ouan-Kasa Formation; *lindlarensis-sextantii* to *svalbardiae-eximius* zones.

Previous record. Similar specimens have been found from upper Emsian–lower Eifelian of Paraná Basin, Brazil (PB, pers. obs.).

Synorisporites papillensis McGregor, 1973
Figure 42G–H

1966 Unidentified McGregor and Owens, pl. 2, fig. 32.
? 1968 Spore trilète à papilles proximales sp. 5 Jardiné and Yapaudjian, pl. 1, fig. 7.
1969 *Synorisporites* sp. A Richardson and Lister, p. 234, pl. 41, figs 1–2.
1973 *Synorisporites papillensis* McGregor, p. 51, pl. 6, figs 26–29.
1976 *Synorisporites* ?*papillensis* McGregor; McGregor and Camfield, p. 28, pl. 1, figs 24, 27.

Dimensions. 24(32)43 μm; 55 specimens measured.

Comparison. *Synorisporites tripapillatus* Richardson and Lister, 1969 is smaller and has a distal sculpture consisting of convolute and anastomosing muri. Jardiné and Yapaudjian (1968) figure an undescribed species resembling *S. papillensis*. ?*Knoxisporites riondae* Cramer and Díez, 1975 is characterized by a distal annulus which may be very faint. The taxonomic attribution of intermediary specimens is difficult because these two species intergrade and are placed in the *S. papillensis* Morphon defined here (Table 1).

Occurrence. WELL-1, BAQA-1, BAQA-2, WELL-2, WELL-3, JNDL-1, JNDL-3, JNDL-4, WELL-4 and WELL-7; Jauf and Jubah Formation; *papillensis-baqaensis* to *svalbardiae-eximius* zones. A1-69; Ouan-Kasa and Awaynat Wanin I formations; *lindlarensis-sextantii* to *svalbardiae-eximius* zones. MG-1; Ouan-Kasa and Awaynat Wanin I formations; *annulatus-protea* to *rugulata-libyensis* zones but the highest occurrences are probably reworked.

Previous records. From lower Lochkovian – upper Pragian of Belgium (Steemans 1989); upper Lochkovian – lower Emsian of Paraná and Solimões basins, Brazil (Rubinstein *et al.* 2005; Mendlowicz Mauller *et al.* 2007); Lochkovian–Emsian of Canada (McGregor and Owens 1966; McGregor 1973; McGregor and Camfield 1976); Pragian of Armorican Massif, France (Le Hérissé 1983); middle Přídolí of Libya (Rubinstein and Steemans 2002); Emsian of Saudi Arabia (Al-Ghazi 2007); and Lochkovian of Wales (Richardson and Lister 1969).

Synorisporites verrucatus Richardson and Lister, 1969
Figure 42I–J

1969 *Synorisporites verrucatus* Richardson and Lister, p. 233, pl. 40, figs 10–12.
1981 *Synorisporites* cf. *verrucatus* Richardson and Lister; Steemans, pl. 1, figs 12–13.

Dimensions. 28(33)40 µm; seven specimens measured.

Comparison. Synorisporites tripapillatus Richardson and Lister, 1969 is smaller and has a distal sculpture of convolute and anastomosing muri, and three proximal interradial papillae. *S. papillensis* McGregor, 1973 has also proximal papillae. *Synorisporites? libycus* Richardson and Ioannides, 1973 have larger verrucae.

Occurrence. A1-69; Ouan-Kasa and Awaynat Wanin I formations; *annulatus-protea* to *svalbardiae-eximius* zones. Ouan-Kasa and Awaynat Wanin I formations; *lindlarensis-sextantii* to *svalbardiae-eximius* zones.

Previous records. From Silurian–Pragian of Algeria (Boumendjel *et al.* 1988); lower Lochkovian of Belgium (Steemans 1989); Lochkovian of Bolivia (McGregor 1984); upper Lochkovian – lower Emsian of Paraná and Solimões basins, Brazil (Rubinstein *et al.* 2005; Mendlowicz Mauller *et al.* 2007); Emsian of Canada (McGregor 1973), Lochkovian–Pragian of Armorican Massif, France (Le Hérissé 1983; Steemans 1989); ?Ludlow– middle Přídolí of Libya (Rubinstein and Steemans 2002); Lower Silurian–Pragian of Saudi Arabia (Steemans 1995); Wenlock–lower Pragian of Spain (Rodriguez 1978*b*); and Přídolí of Wales (Richardson and Lister 1969).

Genus VERRUCIRETUSISPORA Owens, 1971

Type species. Verruciretusispora dubia (Eisenack) Richardson and Rasul, 1978.

Verruciretusispora dubia (Eisenack) Richardson and Rasul, 1978
Figure 42K

1978 *Verruciretusispora dubia* (Eisenack) Richardson and Rasul, p. 443 (cum syn.), pl. 1, fig. 6.

Dimensions. 52(56)58 µm; four specimens measured.

Occurrence. JNDL-1, JNDL-4 and WELL-7; Jauf (Subbat and Hammamiyat members) and Jubah formations; *lindlarensis-sextantii* to *svalbardiae-eximius* zones.

Previous records. From Emsian–lower Givetian of Australia (Hashemi and Playford 2005); Emsian–lower Givetian of Canada (McGregor and Owens, 1966; Owens 1971; McGregor and Uyeno 1972; McGregor 1973; McGregor and Camfield 1982); Emsian of Germany (Lanninger 1968); middle–upper Emsian of Luxembourg (Steemans *et al.* 2000*a*); Emsian–Givetian of Poland (Turnau 1986, 1996; Turnau and Racki 1999; Turnau *et al.* 2005); and upper Eifelian – lower Frasnian of Scotland (Marshall 1988, 2000; Marshall *et al.* 1996; Marshall and Fletcher 2002).

Genus VERRUCISPORITES Chi and Hills, 1976

Type species. Verrucisporites medius Chi and Hills, 1976.

Verrucisporites ellesmerensis (Chaloner) Chi and Hills, 1976
Figures 42L, 52AE–AG

1959 *Biharisporites ellesmerensis* Chaloner, p. 322, pl. 55, fig. 2, text-fig. 1.
1976 *Verrucisporites ellesmerensis* (Chaloner) Chi and Hills, p. 701, pl. 2, figs 9–16; pl. 3, figs 1–6.

FIG. 42. Each figured specimen is identified by borehole, sample, slide number and England Finder Co-ordinate location. All figured specimens are at magnification ×1000 except where mentioned otherwise. A–B, *Squamispora arabica* Breuer *et al.*, 2007*c*. A, JNDL-1, 174.6 ft, 60848, B43/3. B, JNDL-1, 172.7 ft, PPM007, F33/3. C–D, *Stellatispora multicostata* Breuer *et al.*, 2007*c*. C, JNDL-4, 448.6 ft, 68693, R38. D, JNDL-4, 52.3 ft, 03CW188, J38/2. E–F, *Synorisporites* cf. *S. lobatus* Rodriguez, 1978*a*. E, MG-1, 2631.2 m, 62551, F37/4. F, MG-1, 2728 m, 62855, R49. G–H, *Synorisporites papillensis* McGregor, 1973. G, BAQA-1, 222.5 ft, 03CW108, W51/3. H, BAQA-1, 222.5 ft, 03CW108, M30/2. I–J, *Synorisporites verrucatus* Richardson and Lister, 1969. I, A1-69, 2039–2041 ft, 27279, J53/4. J, MG-1, 2631.2 m, 62552, K34-35. K, *Verruciretusispora dubia* (Eisenack) Richardson and Rasul, 1978. JNDL-1, 177.0 ft, PPM009, J39. L, *Verrucisporites ellesmerensis* (Chaloner) Chi and Hills, 1976, magnification ×250. S-462, 1710–1715 ft, 63222, H31. M–P, *Verrucosisporites nafudensis* sp. nov. M, JNDL-4, 331.9 ft, 68669, Q26/3. N, Paratype, JNDL-4, 495.2 ft, 68702, Y40/2. O, Holotype, BAQA-2, 50.8 ft, 03CW127, H30/4. P, BAQA-1, 227.1 ft, 03CW110, N35.

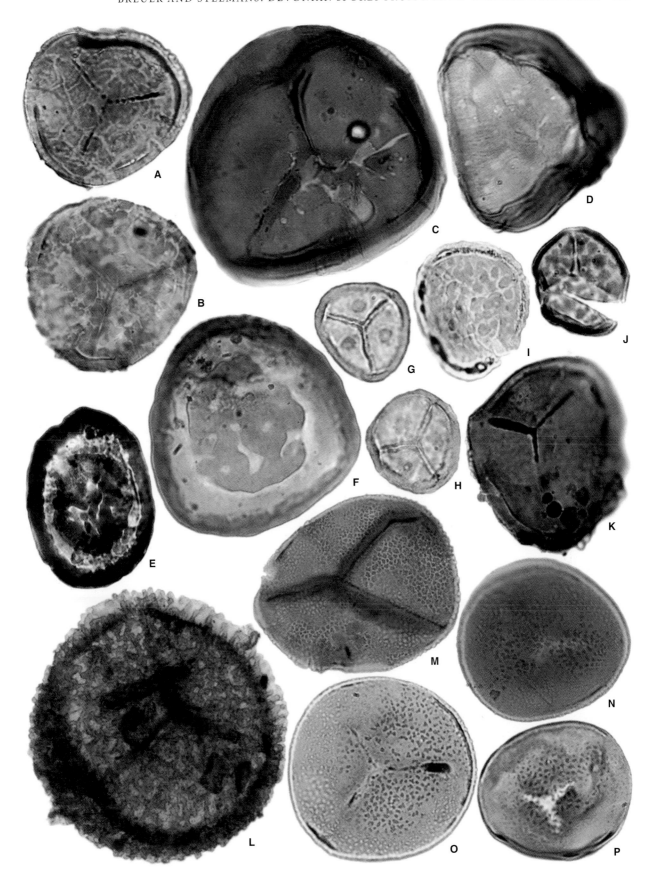

2007 *Verrucisporites yabrinensis* Marshall *et al.*, p. 77, pl. 1, figs 1–11, pl. 2, figs 1–2, 9–10, text- fig. 2A.

Dimensions. 265(299)350 μm; 26 specimens measured.

Remarks. Chi and Hills (1976) define three varieties of *V. ellesmerensis* which are slightly different in detail, but the dimensions of ornamentation intergrade. As the specimens described here are very variable, they may belong to the different defined varieties. Consequently, they were not separated into varieties because the morphological changes seem to be continuous.

Comparison. Verrucisporites yabrinensis Marshall *et al.*, 2007 was erected because its ornament does not correspond exactly to one of the three varieties of *V. ellesmerensis* defined by Chi and Hills, 1976. Indeed, Marshall *et al.* (2007) compare their population only with *V. ellesmerensis* (Chaloner) var. *conatus* Chi and Hills, 1976 which have longer sculptural elements. According to J. E. A. Marshall (pers. comm. 2007), the dimensions of the ornament of *V. yabrinensis* correspond rather to those of *V. ellesmerensis* var. *parvus* but the sculpture is fused like that of *V. ellesmerensis* var. *ellesmerensis*. Thus, the characteristics of ornament of *V. yabrinensis* are included at least in the original definition of *V. ellesmerensis*, i.e. at the specific level. In addition, Marshall *et al.* (2007) also erected a Saudi species different from Arctic Canada because they thought that heterosporous plants are geographically limited by the large size of the megaspore (J. E. A. Marshall, pers. comm. 2007). Although heterosporous plants have little potential to migrate, presence of some identical megaspore species on Gondwana and Euramerica suggests that there was no palaeogeographical barriers to prevent their migration during the Middle Devonian (Steemans *et al.* 2011*b*). Consequently, *V. yabrinensis*, which shows a certain morphological variability, needs to be objectively considered here as a junior synonym of *V. ellesmerensis*.

Occurrence. S-462 and WELL-8; Jubah Formation; *triangulatus-catillus* to *langii-concinna* zones (Marshall *et al.* 2007).

Previous records. From Givetian–Frasnian of Canada (Chi and Hills 1976); and middle Givetian of Libya (Moreau-Benoit 1989).

Genus VERRUCOSISPORITES Ibrahim emend. Smith, 1971

Type species. Verrucosisporites verrucosus (Ibrahim) Ibrahim, 1933.

Verrucosisporites nafudensis sp. nov.
Figure 42M–P

Derivation of name. From *nafudensis* (Latin) meaning from the Nafud Basin; refers to its geographical occurrence.

Holotype. EFC H30/4 (Fig. 42O), slide 03CW127.

Paratype. EFC Y40/2 (Fig. 42N), slide 68702; JNDL-4 core hole, sample 495.2 ft.

Type locality and horizon. BAQA-2 core hole, sample 50.8 ft; Jauf Formation at Baq'a, Saudi Arabia.

Diagnosis. A *Verrucosisporites* sculptured with small flat-topped verrucae and coni, sub-circular to angular and irregular in plan view.

Description. Amb is sub-circular. Laesurae are straight, simple, one-half to four-fifths of the amb radius in length. Curvaturae are barely visible. Exine is 0.5–2 μm thick. Proximal surface is laevigate. Equatorial and distal regions are sculptured with flat-topped verrucae and coni, irregular in plan view, 0.5–3 μm wide at base, 0.5–1 μm high and 0.5–2.5 μm apart. Sculptural elements are discrete or locally fused at base.

Dimensions. 43(54)68 μm; 15 specimens measured.

Comparison. Verrucosisporites polygonalis Lanninger, 1968 has an ornamentation polygonal to more rounded in plan view and closely spaced resulting in a polygonal pattern. *Cyclogranisporites retisimilis* Riegel, 1968 has a darker apical area, sub-triangular with concave sides.

Occurrence. BAQA-1, BAQA-2 and JNDL-4; Jauf Formation (Sha'iba to Subbat members); *papillensis-baqaensis* to *ovalis-biornatus* zones.

Verrucosisporites onustus sp. nov.
Figure 43A–E

Derivation of name. From *onustus* (Latin), meaning swollen; refers to the apices of the proximo-equatorial sculptural elements.

Holotype. EFC J34/2 (Fig. 43E), slide 03CW128.

FIG. 43. Each figured specimen is identified by borehole, sample, slide number and England Finder Co-ordinate location. All figured specimens are at magnification ×1000 except where mentioned otherwise. A–E, *Verrucosisporites onustus* sp. nov. A, Paratype, BAQA-2, 54.8 ft, 03CW129, H50. B, BAQA-1, 345.5 ft, 03CW114, R41/2. C, BAQA-2, 57.2 ft, 66817, D50/4. D, BAQA-2, 56.0 ft, 03CW130, G27-28. E, Holotype, BAQA-2, 52.0 ft, 03CW128, J34/2. F–I, *Verrucosisporites polygonalis* Lanninger, 1968. F, MG-1, 2713 m, 62811, Q40. G, BAQA-1, 371.1 ft, 03CW117, R25/4. H, JNDL-1, 495.0 ft, PPM014, H30/4. I, BAQA-1, 169.1 ft, 03CW103, Q29. J–K, *Verrucosisporites premnus* Richardson, 1965. J, MG1, 2405 m, 62821, V51/1. K, MG-1, 2456 m, 62737, Y40. L–N, *Verrucosisporites scurrus* (Naumova) McGregor and Camfield, 1982. L, MG-1, 2161.8 m, 62528, M45/4. M, A1-69, 1483 ft, 26994, H31/2. N, A1-69, 1334 ft, 27127, K40.

Paratype. EFC H50 (Fig. 43A), slide 03CW129; BAQA-2 core hole, sample 54.8 ft.

Type locality and horizon. BAQA-2 core hole, sample 52.0 ft; Jauf Formation at Baq'a, Saudi Arabia.

Diagnosis. A *Verrucosisporites* sculptured with unevenly distributed verrucae and pila, sub-polygonal to sub-circular to in plan view and with explaned apices.

Description. Amb is sub-circular. Laesurae are straight and simple, seven-tenths to three-fourths of the amb radius in length. Exine is 1–2.5 μm thick equatorially. Proximo-equatorial and distal regions are sculptured with verrucae and pilae commonly 0.5–2.5 μm at base, 1–4.5 μm wide at apice, 1–3.5 μm high. Basal width of ornament is smaller than height. Sculptural elements usually have expanded apices, 1–4.5 μm wide, slightly rounded or flat-topped in profile, sub-polygonal to more rounded in plan view and unevenly distributed. Proximal region granulate.

Dimensions. 44(61)75 μm; 12 specimens measured.

Comparison. *Verrucosisporites polygonalis* Lanninger, 1968 is slightly different by having regularly less high, parallel-sided verrucae, which are evenly distributed and closely spaced, resulting in a polygonal pattern.

Occurrence. BAQA-1 and BAQA-2; Jauf Formation (Sha'iba to Subbat members); *papillensis-baqaensis* to *ovalis* zones.

Verrucosisporites polygonalis Lanninger, 1968
Figure 43F–I

1968 *Verrucosisporites polygonalis* Lanninger, p. 128, pl. 22, fig. 19.
1973 *Verrucosisporites ?polygonalis* Lanninger; McGregor, p. 37 (*cum syn.*), pl. 4, figs 15, 21– 22.

Dimensions. 31(50)90 μm; 21 specimens measured.

Occurrence. BAQA-1, BAQA-2, JNDL-1, JNDL-3, JNDL-4, WELL-2, WELL-3, WELL-4, WELL-6 and WELL-7; Jauf Formation; *papillensis-baqaensis* to *annulatus-protea* zones. A1-69; Ouan-Kasa Formation; *lindlarensis-sextantii* Zone. MG-1; Ouan-Kasa Formation; *annulatus-protea* to *svalbardiae-eximius* zones.

Previous records. From Emsian of Algeria (Moreau-Benoit *et al.* 1993) and Australia (Hashemi and Playford 2005); lower Pragian–Emsian of Belgium (Steemans 1989); upper Pragian – lower Emsian of the Paraná and Parnaíba basins, Brazil (Grahn *et al.* 2005; Mendlowicz Mauller *et al.* 2007); Pragian–Emsian of Canada (McGregor and Owens 1966; McGregor 1973; McGregor and Camfield 1976), Iran (Ghavidel-Syooki 2003), Poland (Turnau 1986; Turnau *et al.* 2005) and Saudi Arabia (Steemans 1995;

Al-Ghazi 2007); lower Pragian–Emsian of Germany (Steemans 1989); upper Pragian of Armorican Massif, France (Le Hérissé 1983); Emsian–lowermost Eifelian of Libya (Moreau-Benoit 1989); and upper Pragian – upper Emsian of Luxembourg (Steemans *et al.* 2000*a*).

Verrucosisporites premnus Richardson, 1965
Figure 43J–K

1965 *Raistrikia* sp. A Richardson, p. 574, pl. 90, fig. 3.
1965 *Raistrikia* cf. *clavata* Hacquebard; Richardson, p. 575, pl. 90, fig. 5.
1965 *Verrucosisporites premnus* Richardson, p. 572, pl. 90, figs 1–2.
non 1968 *Verrucosisporites premnus* Richardson; Lanninger, p. 128, pl. 22, fig. 20.

Dimensions. 30(53)85 μm; 20 specimens measured.

Remarks. *Verrucosisporites premnus* is included in the *V. scurrus* Morphon (Table 1) defined by McGregor and Playford (1992). The latter comprises species characterized by varied, closely spaced or fused, evenly or asymmetrically distributed coni, bacula or verrucae. These taxa have simple laesurae and a sub-circular amb. Consequently, some transitional specimens are difficult to assign to species.

Comparison. *Raistrikia* sp. A and *Raistrickia* cf. *clavata* Hacquebard, 1957 *in* Richardson (1965) have most features in common with *V. premnus*. The major point of difference between these forms and *V. premnus* is that they may bear spatulate or club-shaped sculptural elements in addition to bacula. The variation of ornamentation allowed by the diagnosis of *V. premnus* makes separation of *Raistrikia* sp. A and *Raistrickia* cf. *clavata* Hacquebard, 1957 *in* Richardson (1965) difficult, unrealistic and thus they are considered as synonymous (McGregor and Camfield 1982). *V. scurrus* (Naumova) McGregor and Camfield, 1982 has smaller sculptural elements that are commonly tapered in profile.

Occurrence. WELL-1 and S-462; Jubah Formation; *rugulata-libyensis* to *langii-concinna* zones, some specimens from S-462 may be caved in older strata. A1-69; Awaynat Wanin II Formation; *lemurata-langii* to *triangulatus-catillus* zones. MG-1; Awaynat Wanin I and Awaynat Wanin II formations; *rugulata-libyensis* to *langii-concinna* zones.

Previous records. From middle Givetian of Algeria (Moreau-Benoit *et al.* 1993), Australia (Grey 1991) and Belgium (Gerrienne *et al.* 2004); upper Givetian–Famennian of Bolivia (Perez-Leyton 1990); upper Eifelian – lowermost Famennian of the Amazon and Parnaíba basins, Brazil (Loboziak *et al.* 1992*b*, Melo and Loboziak 2003); middle Eifelian – lower Givetian of Canada (McGregor and Camfield 1982); upper Givetian – lower Frasnian of France (Brice *et al.* 1979; Loboziak and Streel 1988); Eifelian–

Givetian of Germany (Loboziak *et al.* 1990); Eifelian–lower Givetian of Iran (Ghavidel-Syooki 2003); lower Eifelian – middle Givetian of Libya (Moreau-Benoit 1989); upper Eifelian–Givetian of Poland (Turnau 1996; Turnau and Racki 1999); upper Givetian – lower Frasnian of Portugal (Lake *et al.*, 1988); and Givetian–lower Frasnian of Scotland (Richardson 1965; Marshall *et al.* 1996; Marshall 2000).

Verrucosisporites scurrus (Naumova) McGregor and Camfield, 1982
Figure 43L–N

1965 *Raistrickia aratra* Allen, p. 701, pl. 96, figs 3–4.
1982 *Verrucosisporites scurrus* (Naumova) McGregor and Camfield, p. 61 (*cum syn.*), pl. 18, figs 10–17, 22; text-fig. 96.

Dimensions. 35(51)90 μm; 19 specimens measured.

Remarks. Specimens herein assigned to this species belong to a more or less intergrading series from those with predominantly conate and small verrucose sculpture (see *Dibolisporites farraginis* McGregor and Camfield, 1982 and *D. uncatus* (Naumova) McGregor and Camfield, 1982) to those with large verrucate sculptural elements, and thus conform rather closely to the diagnosis of *V. scurrus* and *V. premnus* Richardson, 1965. All these forms are included in the *V. scurrus* Morphon (Table 1).

Comparison. Extreme forms of *V. scurrus* (Naumova) McGregor and Camfield, 1982 intergrade with *D. uncatus* (Naumova) McGregor and Camfield, 1982, *V. premnus*, *V. tumulentis* Clayton and Graham, 1974 and possibly *Chelinospora timanica* (Naumova) Loboziak and Streel, 1989. Typically *D. uncatus* has predominantly somewhat smaller and less crowded conate spinose sculpture. *V. premnus* has larger, predominantly flat-topped, baculate-spatulate sculpture. *C. timanica* has predominantly convolute sculpture and a thicker wall. *Raistrickia nigra* Love, 1960 bears more regularly spaced ornamentation of about the same size on each specimen. *R. commutata* sp. nov. has more widely spaced, and more elongate baculate sculptural elements.

Occurrence. S-462, WELL-1 and WELL-8; Jubah Formation; *rugulata-libyensis* to *langii-concinna* zones. A1-69; Awaynat Wanin II Formation; *rugulata-libyensis* to *langii-concinna* zones. MG-1; Ouan-Kasa, Awaynat Wanin I, Awaynat Wanin II and Awaynat Wanin III formations; *annulatus-protea* to *langii-concinna* zones.

Previous records. From upper Givetian – lower Frasnian of Argentina (Ottone 1996); lower Givetian–lower Frasnian of Australia (Grey 1991; Hashemi and Playford 2005); upper Eifelian–Givetian of Russian Platform (Avkhimovitch *et al.* 1993; Arkhangelskaya and Turnau 2003); middle Givetian of Belgium

(Gerrienne *et al.* 2004); upper Eifelian – lower Frasnian of Bolivia (Perez-Leyton 1990); Givetian–lowermost Famennian of Brazil (Loboziak *et al.* 1988, 1992*b*; Melo and Loboziak 2003); Eifelian–lower Givetian of Canada (McGregor and Camfield 1982); Eifelian–Givetian of Germany (Loboziak *et al.* 1990); lower Eifelian – upper Frasnian of Libya (Paris *et al.* 1985; Streel *et al.* 1988; Moreau-Benoit 1989); Givetian–Frasnian of Spistbergen, Norway (Vigran 1964; Allen 1965); Givetian of Poland (Turnau 1996; Turnau and Racki 1999); and upper Eifelian – lower Frasnian of Scotland (Richardson 1965; Marshall 1988, 2000; Marshall *et al.* 1996).

Verrucosisporites stictus sp. nov.
Figure 44A–G

Derivation of name. From *stictus* (Latin), meaning spotted; refers to the character of the verrucate ornamentation.

Holotype. EFC V41 (Fig. 44B), slide 62272.

Paratype. EFC O28/1 (Fig. 44E), slide 03CW121; BAQA-1 core hole, sample 395.2 ft.

Type locality and horizon. BAQA-1 core hole, sample 395.2 ft; Jauf Formation at Baq'a, Saudi Arabia.

Diagnosis. A *Verrucosisporites* densely sculptured by verrucae, closely appressed or fused at base, surmounted by 1 or more minute coni. Proximal region bearing a kyrtome on each interradial area.

Description. Trilete spores with sub-circular to sub-triangular amb. Laesurae straight, simple and extending almost to the equator. Proximal region laevigate and bearing a kyrtome on each interradial area. Equatorial and distal regions sculptured with densely distributed verrucae, 1.5–4 μm wide at base, 1–2.5 μm high, generally less than 0.5 μm apart. Sculptural elements discrete, sub-circular or polygonal in plan view, rounded or slightly tapered with rounded or more or less flat apices in profile view, surmounted by 1 or more minute coni, *c.* 0.5 μm wide and high, at each apex.

Dimensions. 29(44)53 μm; 18 specimens measured.

Remarks. On well-preserved specimens, the apex of verrucae, where the coni are present, appears as a paler spot in polar view. This phenomenon is interpreted to be the result of light reflection. On more badly preserved specimens, this phenomenon is not observed.

Comparison. *Verrucosisporites* sp. 1 does not possess kyrtomes on its proximal face and is less densely sculptured.

Occurrence. BAQA-1, BAQA-2, JNDL-3 and WELL-4; Jauf Formation (Sha'iba to Subbat members); *papillensis-baqaensis* to *lindlarensis-sextantii* zones. A1-69; Ouan-Kasa Formation; *lindlarensis-sextantii* Zone.

Verrucosisporites sp. 1
Figure 44H–J

Description. Amb is sub-circular to sub-triangular. Laesurae are straight, simple and extend to the equator. Exine 1–2 μm thick. Proximal region is laevigate. Equatorial and distal regions are sculptured with evenly distributed verrucae or truncated coni, 1.5–5 μm wide at base, 2–5 μm high and 0.5–3 μm apart. Sculptural elements are circular or sub-circular in plan view, parallel-sided or slightly tapered with flat or rounded apices in profile view, sometimes surmounted by 1 or 2 minute coni at each apex.

Dimensions. 34(39)45 μm; four specimens measured.

Comparison. *Cymbosporites dammamensis* Steemans, 1995 is patinate and bears small bacula generally with bifurcate-shaped apices. *V. stictus* sp. nov. possesses a kyrtome in each interradial area.

Occurrence. BAQA-1, JNDL-4 and WELL-4; Jauf Formation (Subbat and Hammamiyat members); *ovalis-biornatus* to *lindlarensis-sextantii* zones.

Genus ZONOTRILETES Luber and Waltz, 1938

Type species. None designated.

Zonotriletes armillatus Breuer et al., 2007c
Figure 44K–M

   2007c *Zonotriletes armillatus* Breuer *et al.*, p. 51, pl. 11, figs 9–12, pl. 12, figs 1–2.

Dimensions. 43(64)109 μm; 35 specimens measured.

Occurrence. JNDL-1, Jubah Formation, *svalbardiae-eximius* Zone. MG-1, Awaynat Wanin I Formation, *rugulata-libyensis* Zone.

Previous record. From upper Eifelian – lower Givetian of Parnaíba Basin, Brazil (Breuer and Grahn 2011).

Zonotriletes brevivelatus sp. nov.
Figure 44N–Q, 45A–B

   1968 *Zonotriletes* sp. 1 Jardiné and Yapaudjian, pl. 2, fig. 6.
   1976 *Perotriletes* sp. Massa and Moreau-Benoit, pl. 1, fig. 5.

Derivation of name. From *brevi-* and *velatus* (Latin), meaning bearing a narrow flange; refers to the flange.

Holotype. EFC Q46/3 (Fig. 44O), slide 62779.

Paratype. EFC D55/1 (Fig. 44P), slide 26912; A1-69 borehole, sample 2108–2111 ft.

Type locality and horizon. MG-1 borehole, sample 2639 m; Ouan-Kasa Formation at Mechiguig, Tunisia.

Diagnosis. A simple *Zonotriletes* with a narrow proximo-equatorial flange.

Description. Amb is sub-circular to sub-triangular. Laesurae are distinct, straight, simple or sometimes bordered with labra *c.* 0.5 μm wide individually, extending almost to the inner margin of the zona. Curvaturae are often visible. The central body diameter equals is commonly three-quarters to nine-tenths of the total amb diameter. Exine of the central body is 2–5 μm thick equatorially. The proximo-equatorial flange is commonly 2–8 μm wide and is uniformly wide but is often folded back locally. Thin transverse attachment lines of the flange on the central body can sometimes be distinguished on the proximal face. Proximal and distal surfaces are entirely laevigate.

Dimensions. 49(59)71 μm; 22 specimens measured.

Comparison. This species differs from other species of *Zonotriletes* Luber and Waltz, 1938 by its regular narrow zona. *Z. simplicissimus* Breuer *et al.*, 2007c is more robust and possesses wide labra bordering laesurae. Its flange is generally narrower opposite the laesurae.

Occurrence. BAQA-1, BAQA-2, JNDL-1, JNDL-4 and WELL-7, Jauf (Sha'iba to Hammamiyat members) and Jubah formations; *ovalis-biornatus* to *svalbardiae-eximius* zones. A1-69; Ouan-Kasa Formation; *lindlarensis-sextantii* Zone. MG-1, Ouan-Kasa and Awaynat Wanin I formations; *svalbardiae-eximius* to *rugulata-libyensis* zones.

FIG. 44. Each figured specimen is identified by borehole, sample, slide number and England Finder Co-ordinate location. All figured specimens are at magnification ×1000 except where mentioned otherwise. A–G, *Verrucosisporites stictus* sp. nov. A, BAQA-2, 50.8 ft, 66813, E41. B, Holotype, BAQA-1, 395.2 ft, 62272, V41. C, BAQA-1, 227.1 ft, 03CW110, K39/1. D, BAQA-1, 395.2 ft, 03CW121, X45. E, Paratype, BAQA-1, 395.2 ft, 03CW121, O28/1. F, BAQA-1, 395.2 ft, 62275, Q29-30. G, BAQA-2, 134.4 ft, 03CW137, N27/4. H–J, *Verrucosisporites* sp. 1. H, JNDL-4, 419.3 ft, 03CW261, M35/3. I, JNDL-4, 163.7 ft, 03CW212, V33/4. J, JNDL-4, 411.5 ft, 03CW259, T45/2. K–M, *Zonotriletes armillatus* Breuer *et al.*, 2007c. K, JNDL-1, 155.6 ft, 60837, O34. L, JNDL-1, 172.7 ft, 60846, T32. M, JNDL-1, 155.6 ft, PPM003, T42/4. N–Q, *Zonotriletes brevivelatus* sp. nov. N, MG-1, 2639 m, 62780, R37/1. O, Holotype, MG-1, 2639 m, 62779, Q46/3. P, Paratype, A1-69, 2108–2111 ft, 26912, D55/1. Q, A1-69, 2108–2111 ft, 26913, J44/4.

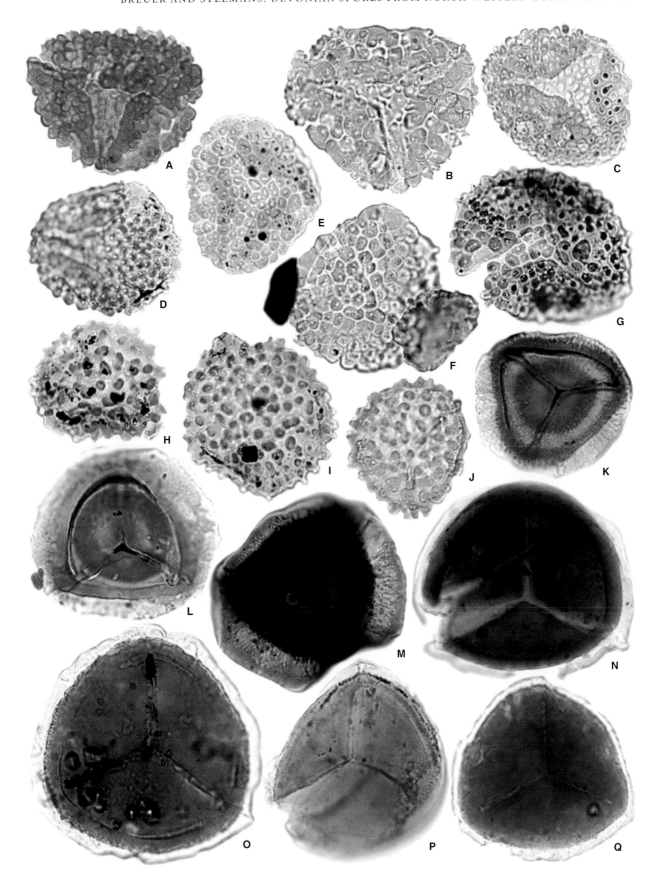

Previous records. From Lochkovian–Pragian of Algeria (Jardiné and Yapaudjian 1968); and Pragian–Emsian of Libya (Massa and Moreau-Benoit 1976).

Zonotriletes rotundus sp. nov.
Figure 45C–G

Derivation of name. From *rotundus* (Latin), meaning rounded; refers to the distal sculpture.

Holotype. EFC O41/2 (Fig. 45C), slide 62781.

Paratype. EFC H51/3 (Fig. 45F), slide 26976; A1-69 borehole, sample 1486 ft.

Type locality and horizon. MG-1 borehole, sample 2315 m; Awaynat Wanin II Formation at Mechiguig, Tunisia.

Diagnosis. A sub-circular *Zonotriletes* with a flange of the same width along the amb. Distal surface contains a distinct annular thickening.

Description. Amb is sub-circular to sub-triangular. Laesurae are distinct, straight to slightly sinuous, simple and rarely elevated up to 2 μm, extending or almost to the inner margin of the zona. Curvaturae are not visible. The central body diameter equals commonly seven-tenths to three-quarters of the total amb diameter. Exine of the central body is commonly 1–3 μm thick equatorially. The thin proximo-equatorial flange is generally 5–20 μm wide but up to 24 μm for the larger specimens. The flange can be folded back opposite the laesurae and radially folded. Thin transverse attachment lines of the flange on the central body may be distinguished on the proximal face. An annulus is present on the distal face and is 6–10 μm wide. The annulus diameter equals two-fifths to three-fifths of the central body diameter. In some specimens, the annulus is barely perceptible and may be represented by a rounded darker area. Proximal and distal surfaces are laevigate.

Dimensions. 55(83)110 μm; 20 specimens measured.

Comparison. *Zonotriletes armillatus* Breuer *et al.*, 2007c has a more triangular appearance and a generally less wide flange which is always narrower opposite the laesurae.

Occurrence. JNDL-1, JNDL-3 and JNDL-4; Jauf (Hammamiyat and Murayr members) and Jubah formations; *lindlarensis-sextantii* to *svalbardiae-eximius* zones. A1-69; Awaynat Wanin I and Awaynat Wanin II formations; *svalbardiae-eximius* to *triangulatus-catillus* zones. MG-1; Ouan-Kasa, Awaynat Wanin I and Awaynat Wanin II formations; *annulatus-protea* to *triangulatus-catillus* zones.

Previous record. From upper Eifelian – lower Givetian of Parnaíba Basin, Brazil (Breuer and Grahn 2011).

Zonotriletes simplicissimus Breuer et al., 2007c
Figure 45H–I

```
? 1968    Zonotriletes sp. 3 Jardiné and Yapaudjian, pl. 2,
              fig. 2.
  2007c   Zonotriletes simplicissimus Breuer et al., p. 52,
              pl. 12, figs 3–7.
```

Dimensions. 38(59)82 μm; 26 specimens measured.

Occurrence. JNDL-1 and S-462, Jubah Formation, *svalbardiae-eximius* to *rugulata-libyensis* zones. A1-69; Awaynat Wanin II Formation; *rugulata-libyensis* Zone.

Previous record. From upper Eifelian – lower Givetian of Parnaíba Basin, Brazil (Breuer and Grahn 2011).

Zonotriletes venatus sp. nov.
Figure 46A–C

```
  1968   Zonotriletes sp. 2 Jardiné and Yapaudjian, pl. 2,
             fig. 1.
  1988   Perotrilites sp. cf. Zonotriletes sp. 2 in Jardiné and
             Yapaudjian; Boumendjel et al., pl. 1, fig. 1.
```

Derivation of name. From *venatus* (latin), meaning veined; refers to the flange.

Holotype. EFC T33 (Fig. 46B), slide 62552.

Paratype. EFC Q36/3 (Fig. 46A), slide 62552; MG-1 borehole, sample 2631.2 m.

Type locality and horizon. MG-1 borehole, sample 2631.2 m; Ouan-Kasa Formation at Mechiguig, Tunisia.

Diagnosis. A large *Zonotriletes* with a very wide veined, striated, equatorial flange.

FIG. 45. Each figured specimen is identified by borehole, sample, slide number and England Finder Co-ordinate location. All figured specimens are at magnification ×1000 except where mentioned otherwise. A–B, *Zonotriletes brevivelatus* sp. nov. A, MG-1, 2639 m, 62780, L39. B, JNDL-4, 328.3 ft, 03CW245, P35/4. C–G, *Zonotriletes rotundus* sp. nov. C, Holotype, MG-1, 2315 m, 62781, O41/2. D, JNDL-1, 155.6 ft, 60837, C50/1. E, A1-69, 1950 ft, 27276, R41. F, Paratype, A1-69, 1486 ft, 26976, H51/3. G, JNDL-1, 155.6 ft, 60837, Q34/4. H–I, *Zonotriletes simplicissimus* Breuer *et al.*, 2007c. H, JNDL-1, 172.7 ft, PPM007, D31; the transverse attachment lines of the flange on the central body are distinguishable. I, JNDL-1, 177.0 ft, 60849, G35/4.

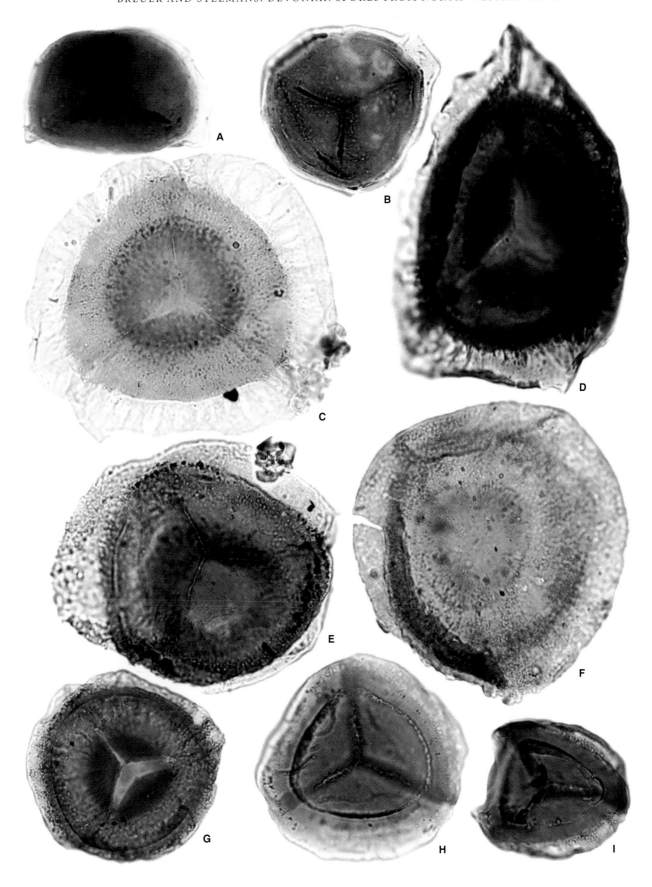

Description. Zonate trilete spores with sub-circular to sub-triangular amb. Laesurae distinct, straight to slightly sinuous, sometimes elevated, extending to the outer margin of the zona. Curvaturae not visible. The central body diameter equals two-fifths to three-fifths of the total amb diameter. Exine of the central body relatively thin equatorially. The equatorial flange is finely veined and striated. Proximal and distal surfaces entirely laevigate.

Dimensions. 77(130)176 μm; 10 specimens measured.

Comparison. *Zonotriletes* sp. 1 is usually smaller and has a narrower zona.

Remarks. The equatorial flange is thin and very large; as a result, it is often broken and/or folded.

Occurrence. BAQA-1 and JNDL-4; Jauf Formation (Qasr to Subbat members); *ovalis-biornatus* to *lindlarensis-sextantii* zones. MG-1; Ouan-Kasa Formation; *svalbardiae-eximius* Zone.

Previous records. From Lochkovian–Pragian of Algeria (Jardiné and Yapaudjian 1968; Boumendjel *et al.* 1988); lower Lochkovian – lower Emsian of Brazil (Grahn *et al.* 2005; Mendlowicz Mauller *et al.* 2007; Steemans *et al.* 2008).

Zonotriletes sp. 1
Figure 46D–E

Description. Amb is sub-circular to sub-triangular. Laesurae are distinct, straight to slightly sinuous, simple or bordered with labra up to 3 μm wide individually and extending to the outer margin of the zona. Curvaturae are not visible. The central body diameter equals three-fifths to four-fifths of the total amb diameter. Exine of the central body is 2–5 μm thick equatorially. Zona is divided entirely or partially into three individual proximo-equatorial flanges, the maximum width (commonly 14–25 μm) of which is opposite the laesurae. Zona laevigate to infragranular, generally folded back opposite the laesurae, resulting in a tri-lobed appearance. Thin transverse attachment lines of the flange on the central body can be distinguished on the proximal face. Proximal and distal surfaces of the central body are laevigate.

Dimensions. 65(82)102 μm; five specimens measured.

Comparison. *Alatisporites*? *trisacculus* sp. nov. is more rounded and possess three individual sacci. Furthermore, the maximum width of the zona is opposite the laesurae.

Occurrence. JNDL-1; Jubah Formation; *svalbardiae-eximius* Zone. A1-69; Awaynat Wanin I Formation; *svalbardiae-eximius* Zone.

BIOSTRATIGRAPHY

The two main Devonian spore zonations defined by Richardson and McGregor (1986) and Streel *et al.* (1987) from Euramerican material are commonly used in most of the palynological studies (see above). Loboziak and Melo (2002) simplified the Western European biozonation of Streel *et al.* (1987) to apply it to western Gondwanan localities. In addition, they associated the ranges of some endemic spores which could represent useful zonal markers for the regional biozonation (Loboziak and Streel 1995a). In north-western Gondwana, Massa and Moreau-Benoit (1976) and Moreau-Benoit (1989) established also a unique detailed biozonation (see above). The palynological analyses show here that the reference spore zones usually used in Euramerica are not all recognized in the Gondwanan coeval sections. It is due to the absence or the rarity of several index species.

In this work, about 88 per cent of species from the North African assemblages are also found in Saudi Arabia. Among the 205 different taxa described herein, about 48 per cent are found elsewhere outside Gondwana (cosmopolitan species) and the remaining 52 per cent are restricted to Saudi Arabia, North Africa or more generally to Gondwana (endemic species). In terms of number of specimens, these endemic taxa may represent a percentage greater again of the whole spore assemblages. Almost since their origin, land plant taxa have not been cosmopolitan, and therefore the lateral extension of correlation by spores is risky (Traverse 2007). Precise transcontinental correlation is sometimes difficult, intercontinental correlation is again more difficult and sometimes impossible (except in very broad terms). As one of the main problems for spore-based palynostratigraphy and biostratigraphy in general is provincialism, a new biozonation based on the own characteristics of the spore assemblages described here could allow more accurate local/regional correlations and also tentative intercontinental correlations.

New biozonation for north-western Gondwana

The known ranges of the nominal species and main characteristic species occurring in North Africa and Saudi Arabia are plotted in a composite stratigraphical chart in Figures 53 and 54. The plotted stratigraphical ranges of these characteristic taxa do not represent a mean based on their local range in each section but is their total

---

FIG. 46. Each figured specimen is identified by borehole, sample, slide number and England Finder Co-ordinate location. All figured specimens are at magnification ×1000 except where mentioned otherwise. A–C, *Zonotriletes venatus* sp. nov., magnification ×750. A, Paratype, MG-1, 2631.2 m, 62552, Q36/3. B, Holotype, MG-1, 2631.2 m, 62552, T33. C, BAQA-1, 406.0 ft, 03CW123, L23/3. D–E, *Zonotriletes* sp. 1. D, JNDL-1, 156.0 ft, PPM004, K38/1. E, JNDL-1, 162.3 ft, 60844, L34.

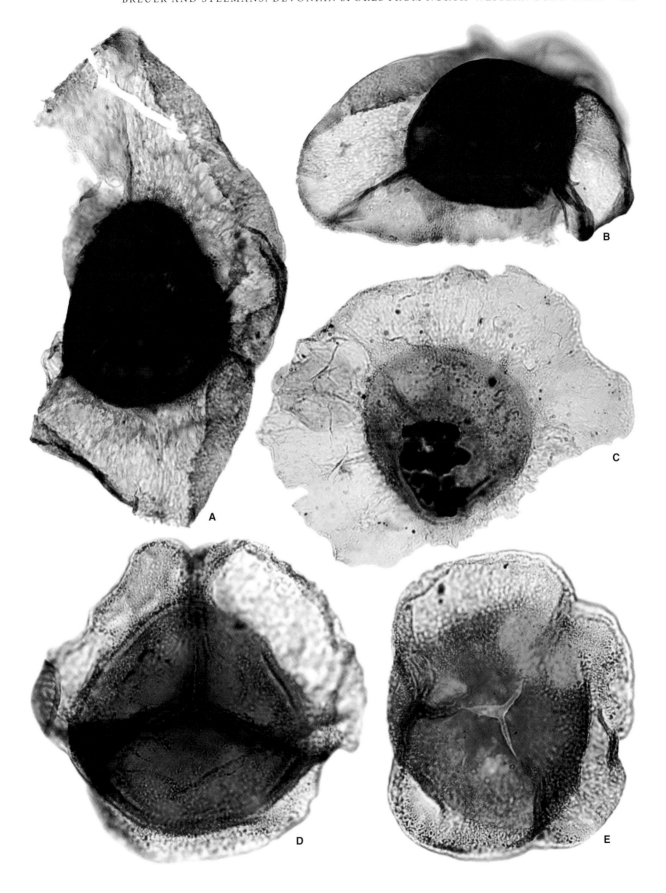

stratigraphical range. The first and last occurrences of each taxon are taken into account in all cored sections. By studying, analyzing and comparing stratigraphical range of spores species and composition of different assemblages of the studied sections, a new biozonation can be defined for north-western Gondwana. As each type of biozones has different advantages and disadvantages, different types of biozones are combined to establish the most useful biozonation. The assemblage zones are used to subdivide the stratigraphical column on a large scale. Firstly, their advantage is that they are probably the most easily applicable for long-distance (intercontinental) correlation to other studies. Secondly, some general characteristics or trends of the assemblages may be included in order to be useful also in a local/regional use. Finally, several index species may be designated for the recognition of the zone. Although the assemblage zones boundaries defined here are based on one single index species and consequently correspond in theory to an interval zone, each biozone is described as an assemblage zone because stratigraphically useful species and general characteristics of the assemblages are included in the characterization of these biozones. Therefore, in spite of the eventual absence of the index species in a studied section, the assemblage zone might be recognizable. The assemblage zones are themselves divided into subzones when it is possible for a more accurate stratigraphy. These subzones are either interval or acme zones. The interval zones are not useful if their index taxon is absent for any reason, but they can be very useful for local/regional correlations. However, a note of caution is expressed regarding the use of interval zones because it is quite possible that the age of first occurrences may vary spatially due to immigration/migration or subtle biogeographical, ecological or facies effects (Wellman 2006). Nevertheless, the use of interval zones is still the only way to correlate spores with other palaeontological groups such as conodonts and foraminifera. As regards the acme or abundance zones, their use is tenuous and probably more of ecological significance than anything else but they are above all useful for local correlation.

The new proposed scheme based on a combination of diverse types of biozones consists of nine assemblage zones, nine interval zones and one acme zone spanning from upper Pragian to lower Frasnian. Thanks to this new biozonation, the studied sections from the three distinct regions are correlated (Fig. 55).

From base to top, the assemblages are defined and compared to biozones defined in the literature (Fig. 56), especially the two main Devonian spore zonations defined on Euramerica by Richardson and McGregor (1986) and Streel *et al.* (1987) and the operational palynological zonation used by Saudi Aramco (Al-Hajri *et al.* 1999). The age of each assemblage is discussed, but the biostratigraphical correlations remain approximative (Fig. 56), as there are any outcrops or sections that allow to correlate the different biozones. Indeed few precise independent palaeontological dating is available for the studied areas; therefore, we have to keep in mind that the suggested ages are only based on the comparisons with reference spore biozones, which themselves are dated independently. The use of the high-resolution conodont-based international Devonian stratigraphy allows to date the spore zonation from Western Europe of Streel *et al.* (1987; Streel and Loboziak 1996; Streel *et al.* 2000).

Description of the palynological zones and subzones

The name given to each spore zones comprises (1) the names of one or two characteristic taxa referred to as 'nominal species', and (2) a statement of the kind of zone. The nominal species are chosen for various reasons: lateral widespread and abundant occurrence (acme zone), restriction to a definite stratigraphical interval, characteristic first appearance or extinction. Stratigraphical range of nominal species is given in Figure 53. On its first citation, the name of the zone is given in full, for example the *Acinosporites lindlarensis–Camarozonotriletes sextantii* Assemblage Zone. In subsequent citations the name of the zone is abbreviated: the *lindlarensis-sextantii* Zone.

FIG. 47. Each figured specimen is identified by borehole, sample, slide number and England Finder Co-ordinate location. All figured specimens are at magnification ×1000 except where mentioned otherwise. A–C, *Contagisporites optivus* (Chibrikova) Owens, 1971. A, S-462, 2110–2115 ft, 63273, N40/1. B, S-462, 2260–2265 ft, 63281, Q28/2. C, S-462, 2110–2115 ft, 63272, O31. D–F, *Corystisporites collaris* Tiwari and Schaarschmidt, 1975. D, A1-69, 1109 ft, 27274, O45/1. E, S-462, 1760–1765 ft, 63255, P36/1. F, A1-69, 971 ft, 62639, T32. G–L, *Corystisporites undulatus* Turnau, 1996. G, A1-69, 1109 ft, 27273, E50/4. H, A1-69, 971 ft, 62641, K46/2. I, A1-69, 1277 ft, 62636, R29/2. J, MG-1, 2205 m, 62595, G30/2. K, MG-1, 2295 m, 63005, X51/3. L, A1-69, 1277 ft, 62636, E35/3. M–O, *Craspedispora ghadamesensis* Loboziak and Street, 1989. M, A1-69, 1322 ft, 27125, K54/2. N, A1-69, 1596 ft, 26990, D44. O, A1-69, 1700 ft, 62634, M37/2. P–R, *Craspedispora paranaensis* Loboziak *et al.*, 1988. P, A1-69, 1700 ft, 62632, R50. Q, A1-69, 1700 ft, 62633, T43/2. R, A1-69, 1596 ft, 26989, T39. S–U, *Cristatisporites (Calyptosporites) reticulatus* (Tiwari and Schaarschmidt) comb. nov. S, MG-1, 2713 m, 62811, H36. T, MG-1, 2264 m, 62951, L42/3. U, MG-1, 2285 m, 62846, L46. V–X, *Cristatisporites streelii* sp. nov. V, Paratype, MG-1, 2241 m, 62964, H44/3. W, Holotype, MG-1, 2270 m, 62849, R42/3. X, MG-1, 2375 m, 62773, L33.

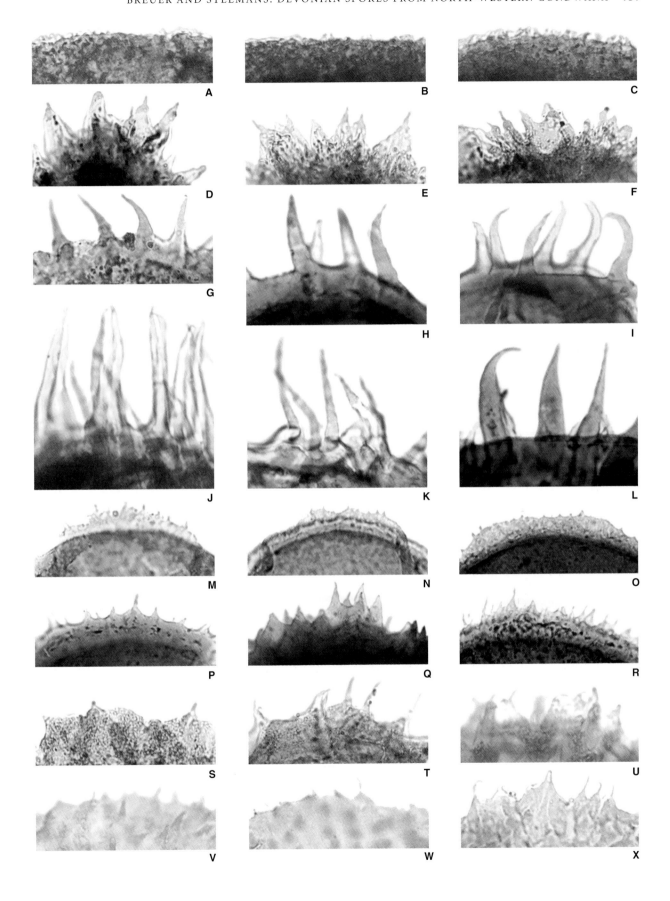

Synorisporites papillensis–Cymbohilates baqaensis *Assemblage Zone*

Reference section. Borehole BAQA-2: samples 133.0 and 134.4 ft.

Distribution. Lower part of the Sha'iba Member, Jauf Formation, Saudi Arabia. This biozone is only recognized in two samples. Its lower stratigraphical limit of this assemblage is presently unknown, as it occurs at the lowest depth sampled.

Description. The most characteristic trilete spores are: *Ambitisporites eslae*, *Apiculiretusispora plicata*, *Biornatispora elegantula*, *Camarozonotriletes filatoffii*, *Chelinospora carnosa*, *C. vulgata*, *Cirratriradites? diaphanus*, *Clivosispora verrucata*, *Cymbosporites dammamensis*, *C. rarispinosus*, *C. wellmanii*, *Dictyotriletes emsiensis*, *D. subgranifer*, *Lycospora culpa*, *Raistrickia jaufensis*, *Scylaspora costulosa*, *Synorisporites papillensis*, *Verrucosisporites onustus* and *V. polygonalis*. Cryptospores are rather well diversified and common including above all *Cymbohilates baqaensis*, *C. comptulus* and *Gneudnaspora divellomedia* var. *minor*). Although all these species persist upwards in the succeeding *Latosporites ovalis–Dictyotriletes biornatus* Assemblage Zone, the *papillensis-baqaensis* is primarily distinguished by the absence of laevigate monolete spores. This assemblage can also be recognized by the occurrence of *Cymbosporites wellmanii* and *Dictyotriletes granulatus*, which are exclusive to it. The spores of this section are mainly simple laevigate spores, simple spores sculptured with discrete elements and proximal interradial papillae are common.

Comparison with reference biozones. The co-occurrence of *Verrucosisporites polygonalis* and *Dictyotriletes emsiensis* is the criterion to correlate this assemblage to the *polygonalis-emsiensis* Assemblage Zone of Richardson and McGregor (1986). The presence of *Dictyotriletes subgranifer* indicates more precisely the Su Interval Zone of Steel *et al.* (1987).

Stage. Upper Pragian.

Latosporites ovalis–Dictyotriletes biornatus *Assemblage Zone*

Reference section. No section recorded entirely the *ovalis-biornatus* Zone but the combination of core holes BAQA-2 (from 64.5 ft) and BAQA-1 (to 161 ft) can serve as a reference section since these sections are easily correlated by lithostratigraphic and sedimentologic data.

Distribution. Upper Sha'iba, Qasr, lower and middle Subbat members, Jauf Formation, Saudi Arabia.

Zone base definition. Its lower boundary is based on the first occurrence of *Latosporites ovalis*, the first laevigate monolete spore. Although this species is not a main component of the spore assemblages, it is easily distinguishable even in a poorly preserved material.

Description. The most characteristic species to appear through the assemblage are: *Artemopyra recticosta*, *Biornatispora dubia*, *Brochotriletes crameri*, *B. hudsonii*, *B. tenellus*, *Coronaspora inornata*, *Cymbosporites asymmetricus*, *Diaphanospora milleri* Morphon, *Dibolisporites bullatus*, *Dictyotriletes biornatus* Morphon, *D. favosus*, *Emphanisporites schultzii*, *Latosporites ovalis*, *Leiozosterospora* cf. *L. andersonii*, *Reticuloidosporites antarcticus Retusotriletes tenerimedium*, *Rhabdosporites minutus* and *Stellatispora multicostata*. *Apiculiretusispora brandtii* becomes more common. The trilete spores are still dominated by simple forms but are more and more diversified. Spores become larger (up to more than 130 μm). Cavate spores are present from the base of the assemblage with *Leiozosterospora* cf. *L. andersonii*. Although a unique specimen of sculptured monolete spore (*Devonomonoletes* sp. 1), which is a poorly-known species, is recorded in the underlying the *papillensis-baqaensis* Zone, the *ovalis-biornatus* Zone is characterized by the first significant inception of monolete spores (*Latosporites ovalis*) and followed later by the short-ranged reticulate form *Reticuloidosporites antarcticus* only known in Gondwana. The cryptospores still constitute a significant component.

FIG. 48. Each figured specimen is identified by borehole, sample, slide number and England Finder Co-ordinate location. All figured specimens are at magnification ×1000 except where mentioned otherwise. A–C, *Cristatisporites streelii* sp. nov. A, MG-1, 2241 m, 62964, T30/1. B, MG-1, 2285 m, 62845, J27. C, A1-69, 1322 ft, 27126, O37/1. D–I, *Densosporites devonicus* Richardson, 1960. D, MG-1, 2292 m, 63023, T29. E, MG-1, 2483 m, 62802, C37/4. F, MG-1, 2315 m, 62783, Q42. G, MG-1, 2456 m, 62739, G44. H, MG-1, 2295 m, 63007, E47/1. I, MG-1, 2527 m, 63003, G34/2. J–O, *Grandispora cassidea* (Owens) Massa and Moreau-Benoit, 1976. J, MG-1, 2465 m, 62852, R43/2. K, MG-1, 2518 m, 62805, N27/4. L, MG-1, 2161.8 m, 62529, J49/1. M, MG-1, 2536 m, 62740, L40. N, MG-1, 2161.8 m, 62528, X52/2. O, MG-1, 2639 m, 62778, T42. P–X, *Grandispora douglastownensis* McGregor, 1973. P, A1-69, 1962 ft, 27278, O39. Q, A1-69, 1962 ft, 27278, P44/3. R, A1-69, 1962 ft, 27277, U54/3. S, JNDL-1, 174.6 ft, PPM008, T32. T, JNDL-1, 177.0 ft, 60850, P45/4. U, JNDL-1, 162.3 ft, 60841, K44. V, JNDL-1, 156.0 ft, PPM004, P36/3. W, JNDL-1, 167.8 ft, PPM006, N37. X, WELL-1, 16327.6 ft, 61944, E49. Y–AA. *Grandispora fibrilabrata* Balme, 1988. Y, S-462, 1810–1815 ft, 63257, R48. Z, S-462, 1860–1865 ft, 63258, P27/1. AA, S-462, 2010–2015 ft, 63266, S34.

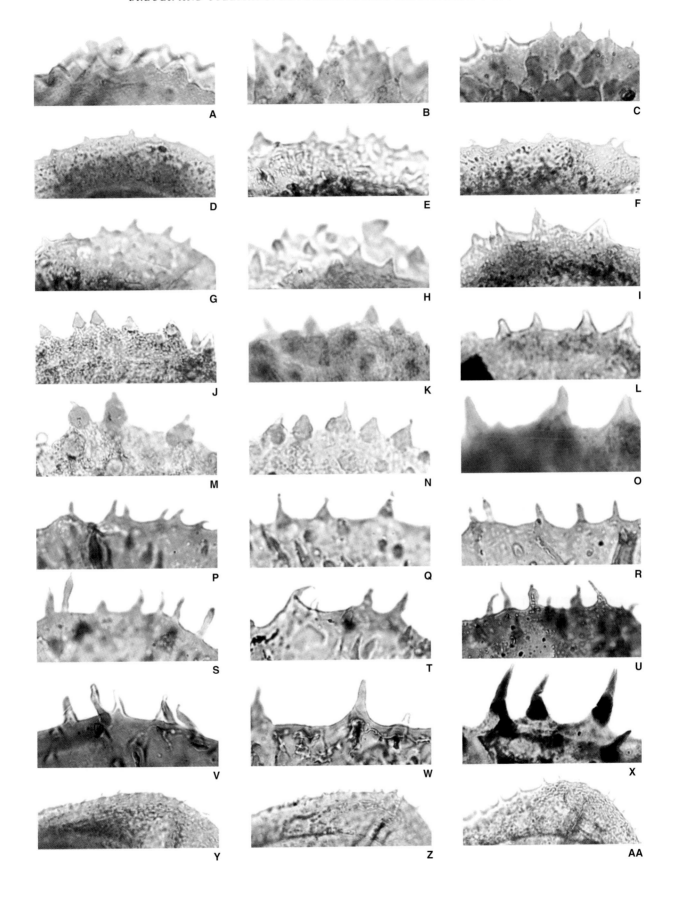

Comparison with reference biozones. The index species used in Euramerica and recognized all throughout the *ovalis-biornatus* Zone are *Dictyotriletes emsiensis, D. subgranifer* and *Verrucosisporites polygonalis*. The diversified composition of the *ovalis-biornatus* Zone seems, however, to indicate a younger biozone than the *polygonalis-emsiensis* Assemblage Zone of Richardson and McGregor (1986) and the Su Interval Zone of Streel *et al.* (1987). The occurrence of first monolete spores is interesting as they are not older than Emsian according to Traverse, 2007. However, Wellman (2006) found a unique specimen of monolete spores in an assemblage corresponding to the Su Interval Zone. In Antarctica, *Reticuloidosporites antarcticus* was found in the lower Emsian (Kemp 1972), the age of which is recently reinterpreted as Pragian by Troth *et al.* (2011). In Euramerica, first rare monolete spores occur from the Emsian *annulatus-sextantii* Assemblage Zone of Richardson and McGregor (1986). *Dibolisporites echinaceus sensu stricto* and *Emphanisporites schultzii*, the first appearance of which is known to be Emsian, combined with the last occurrences of *Chelinospora retorrida* and *Iberoespora cantabrica*, which are usually not younger than upper Pragian, suggest that the Pragian/Emsian boundary lies within the *Latosporites ovalis* Interval Zone. The occurrence of *Rhabdosporites minutus* allows to partly correlate the *Cymbosporites asymmetricus* Interval Zone to the Min Interval Zone of Streel *et al.* (1987) that is comprised in the FD Oppel Zone.

Stages. Upper Pragian – lower Emsian.

Latosporites ovalis *Interval Zone*

Reference section. No section recorded the *ovalis* Zone in its entirety, but the combination of core holes BAQA-2 (from 64.5 ft) and the lower part of BAQA-1 (to 345.5 ft) can serve as a reference section.

Distribution. Upper Sha'iba to the lowermost part of the Subbat members, Jauf Formation, Saudi Arabia.

Zone base definition. Its lower boundary corresponds to that of the *ovalis-biornatus* Zone.

Description. The *ovalis* Zone is also characterized by very common specimens of *Gneudnaspora divellomedia* var. *minor* and the first occurrence of the different members of the *Dictyotriletes biornatus* Morphon. Last occurrences of *Chelinospora carnosa, Lycospora culpa* and *Verrucosisporites onustus* and entire ranges of *Biornatispora microclavata, Chelinospora densa* and *Dictyotriletes ?gorgoneus* are restricted to this subzone.

Diaphanospora milleri *Interval Zone*

Reference section. BAQA-1 (from 345.5 to 227.1 ft).

Distribution. Lower Subbat Member, Jauf Formation, Saudi Arabia.

Zone base definition. Its lower boundary is based on the first occurrence of *Diaphanospora milleri*.

Remarks. The *milleri* Zone is also characterized by the first occurrence of *Dibolisporites bullatus, Perotrilites caperatus, Retusotriletes celatus* and *Reticuloidosporites antarcticus*.

Cymbosporites asymmetricus *Interval Zone*

Reference section. BAQA-1 (from 227.1 to 161 ft).

Distribution. Middle Subbat Member, Jauf Formation, Saudi Arabia.

Zone base definition. Its lower boundary is based on the first occurrence of *Cymbosporites asymmetricus*.

Remarks. The *asymmetricus* Zone is also characterized by a significant increase of *Cymbosporites senex* specimens. *Artemopyra recticosta, Brochotriletes crameri, Rhabdospor-*

FIG. 49. Each figured specimen is identified by borehole, sample, slide number and England Finder Co-ordinate location. All figured specimens are at magnification ×1000 except where mentioned otherwise. A–C, *Grandispora gabesensis* Loboziak and Streel, 1989. A, A1-69, 1596 ft, 26990, P39/4. B, A1-69, 1962 ft, 27278, H51. C, A1-69, 1483 ft, 26994 ft, S46/4. D–L, *Grandispora incognita* (Kedo) McGregor and Camfield, 1976. D, WELL-8, 16684.3 ft, 62427, D43. E, A1-69, 1416 ft, 26992, Q38/4. F, A1-69, 1596 ft, 26990, H36. G, A1-69, 1540 ft, 26988, Q37/2. H, A1-69, 1540 ft, 26988, N40/3. I, A1-69, 1540 ft, 26987, H37/1. J, A1-69, 1277 ft, 62636, K39. K, A1-69, 1277 ft, 62636, O41/4. L, A1-69, 1277 ft, 62636, M33. M, *Grandispora inculta* Allen, 1965, S-462, 2060–2065 ft, 63270, N30/3. N–Y, *Grandispora libyensis* Moreau-Benoit, 1980*b*. N, A1-69, 1322 ft, 27126, H36. O, A1-69, 1277 ft, 62635, Y31/2. P, A1-69, 1416 ft, 26993, G31/3. Q, A1-69, 1277 ft, 62637, M31. R, A1-69, 1416 ft, 26993, R40. S, A1-69, 1416 ft, 26992, C37/2. T, A1-69, 1416 ft, 26993, S55/2. U, A1-69, 1416 ft, 26993, O43. V, A1-69, 1277 ft, 62636, O31/3. W, A1-69, 1277 ft, 62637, L51. X, A1-69, 1174 ft, 62673, O32/3. Y, A1-69, 1416 ft, 26993, K31/3.

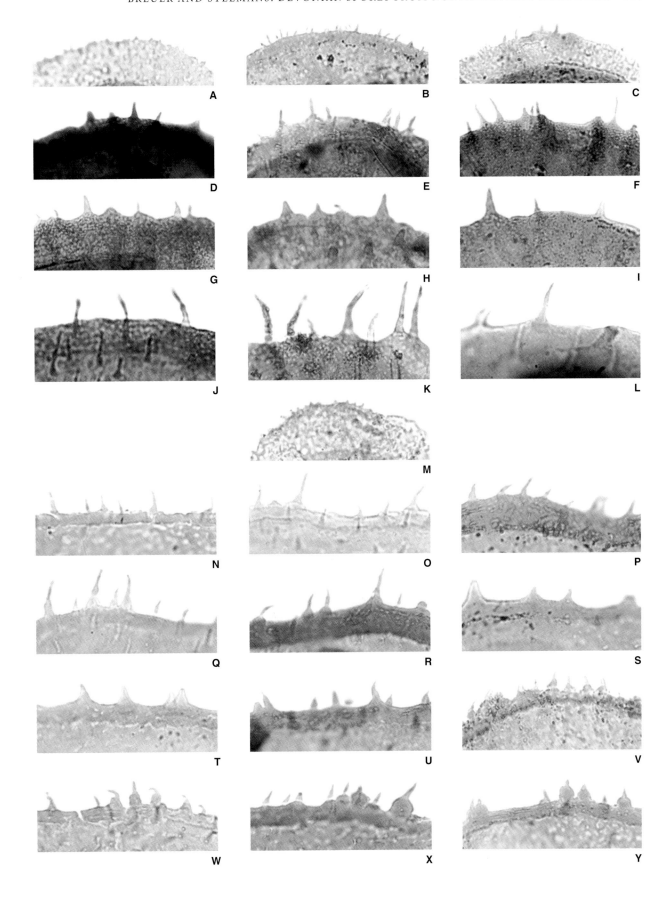

ites minutus and *Stellatispora multicostata* first occur within this subzone, whereas *Ambitisporites eslae*, *Biornatispora elegantula*, *Chelinospora condensata*, *C. laxa*, *C. vulgata*, *Raistrickia jaufensis*, *Verrucosisporites nafudensis* and *V. stictus* disappear.

Acinosporites lindlarensis–Camarozonotriletes sextantii Assemblage Zone

Reference section. The reference section is the combination of core holes JNDL-4 (from 404.8 ft) and JNDL-3 (to 222.7 ft). As the core holes are distant from about 15 km, they are easily correlated by lithostratigraphic and sedimentologic data.

Distribution. Upper Subbat, Hammamiyat and most of Muray members, Jauf Formation, Saudi Arabia. Ouan-Kasa Formation, Libya.

Zone base definition. Its lower boundary is based on the first occurrence of *Acinosporites lindlarensis*.

Description and regional variation. The most characteristic spore species to appear in the *lindlarensis-sextantii* Zone are *Camarozonotriletes sextantii*, *Cymbohilates heteroverrucosus*, *Dibolisporites gaspiensis*, *D. tuberculatus*, *Dictyotriletes marshallii*, *Emphanisporites erraticus* and *E. plicatus*. The present spore assemblage comprises mainly simple laevigate spores, simple sculptured spores and radial-patterned spores. The zonate spores become more common. *Dibolisporites* spp. increases in term of specimens in comparisons with the previous assemblages while *Chelinospora* spp. representatives are no longer present. Entire range of *Dictyotriletes marshallii*, *Iberoespora* cf. *guzmani* and *Emphanisporites erraticus* are restricted to the *lindlarensis-sextantii* Zone. Lots of species disappear within this zone, the most significant of which are *Ambitisporites asturicus*, *Amicosporites jonkeri*, *Brochotriletes hudsonii*, *Cirratriradites? diaphanus*, *Cymbohilates comptulus*, *Dictyotriletes biornatus* Morphon, *D. favosus*, *Emphanisporites schultzii*, *?Knoxisporites riondae* and *Stellatispora multicostata*.

In Saudi Arabia, the *lindlarensis-sextantii* Zone includes the D3B Subzone of Al-Hajri *et al.* (1999), which represents a monospecific algal bloom. These leiospherid-rich, low-diversity assemblages occur in north-western Saudi Arabia not as a unique event but a series of marine pulses that extends on a 400-ft thickness (Breuer *et al.* 2007a, b, c). The D3B Subzone is much thicker than in eastern Saudi Arabia where it is normally several tens of feet. These leiospherid-rich assemblages alternate with normal spore-dominated assemblages.

Comparison with reference biozones. The presence notably of *Acinosporites lindlarensis*, *Camarozonotriletes sextantii* and *Rhabdosporites minutus* allows to partly correlate the *lindlarensis-sextantii* Zone to the *annulatus-sextantii* Assemblage Zone of Richardson and McGregor (1986) and the Min Interval Zone of Streel *et al.* (1987) that is comprised in the FD Oppel Zone.

Stage. Middle–upper Emsian.

Emphanisporites annulatus–Grandispora protea Assemblage Zone

Reference section. Borehole A1-69 (sample 2040 ft).

Distribution. Uppermost part of Murayr Member (Jauf Formation) and the lowermost part of the Jubah Formation, Saudi Arabia. Ouan-Kasa Formation, North Africa.

Zone base definition. Its lower boundary is marked by the first occurrence of *Emphanisporites annulatus*.

Description and regional variation. The present assemblage zone is primarily characterized by the first occurrence of the large apiculate and spinose zonate-pseudosaccate spores (*Grandispora/Samarisporites* complex). The most characteristic spore species to appear in the *annulatus-protea* Zone are *Craspedispora ghadamesensis*, *Geminospora convoluta*, *Grandispora douglastownensis*, *G. protea* and *Granulatisporites concavus*. Several typical Early Devonian

---

FIG. 50. Each figured specimen is identified by borehole, sample, slide number and England Finder Co-ordinate location. All figured specimens are at magnification ×1000 except where mentioned otherwise. A–I, *Grandispora libyensis* Moreau-Benoit, 1980b. A, A1-69, 1174 ft, 62673, O32/3. B, A1-69, 1334 ft, 27128, H39/4. C, A1-69, 1334 ft, 27128, H39/4. D, A1-69, 1296 ft, 62644, U29/1. E, A1-69, 1296 ft, 62644, U29/1. F, A1-69, 1296 ft, 62645, F36. G, A1-69, 1296 ft, 62645, G47/2. H, A1-69, 1296 ft, 62644, T35/1. I, A1-69, 1296 ft, 62645, R43. J–L, *Grandispora maura* sp. nov. J, MG-1, 2241 m, 62966, S39/1. K, Holotype, MG-1, 2247 m, 62942, K37/3. L, MG-1, 2465 m, 62852, H35. M–U, *Grandispora naumovae* (Kedo) McGregor, 1973. M, S-462, 1710–1715 ft, 63222, G44/3. N, S-462, 1710–1715 ft, 63222, V40. O, S-462, 1710–1715 ft, 63224, F49-50. P, S-462, 1660–1665 ft, 63219, L38/4. Q, S-462, 1760–1765 ft, 63254, J30. R, S-462, 1470–1475 ft, 63213, J41/3. S–T, A1-69, 1530 ft, 26984, J34/2. U, A1-69, 1334 ft, 27127, O47. V–X, *Grandispora permulta* (Daemon) Loboziak *et al.*, 1999. V, A1-69, 1277 ft, 62637, V49/1. W, A1-69, 1322 ft, 27125, H39/4. X, WELL-1, 16358.5 ft, 61972, W31.

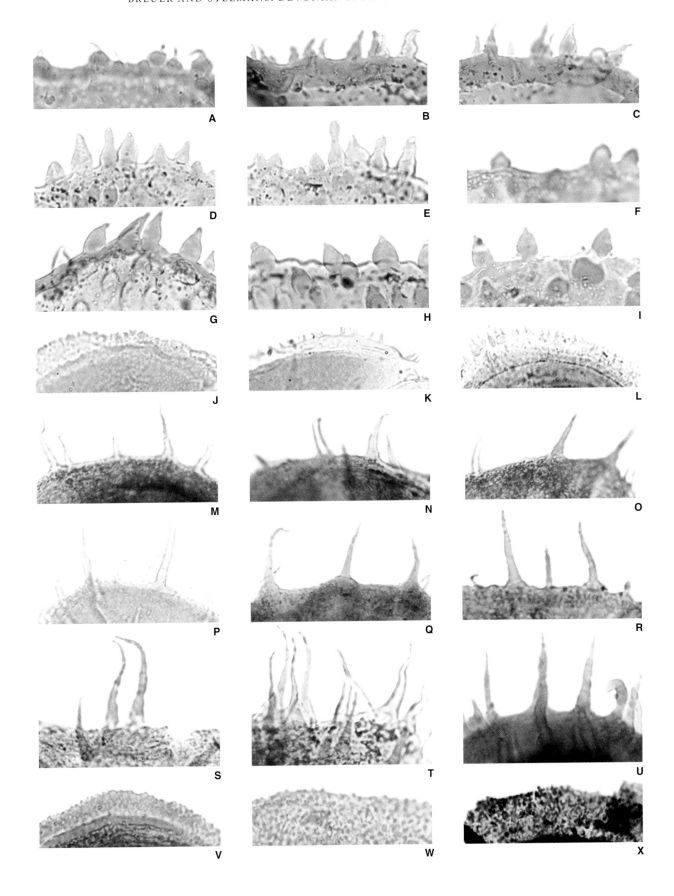

taxa disappear within the *annulatus-protea* Zone; they are *Amicosporites streelii*, *Biornatispora dubia*, *Clivosispora verrucata* var. verrucata, *Dictyotriletes subgranifer*, *Retusotriletes maculatus* and *Verrucosisporites polygonalis*. The cryptospores have decreased in number and seems to be still dominated by *Gneudnaspora* and *Artemopyra*.

Densosporites devonicus is only known in Tunisia. First specimens of *Grandispora permulta* and the spinose-verrucate representatives of *Verrucosisporites scurrus* Morphon are already present in Tunisia, whereas they occur later in Libya and Saudi Arabia from the *Scylaspora rugulata–Grandispora libyensis* Assemblage Zone.

Comparison with reference biozones. The first inception of the large zonate-pseudosaccate spores is known elsewhere all over the world and corresponds in Euramerica to the *douglastownensis-eurypterota* Assemblage Zone of Richardson and McGregor (1986) and the AP Oppel Zone of Streel *et al.* (1987), the index species *Acinosporites apiculatus*, *Grandispora douglastownensis* and *G. protea* of which are found in the *annulatus-protea* Zone. Note that *Densosporites devonicus* which is a nominal species of the middle Eifelian–early Givetian *devonicus-naumovae* Assemblage Zone of Richardson and McGregor (1986) and the AD Oppel Zone of Streel *et al.* (1987) seems to appear earlier in North Africa than in Euramerica. M. Streel (pers. comm. 2007), after many years of palynological consulting, considers now the first inception of *D. devonicus* as unpredictable and consequently is not a reliable criterion to mark the base of the AD Oppel Zone. Besides, Streel and Loboziak (1996) do not consider anymore this species as defining the base of the latter.

Stage. Uppermost Emsian.

Geminospora svalbardiae–Samarisporites eximius Assemblage Zone

Reference section. Borehole A1-69 (from 1962 to 1830 ft).

Distribution. Jubah Formation, Saudi Arabia. Upper part of the Ouan-Kasa and lower part of the Awaynat Wanin I formations, North Africa.

Zone base definition. Its lower boundary is based on the first occurrence of *Geminospora svalbardiae*, which is common.

Description. The most characteristic spore species to appear in the *svalbardiae-eximius* Zone are *Acinosporites acanthomammillatus*, *Ancyrospora nettersheimensis*, *Auroraspora minuta*, *Camarozonotriletes rugulosus*, *Grandispora cassidea*, *G. gabesensis*, *G. velata*, *Samarisporites eximius*, *S. praetervisus*, *Zonotriletes armillatus* and *Z. simplicissimus*. Other ancient species including the last typical Early Devonian holdovers disappear in the *svalbardiae-eximius* Zone; these are *Brochotriletes crameri*, *Camarozonotriletes sextantii*, *Cymbohilates baqaensis*, *Cymbosporites senex*, *Dictyotriletes emsiensis*, *Granulatisporites concavus* and *Synorisporites papillensis*. The large apiculate and spinose zonate-pseudosaccate spores proliferate and become abundant throughout the assemblage. They diversify again more with new forms of *Grandispora* and *Samarisporites*. In addition, other pseudosaccate genera (*Auroraspora* and *Geminospora*) are also well represented. The typical feature of grapnel-tipped spines (*Ancyrospora*) appears for the first time.

In Saudi Arabia, the cryptospores (*Artemopyra*, *Cymbohilates* and *Gneudnaspora*) constitute a significant part of the spore assemblages where they may represent up to about 20 per cent of the whole spore assemblage. *Acinosporites tristratus*, *Dibolisporites Pilatus* and *Squamispora arabica* are restricted to *svalbardiae-eximius* Zone and endemic to Saudi Arabia. Relatively small megaspores (*Jhariatriletes emsiensis*) first appear in this assemblage in North Africa.

Comparison with reference biozones. In Euramerica, *Ancyrospora nettersheimensis* possesses an acme zone (net) in the AP Oppel Zone of Streel *et al.* (1987) and is also known in the *douglastownensis-eurypterota* Assemblage

---

FIG. 51. Each figured specimen is identified by borehole, sample, slide number and England Finder Co-ordinate location. All figured specimens are at magnification ×1000 except where mentioned otherwise. A–F, *Grandispora permulta* (Daemon) Loboziak *et al.*, 1999. A, A1-69, 1277 ft, 62637, U47/1. B, A1-69, 1483 ft, 26995, W43/3. C, A1-69, 1296 ft, 62643, G34/1. D, A1-69, 1596 ft, 26989, U46/3. E, A1-69, 1530 ft, 26984, R46/4. F, A1-69, 1596 ft, 26989, T43. G–R, *Grandispora protea* (Naumova) Moreau-Benoit, 1980*b*. G, JNDL-1, 495.0 ft, 60854, P51. H, A1-69, 1962 ft, 27278, T36/3. I, JNDL-1, 156.0 ft, 60840, Q32. J, JNDL-1, 174.6 ft, 60848, P34. K, JNDL-1, 174.6 ft, 60848, S43/4. L, A1-69, 1530 ft, 26984, T54/1. M, JNDL-1, 174.6 ft, PPM008, Q39/4. N, JNDL-1, 177.0 ft, 60849, L51/2. O, A1-69, 1962 ft, 27278, T44/4. P, JNDL-1, 172.7 ft, 60845, O40/2. Q, A1-69, 1962 ft, 27278, M49/1. R, JNDL-177.0 ft, 60849, K51/1. S–X, *Grandispora rarispinosa* Moreau-Benoit, 1980*b*. S, S-462, 1710–1715, 63222, X40. T, MG-1, 2182.4 m, 62527, Q32. U, MG-1, 2285 m, 62845, L26/3. V, MG-1, 2161.8 m, 62529, E50/4. W, MG-1, 2413 m, 62776, R35/2. X, MG-1, 2160.6 m, 62746, Q27/4. Y–AA, *Grandispora (Calyptosporites) stolidota* (Balme) comb. nov. Y, A1-69, 1322 ft, 27126, Q39/3. Z, A1-69, 1322 ft, 27126, R42. AA, MG-1, 2465 m, 62852, L52/3. AB–AD, *Grandispora velata* (Richardson) McGregor, 1973. AB, A1-69, 1416 ft, 26992, U47/2. AC, A1-69, 1540 ft, 26987, F41. AD, A1-69, 1530 ft, 26984, H40/1.

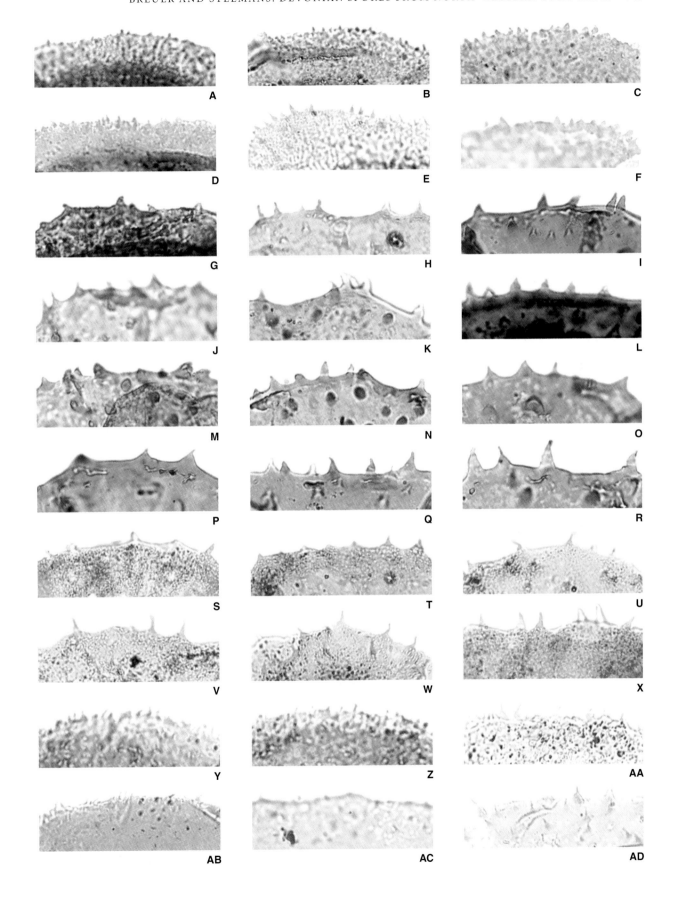

Zone of Richardson and McGregor (1986). *Grandispora velata* and *Acinosporites acanthomammillatus*, present in this assemblage, mark the base of the *velata-langii* Assemblage Zone of Richardson and McGregor (1986), and the former is the nominal species of the Vel Interval Zone of Streel *et al.* (1987).

Stage. Lower Eifelian.

Scylaspora rugulata–Grandispora libyensis *Assemblage Zone*

Reference section. Borehole MG-1 (from 2536 to 2367 m).

Distribution. Jubah Formation, Saudi Arabia. Awaynat Wanin I Formation and the lower part of the Awaynat Wanin II Formations, North Africa.

Zone base definition. Its lower boundary is based on the first occurrence of a series of common species which appear more or less coeval. It comprises *Camarozonotriletes? concavus, Chelinospora timanica, Craspedispora paranaensis, Dictyotriletes hemeri, Grandispora libyensis, G. naumovae, G. stolidota, Scylaspora rugulata* and *Verrucosisporites premnus*. Note that the first specimens of *G. libyensis* are a morphotype with rather slender spines.

Description. In addition to the new species, *Camarozonotriletes asperulus, Elenisporis gondwanensis* and *Grandispora incognita* appear within the *rugulata-libyensis* Zone. In the upper part of the zone, the most common species to appear are *Archaeozonotriletes variabilis* and *Cristatisporites streelii. Grandispora permulta*, and all members of the *Verrucosisporites scurrus* Morphon, some of which already occur in Tunisia, are henceforth present and are particularly characteristic of this interval and constitute a main component of the spore assemblages. The *rugulata-libyen-*sis Zone also marks the vanishing of ancient common species such as *Ancvrospora nettersheimensis* and *Grandispora protea*. The large apiculate and spinose zonate-pseudosaccate spores reach their acme in the *svalbardiae-eximius* Zone. Laevigate zonate spores markedly decrease in comparison with the underlying *svalbardiae-eximius* Zone. The cryptospores are much rarer in the *rugulata-libyensis* Zone; *Gneudnaspora divellomedia* var. *divellomedia* is sometimes found, whereas *Artemopyra* specimens are very rare.

Comparison with reference biozones. Grandispora naumovae and *Verrucosisporites premnus* mark the base of the *devonicus-naumovae* Assemblage Zone of Richardson and McGregor (1986), whereas *Scylaspora rugulata* occur a bit earlier within the *velata-langii* Assemblage Zone in Euramerica. *Archaeozonotriletes variabilis* occurs in the second half of the *devonicus-naumovae* Assemblage Zone. The *rugulata-libyensis* Zone seems to correspond likely to the latter or also the AD-pre Lem Oppel Zone defined in Loboziak and Melo (2002). This last biozone appear undifferentiated here up to the first appearance of *Geminospora lemurata* whose inception marks the base of the Lem Interval Zone of Streel *et al.* (1987).

Stages. Upper Eifelian – lowermost Givetian.

Scylaspora rugulata *Interval Zone*

Reference section. Borehole MG-1 (sample 2536 to 2518 m).

Distribution. Awaynat Wanin I Formation, North Africa.

Zone base definition. Its lower boundary corresponds to that of the *rugulata-libyensis* Zone. It is based on the first occurrence of *Scvlaspora rugulata*.

FIG. 52. Each figured specimen is identified by borehole, sample, slide number and England Finder Co-ordinate location. All figured specimens are at magnification ×1000 except where mentioned otherwise. A–C, *Grandispora velata* (Richardson) McGregor, 1973. A, A1-69, 1867 ft, 26969, R53/4. B, JNDL-1, 177.0 ft, 60849, F32/1. C, A1-69, 1962 ft, 27277, F32/4. D–F, *Hystricosporites brevispinus* sp. nov. D, Holotype, MG-1, 2295 m, 63007, P37. E, Paratype, MG-1, 2375 m, 62772, H34/1. F, MG-1, 2295 m, 63006, L52/3. G–I, *Hystricosporites* sp. 1. G, MG-1, 2160.6 m, 62746, N49. H, MG-1, 2161.8 m, 62528, Q44. I, MG-1, 2160.6 m, 62727, E49. J–L, *Jhariatriletes (Verruciretusispora) emsiensis* (Moreau-Benoit) comb. nov. J, A1-69, 1962 ft, 27277, T45/4. K, A1-69, 1962 ft, 27278, J42. L, A1-69, 1962 ft, 27277, Q40. M–O, *Samarisporites angulatus* (Tiwari and Schaarschmidt) Loboziak and Streel, 1989. M, A1-69, 1596 ft, 26990, K40/2. N, A1-69, 1870 ft, 26973, K41. O, A1-69, 1867 ft, 26969, D37. P–R, *Samarisporites eximius* (Allen) Loboziak and Streel, 1989. P, JNDL-1, 162.3 ft, PPM005, S45/3. Q, A1-69, 1334 ft, 27127, N53/2. R, JNDL-1, 177.0 ft, 60849, O39. S–U, *Samarisporites praetervisus* (Naumova) Allen, 1965. S, A1-69, 1334 ft, 27127, K48. T–U, Holotype, MG-1, 2258 m, 62947, M39/2. V–X, *Samarisporites triangulatus* Allen, 1965. V, A1-69, 1277 ft, 62636, X31/1. W, A1-69, 1277 ft, 62637, T29/1. X, MG-1, 2182.4 m, 62527, Q29/2. Y–AA, *Samarisporites tunisiensis* sp. nov. Y, MG-1, 2741.4 m, 62611, L28/1. Z, Holotype, MG-1, 2741.4 m, 62611, P39/2. AA, MG-1, 2741.4 m, 62611, K41. AB–AD, *Samarisporites* sp. 2. AB, A1-69, 1596 ft, 26989, G38/1. AC, A1-69, 1596 ft, 26989, S55. AD, A1-69, 1596 ft, 26990, N36. AE–AG, *Verrucisporites ellesmerensis* (Chaloner) Chi and Hills, 1976. AE, S-462, 1710–1715 ft, 63222, H31/3. AF, S-462, 1860–1865 ft, 63258, N31/2. AG, S-462, 1860–1865 ft, 63259, H38.

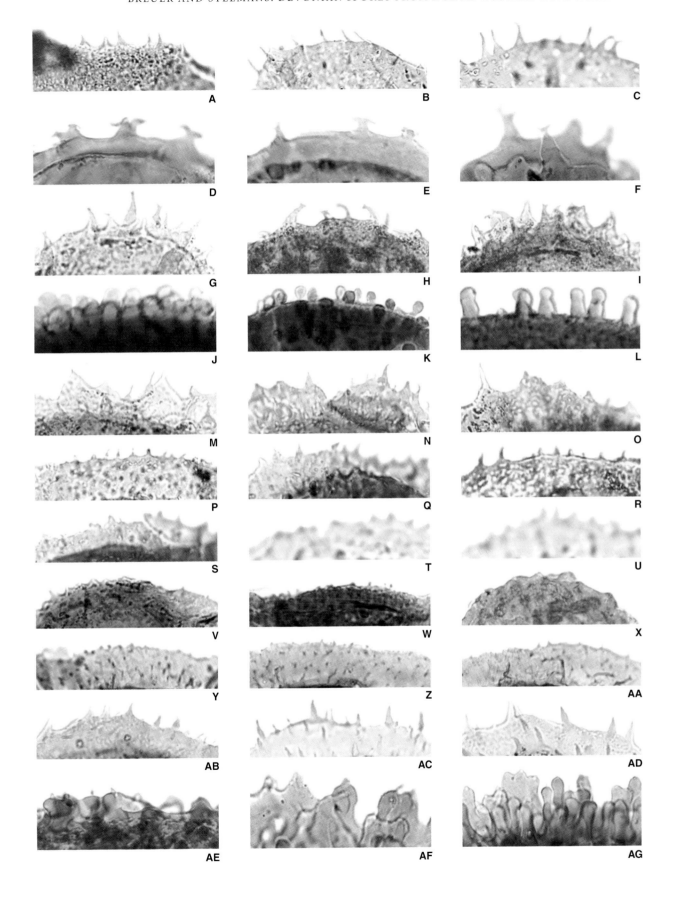

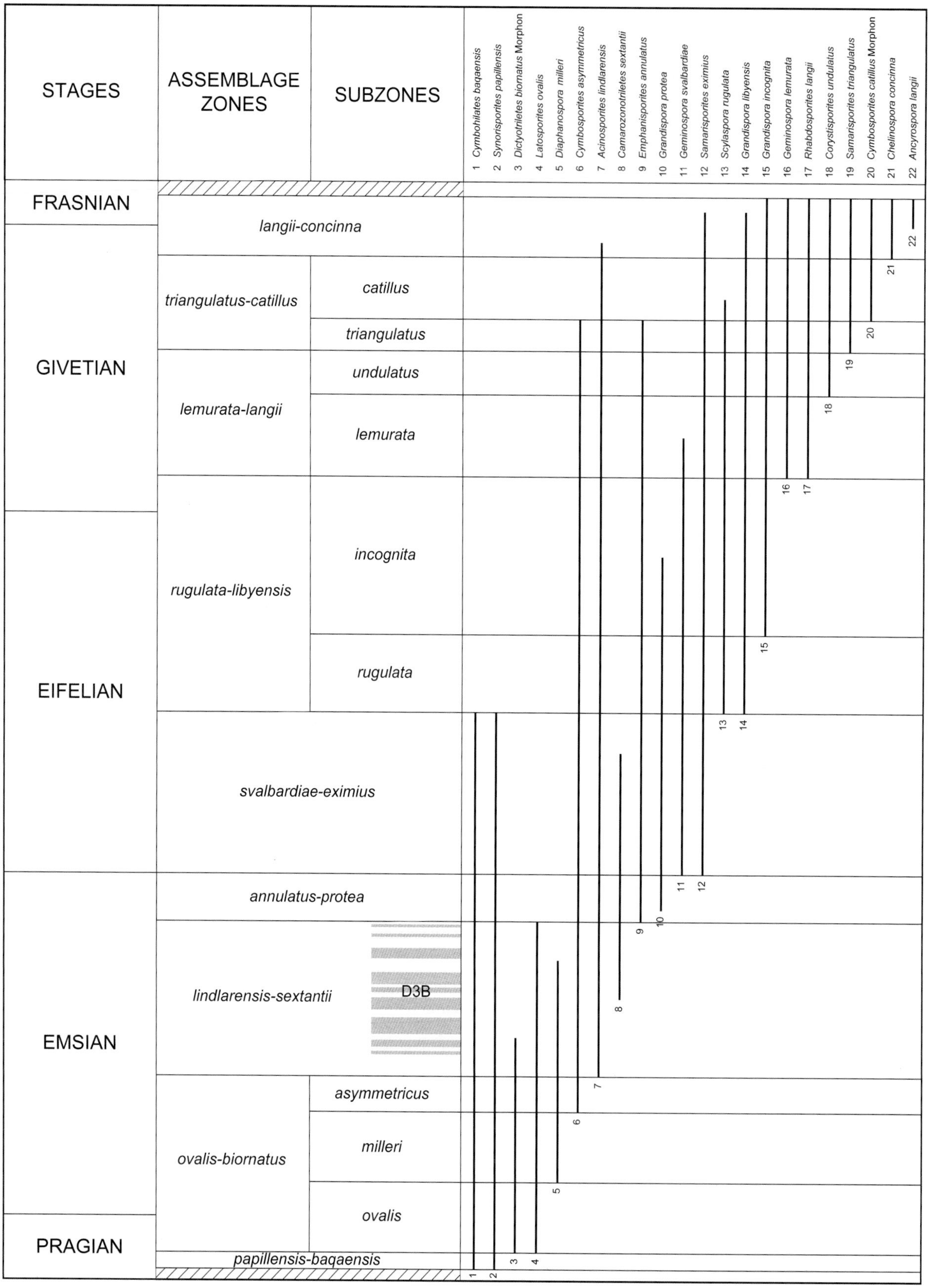

FIG. 53. Stratigraphical ranges of nominal species of the spore assemblage zones. Dashed lines indicate imprecisely defined limits.

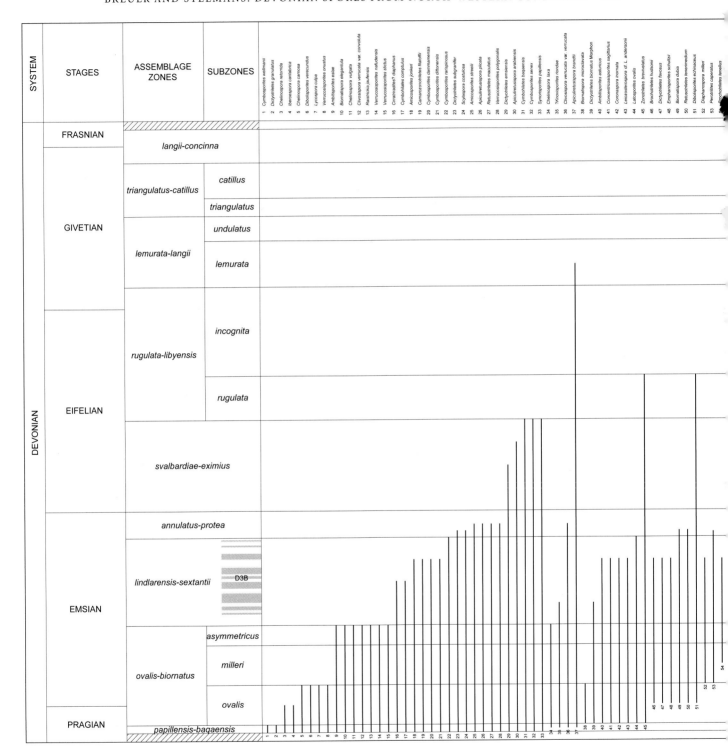

FIG. 54. Stratigraphical ranges of characterizing species of the spore assemblage zones. For ranges of the nominal species only see Figure 53. Dashed lines

northwestern Saudi Arabia

eastern Saudi Arabia

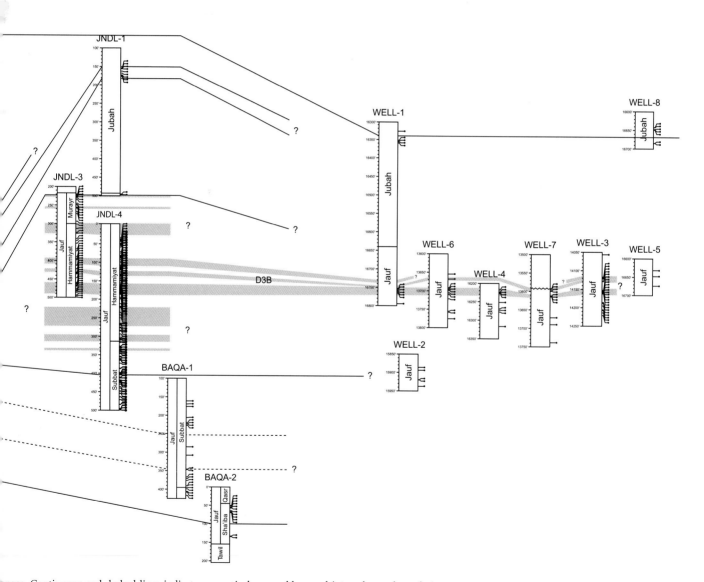

anges. Continuous and dashed lines indicate, respectively, assemblage and interval zone boundaries.

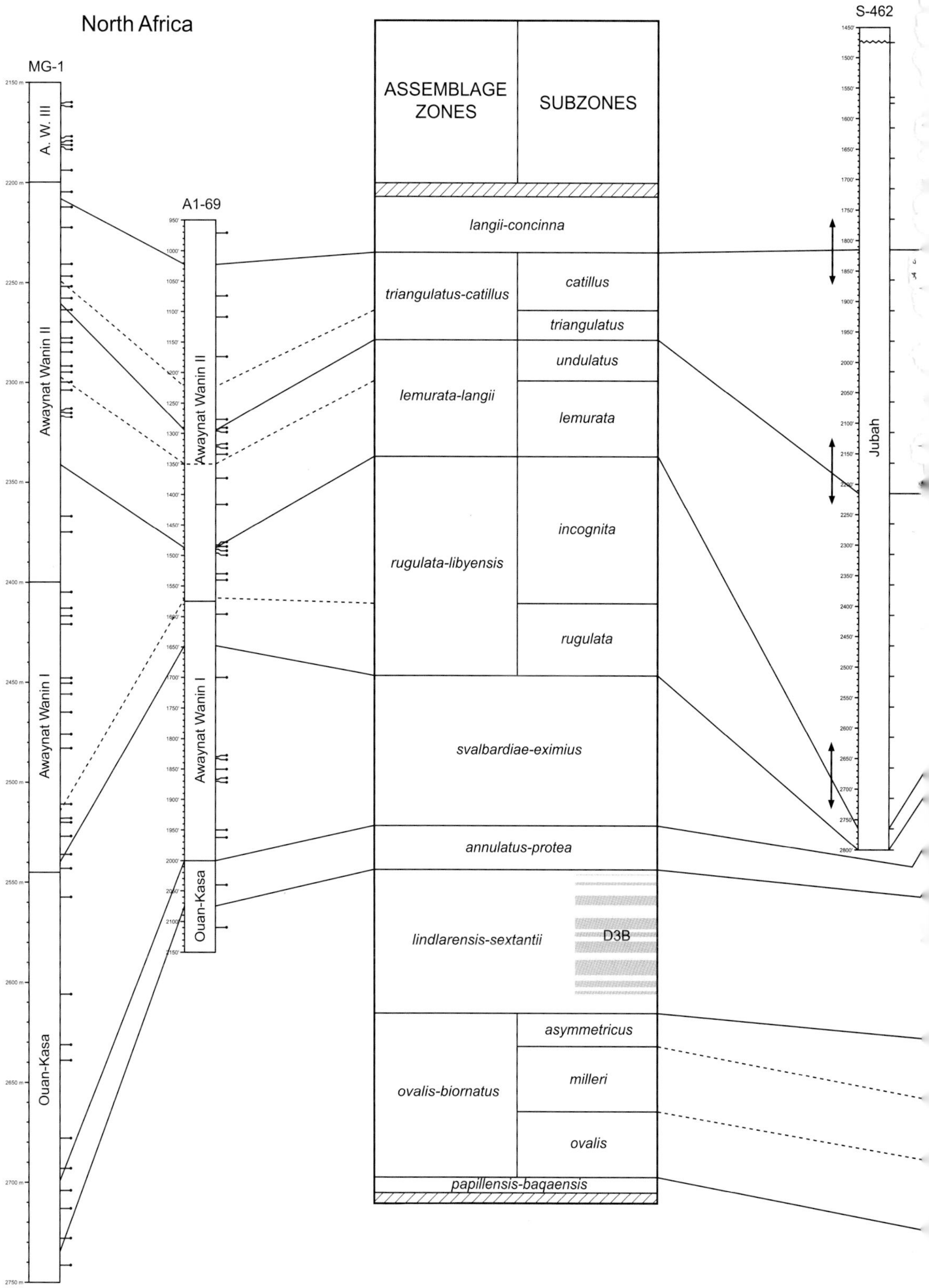

FIG. 55. Spore-based correlation of studied sections. Dots indicate the core samples. Cuttings in S-462 are plotted at tops of composite sample

56	Dibolisporites bullatus
57	Stellatispora multicostata
58	Brochotriletes crameri
59	Rhabdosporites minutus
60	Artemopyra recticosta
61	Cymbosporites asymmetricus
62	Dictyotriletes marshallii
63	Dibolisporites gaspiensis
64	Acinosporites lindlarensis
65	Acinosporites apiculatus
66	Dibolisporites tuberculatus
67	Emphanisporites erraticus
68	Emphanisporites plicatus
69	Camarozonotriletes sextantii
70	Emphanisporites annulatus
71	Geminospora convoluta
72	Grandispora protea
73	Grandispora permulta
74	Verrucosisporites scurrus
75	Grandispora douglastownense
76	Craspedispora ghadamesensis
77	Dibolisporites uncatus
78	Geminospora svalbardiae
79	Grandispora velata
80	Grandispora gabonensis
81	Samarisporites eximius
82	Ancyrospora nettersheimensis
83	Ancyrospora acanthomammillatus
84	Aurorospora minuta
85	Craspedispora paranaensis
86	Scylaspora rugulata
87	Dictyotriletes herreri
88	Verrucosisporites premnus
89	Grandispora libyensis
90	Camarozonotriletes aspervalus
91	Camarozonotriletes ? concavus
92	Chelinospora timanica
93	Grandispora naumovae
94	Grandispora stolidota
95	Eleniospora gondwanensis
96	Dibolisporites farraginis
97	Grandispora incognita
98	Archaeozonotriletes variabilis
99	Cristatisporites strelkii
100	Geminospora lemurata
101	Rhabdosporites langii
102	Cymbosporites variegatus
103	Corystisporites undulatus
104	Dibolisporites turriculatus
105	Lophozonotriletes media
106	Contagisporites optivus
107	Camarozonotriletes parvus
108	Samarisporites triangulatus
109	Verrucosisporites ellesmerensis
110	Cymbosporites catillus Morphon
111	Grandispora floridabala
112	Chelinospora concinna
113	Ancyrospora langii
114	Emphanisporites falcicostatus

ASSEMBLAGE ZONES — **SUBZONES**

Assemblage Zone	Subzone
langii-concinna	
triangulatus-catillus	catillus
	triangulatus
lemurata-langii	undulatus
	lemurata
rugulata-libyensis	incognita
	rugulata
svalbardiae-eximius	
annulatus-protea	
lindlarensis-sextantii	D3B
ovalis-biornatus	asymmetricus
	milleri
	ovalis
papillensis-bagaensis	

ndicate imprecisely defined limits.

Grandispora incognita *Interval Zone*

Reference section. Borehole A1-69 (from 1596 to 1490 ft).

Distribution. Upper part of the Awaynat Wanin I and the lower part of Awaynat Wanin II formations, North Africa.

Biozone boundary definition. Its lower boundary is based on the first occurrence of *Grandispora incognita*.

Geminospora lemurata–Rhabdosporites langii *Assemblage Zone*

Reference section. Borehole Al-69 (from 1486 to 1296 ft).

Distribution. Jubah Formation, Saudi Arabia. Awaynat Wanin II Formation, North Africa.

Zone base definition. The lower boundary is based on the first occurrence of *Geminospora lemurata*.

Description and regional variation. Geminospora punctata and *Rhabdosporites langii* make their first inception more or less in the meantime as *G. lemurata. Cymbosporites variegatus, Camarozonotriletes parvus, Corystisporites undulatus, Dibolisporites turriculatus* and *Lophozonotriletes media* appear within the *lemurata-langii* Zone. First specimens of *Contagisporites optivus* appear near the top of this interval. *Apiculiretusispora brandtii, Geminospora svalbardiae, Grandispora velata* and *Rhabdosporites minutus* are the most characteristic species to vanish in the *lemurata-langii* Zone. The cryptospores are uncommon; only specimens of *Gneudnaspora* and *Arternopyra* are found. The monolete spores seem to reappear inconspicuously after a distinct decline in the underlying assemblages. Monolete *Geminospora lemurata* are not so uncommon and rare large cavate monolete spores (*Archaeoperisaccus* cf. *A. rhacodes*) first occur in Tunisia. More and more megaspore specimens (including *Contagisporites optivus*) appear throughout the *lemurata-langii* Zone. Most of them are unstudied in this work but described in Steemans *et al.* (2011*b*).

Comparison with reference biozones. This assemblage characterized above all by the widespread first occurrence of *Geminospora lemurata* equals the Lem Interval Zone of Streel *et al.* (1987) and the lower part of the *lemurata-magnficus* Assemblage Zone of Richardson and McGregor (1986) in Euramerica. The stratigraphically important species *Rhabdosporites langii* appears later here than in Euramerica. *Lophozonotriletes media*, which is the index

species of the Frasnian BM Oppel Zone of Streel *et al.* (1987), is found earlier than in Euramerica. This difference is probably explained by the use of a larger concept of *L. media* in this work (M. Streel, pers. comm. 2007). As specimens attributed to *L. media* seem to be very variable in time and space, it could be an unreliable key species for precise biostratigraphy.

Stage. Lower Givetian but not the lowermost.

Geminospora lemurata *Interval Zone*

Reference section. Borehole A1-69 (from 1486 to 1416 ft).

Distribution. Jubah Formation, Saudi Arabia. Awaynat Wanin II Formation, North Africa.

Zone base definition. The lower boundary is based on the first occurrence of *Geminospora lemurata*.

Remarks. The lower boundary of the *lemurata* Zone corresponds to that of the *lemurata-langii* Zone as it is based on the first occurrence of *Geminospora lemurata*.

Corystisporites undulatus *Interval Zone*

Reference section. Borehole Al-69 (from 1416 to 1296 ft).

Distribution. Jubah Formation, Saudi Arabia. Awaynat Wanin II Formation, North Africa.

Zonal base definition. Its lower boundary is based on the first occurrence of *Corystisporites undulatus*.

Samarisporites triangulatus–Cymbosporites catillus *Assemblage Zone*

Reference section. Borehole A1-69 (from 1293 to 1074 ft).

Distribution. Jubah Formation, Saudi Arabia. Awaynat Wanin II Formation, North Africa.

Zone base definition. Its lower boundary is only based on the first occurrence of *Samarisporites triangulatus*.

Description. The first inception of *Cymbosporites catillus, C. cyathus, Grandispora fibrilabrata* and *Verrucisporites ellesmerensis* is likely a little bit younger than that of *Samarisporites triangulatus*. The specimens of *G. libyensis* possess rather slender spines as well as more massive bulbous biform elements. The two morphotypes intergrade and coexist mainly in the *triangulatus-catillus* Zone. *Contagisporites optivus*, *Craspedispora ghadamesensis*, *Dictyotriletes hemeri*, *Elenisporis gondwanensis*, *Grandispora douglastownensis*, *Scylaspora rugulata* and *Zonotriletes rotundus* are the most characteristic species to disappear in the *triangulatus-catillus* Zone. The general diversity of the spore populations begins to decrease from the base of this assemblage. The large apiculate and spinose zonate-pseudosaccate spores show a quantitative decrease. They seem to be replaced with numerous small patinate specimens (mainly *Cymbosporites*). *Geminospora lemurata* and *Samarisporites triangulatus* are also abundant. The cryptospores are very uncommon. *Artetmopyra* vanishes at the base of the assemblage and leaves *Gneudnaspora divellomedia* var. *divellomedia* as the last survivor of this group. The megaspores are very diverse in both Saudi Arabia and North Africa; they include notably *Contagisporites optivus* and *Verrucisporites ellesmerensis*. Other megaspores are described in Steemans *et al.* (2011*b*).

Comparison with reference biozones. *Samarisporites triangulatus* is a widespread species and seems to appear almost at the same time everywhere. In Euramerica, the *triangulatus-catillus* Zone corresponds to the TA Oppel Zone of Streel *et al.* (1987) in spite of the absence of *Ancyrospora ancyrea* var. *ancyrea* in the studied sections. The co-occurrence of *Contagisporites optivus* and *Samarisporites triangulatus* could indicate in part the *optivus-triangulatus* Assemblage Zone of Richardson and McGregor (1986). Although it is generally correlated with the TCo Oppel Zone of Streel *et al.* (1987), *C. optivus* appears from the underlying TA Oppel Zone in Belgium (Steemans *et al.* 2011*b*). Nevertheless, the use of megaspore species for correlation is hardly reliable since they are recovered from few types of deposit environments.

Stage. Middle–Upper Givetian.

Samarisporites triangulatus Interval Zone

Reference section. Borehole A1-69 (from 1293.0 to 1277.0 ft).

Distribution. Jubah Formation, Saudi Arabia. Awaynat Wanin II Formation, North Africa.

Zone base definition. Its lower boundary is based on the first occurrence of *Samarisporites triangulatus*.

Cymbosporites catillus Interval Zone

Reference section. MG-I (from 2247 to 2212.5 m).

Distribution. Jubah Formation, Saudi Arabia. Awaynat Wanin II Formation, North Africa.

Zone base definition. Its lower boundary is based on the first occurrence of specimens belonging to the *Cymbosporites catillus* Morphon.

Ancyrospora langii–Chelinospora concinna Assemblage Zone

Reference section. MG-1 (from 2205 to 2160.6 m).

Distribution. Upper Jubah Formation, Saudi Arabia. Uppermost part of the Awaynat Wanin II Formation and the Awaynat Wanin III Formation, North Africa. Its upper stratigraphical limit of this assemblage is presently unknown, as it occurs at the highest depth sampled.

Zone base definition. Its lower boundary is based on the first occurrence of *Chelinospora concinna*.

Description. In addition to the first occurrence of *Chelinospora concinna*, this assemblage is mainly distinguished from the underlying *triangulatus-catillus* Zone by the inception of the common species *Ancyrospora langii* and *Emphanisporites laticostatus*. *Geminospora lemurata*, *Samarisporites triangulatus* and the *Cymbosporites catillus* Morphon are still represented in large amounts. The general diversity of the spore populations is still decreasing. The large apiculate and spinose zonate-pseudosaccate continue their quantitative decline started in the *triangulatus-catillus* Zone. The cryptospores are very uncommon, with a unique species, *Gneudnaspora divellomedia* var. *divellomedia*. The monolete spores are above all present with the monolete form of *Geminospora lemurata*. The megaspores are still present in number, including notably *Verrucisporites ellesmerensis*. The feature of grapnel-tipped spines seems to be more and more frequent with the reappearance of *Ancyrospora* genus and the presence of rarer specimens of *Hystricosporites*.

Comparison with reference biozones. This assemblage is the equivalent of the TCo Oppel Zone of Streel *et al.* (1987) since *Samarisporites triangulatus* and *Chelinospora concinna* are found together. The *langii-concinna* Zone

---

FIG. 56. Chart comparing the new biozonation with the reference spore zonations from Euramerica and Saudi Arabia. As there are any outcrops or sections that allow to correlate the different biozones, biostratigraphical correlations illustrated here are approximative. *Operational palynological zonation of Al-Hajri *et al.* (1999) updated according to PB (pers. obs.).

STAGES	Richardson and McGregor (1986)	Streel *et al.* (1987)			This paper		Al-Hajri *et al.* (1999)*	
FRASNIAN	*ovalis-bulliferus*	(IV)	B β		/////////////		D1	
			B α					
			A					
		BM						
		BJ			?			
GIVETIAN	*optivus-triangulatus* ?	TCo			*langii-concinna*		D2	
	lemurata-magnificus	TA			*triangulatus-catillus*	*catillus*		
						triangulatus		
		AD	Lem		*lemurata-langii*	*undulatus*		
						lemurata		
	devonicus-naumovae ?		Ref		*rugulata-libyensis*	*incognita*		
			Mac			*rugulata*		
EIFELIAN	*velata-langii*	AP	Vel		*svalbardiae-eximius*		D3	A
	douglastownensis -eurypterota		net					
			Pro					
			ked		*annulatus-protea*			
			Cor					
EMSIAN	*annulatus-sextantii*	FD	Min		*lindlarensis-sextantii*	D3B		B
					asymmetricus		D3/D4	
			Pra		*ovalis-biornatus*	*milleri*	D4	A
			Fov					
		AB				*ovalis*		
PRAGIAN	*polygonalis-emsiensis*	PoW	Su		*papillensis-baqaensis* ?			
			Pa β		/////////////			
			Pa α					
			W					
			Po					

also corresponds in part to the *optivus-triangulatus* Assemblage Zone of Richardson and McGregor (1986).

Stages. Upper Givetian – lower Frasnian.

Acknowledgements. We acknowledge the Saudi Arabian Ministry of Petroleum and Mineral Resources and the Saudi Arabian Oil Company (Saudi Aramco) for granting permission to publish this paper. We thank Merrell A. Miller (Saudi Aramco, Dhahran, KSA), John Filatoff (Brisbane, Australia) and Maurice Streel (University of Liège, Belgium) for their encouragement, constructive discussions and critical comments on this work. Marcella Giraldo-Mezzatesta (University of Liège) is acknowledged for processing palynological samples. Thanks are especially expressed to John E. A. Marshall (University of Southampton, UK), Charles H. Wellman (University of Sheffield, UK) and G. Clayton (Trinity College Dublin, Ireland) for comprehensively reviewing the manuscript. This contribution is part of the Saudi Aramco/CIMP joint project on the Palaeozoic of north-western Gondwanan regions.

Editor. Svend Stouge

REFERENCES

AL-GHAZI, A. 2007. New evidence for the Early Devonian age of the Jauf Formation in northern Saudi Arabia. *Revue de Micropaléontologie*, **50**, 59–72.

—— 2009. *Apiculiretusispora arabiensis*, new name for *Apiculiretusispora densa* Al-Ghazi, 2007 (preoccupied). *Revue de Micropaléontologie*, **52**, 193.

AL-HAJRI, S. A. and OWENS, B. (eds) 2000. Stratigraphic palynology of the Palaeozoic of Saudi Arabia. *GeoArabia, Special Publication*, **1**, 231 pp.

—— and PARIS F. 1998. Age and palaeoenvironment of the Sharawra Member (Silurian of northwestern Saudi Arabia). *Geobios*, **31**, 3–12.

—— FILATOFF, J., WENDER, L. E. and NORTON, A. K. 1999. Stratigraphy and operational palynology of the Devonian System in Saudi Arabia. *GeoArabia*, **4**, 53–68.

AL-HUSSEINI, M. and MATTHEWS, R. K. 2006. Devonian Jauf Formation, Saudi Arabia: orbital second-order depositional sequence 28. *GeoArabia*, **11**, 53–70.

ALLEN, K. C. 1965. Lower to Middle Devonian spores of North and Central Vestspitsbergen. *Palaeontology*, **8**, 687–748.

—— 1980. A review of *in situ* late Silurian and Devonian spores. *Review of Palaeobotany and Palynology*, **29**, 253–270.

AMENÁBAR, C. R. 2009. Middle Devonian from the Chiuga Formation, Precordillera region, northwestern Argentina. 177–192. *In* KÖNIGSHOF, P. (ed.). *Devonian change: case studies in palaeogeography and palaeoecology*. Geological Society, London, Special Publications, **314**, 412 pp.

ANDREWS, H. N., GENSEL, P. G. and FORBES, W. H. 1974. An apparently heterosporous plant from the Middle Devonian of New Brunswick. *Palaeontology*, **17**, 387–408.

ARKHANGELSKAYA, A. D. 1976. Basis for Eifelian age of the *Periplecotriletes tortus* zone in central districts of the European part of the USSR. 36–66. *In* BYVSHEVA, T. V. (ed.). *Results of palynological research on Precambrian, Paleozoic and Mesozoic of USSR.* Trudy Vsesoluznogo Naucho-Issledovate skogo Geologoazvedochnogo Neftianogo Instituta, Kama Branch, 192 pp. [In Russian].

—— 1978. Spores from the Lower Devonian of the Lithuanian SSR. *Paleontologicheskii Zhurnal*, **2**, 113–120. [In Russian].

—— 1985. Zonal spore assemblages and stratigraphy of the Lower and Middle Devonian in the Russian Plate. 5–21. *In Atlas of spore and pollen from the Phanerozoic petroleum formations in the Russian and Turanian Plates*. Moscow. [In Russian].

—— and TURNAU E. 2003. New dispersed seed-megaspores from the mid-Givetian of European Russia. *Review of Palaeobotany and Palynology*, **127**, 45–58.

AVKHIMOVITCH, V. I., CHIBRIKOVA, E. V., OBUK-HOVSKAYA, T. G., NAZARENKO, A. M., UMNOVA, V. T., RASKATOVA, L. G., MANTSUROVA, V. N., LOBOZIAK, S. and STREEL, M. 1993. Middle and Upper Devonian spore zonation of Eastern Europe. *Bulletin des Centres de Recherches de Exploration-Production Elf-Aquitaine*, **17**, 79–147.

BALME, B. E. 1962. Upper Devonian (Frasnian) spores from the Carnarvon Basin, Western Australia. *The Palaeobotanist*, **9**, 1–10.

—— 1988. Spores from Late Devonian (early Frasnian) strata, Carnarvon Basin, Western Australia. *Palaeontographica, Abteilung B*, **209**, 109–166.

—— and HASSELL C. W. 1962. Upper Devonian spores from the Canning Basin, Western Australia. *Micropaleontology*, **8**, 1–28.

BÄR, P. and RIEGEL, W. 1974. Les microflores des séries paléozoïques du Ghana (Afrique occidentale) et leurs relations paléofloristiques. *Sciences Géologiques, Bulletin*, **27**, 39–57.

BECK, J. H. and STROTHER, P. 2008. Spores and cryptospores from the Silurian section a Allenport, Pennsylvania, USA. *Journal of Paleontology*, **82**, 857–883.

BECKER, G., BLESS, M. J. M., STREEL, M. and THOREZ, J. 1974. Palynology and ostracode distribution in the Upper Devonian and basal Dinantian of Belgium and their dependence on sedimentary facies. *Mededelingen Rijks Geologische Dienst, Niewe Serie*, **25**, 9–99.

BEJU, D. 1967. Quelques spores, acritarches et chitinozoaires d'âge Dévonien inférieur de la plate-forme Moesienne (Roumanie). *Review of Palaeobotany and Palynology*, **5**, 39–49.

BEN RAHUMA, M. M., PROUST, J.-N. and ESCHARD, R. 2008. *The stratigraphic evolution of the Devonian sequences, Awaynat Wanin area, southern Ghadamis Basin: a fieldguide book*. Libyan Pretroleum Institute, Tripoli, 67 pp.

BERRY, E. W. 1937. Spores from the Pennington Coal, Rhea County, Tennessee. *The American Midland Naturalist*, **18**, 155–160.

BEYDOUN, Z. R. 1991. Arabian Plate hydrocarbon geology and potential – a plate tectonic approach. *American Association of Petroleum Geologists, Studies in Geology*, **33**, 1–77.

BHARADWAJ, D. C. and TIWARI, R. S. 1970. Lower Gondwana megaspores – a monograph. *Palaeontographica, Abteilung B*, **129**, 1–70.

BONAMO, P. 1977. *Rellimia thomsonii* (Progymnospermopsida) from the Middle Devonian of New York State. *American Journal of Botany*, **64**, 1272–1285.

—— and BANKS H. P. 1967. *Tetraxylopteris schmidtii*: its fertile parts and its relationships within the Aneurophytales. *American Journal of Botany*, **54**, 755–768.

BOOTE, D. R. D., CLARK-LOWES, D. D. and TRAUT, M. W. 1998. Palaeozoic petroleum systems of North Africa. 7–68. *In* MCGREGOR, D. S., MOODY, R. T. J. and CLARK-LOWES, D. D. (eds). *Petroleum geology of North Africa*. Geological Society, London, Special Publications, **132**, 442 pp.

BOUCOT, A. J., MCCLURE, H. A., ALVAREZ, F., ROSS, J. P., TAYLOR, D. W., STRUVE, W., SAVAGE, N. N. and TURNER, S. 1989. New Devonian fossils from Saudi Arabia and their biogeographical affinities. *Senckenbergiana Lethaea*, **69**, 535–597.

BOUMENDJEL, K., LOBOZIAK, S., PARIS, F., STEEMANS, P. and STREEL, M. 1988. Biostratigraphie des spores et des chitinozoaires du Silurien supérieur et du Dévonien dans le Bassin d'Illizi (S.E. du Sahara algérien). *Geobios*, **22**, 329–357.

BREUER, P. and GRAHN, Y. 2011. Middle Devonian spore stratigraphy in the eastern outcrop belt of the Parnaíba Basin, northeastern Brazil. *Revista Española de Micropaleontología*, **43**, 1–21.

—— STRICANNE, L. and STEEMANS, P. 2005*a*. Morphometric analysis of proposed evolutionary lineages of Early Devonian land plant spores. *Geological Magazine*, **142**, 241–253.

—— AL-GHAZI, A., FILATOFF, J., HIGGS, K. T., STEEMANS, P. and WELLMAN, C. H. 2005*b*. Stratigraphic palynology of Devonian boreholes from northern Saudi Arabia. 3–9. *In* STEEMANS, P. and JAVAUX, E. (eds). *Pre-Cambrian to Palaeozoic palaeopalynology and palaeobotany*. Carnets de Géologie/Notebooks on Geology, Memoir, **2005/02**, 77 pp.

—— FILATOFF, J. and STEEMANS, P. 2007*a*. Some considerations on Devonian spore taxonomy. 3–8. *In* STEEMANS, P. and JAVAUX, E. (eds). *Pre-Cambrian to Palaeozoic palaeopalynology and palaeobotany*. Carnets de Géologie/Notebooks on Geology, Memoir, **2007/01**, 73 pp.

—— DISLAIRE, G., FILATOFF, J., PIRARD, E. and STEEMANS, P. 2007*b*. Support Vector assisted spore classification based on ornament spatial distribution analysis – an application to Emsian spores from Saudi Arabia. 9–15. *In* STEEMANS, P. and JAVAUX, E. (eds). *Pre-Cambrian to Palaeozoic palaeopalynology and palaeobotany*. Carnets de Géologie/Notebooks on Geology, Memoir, **2007/01**, 73 pp.

—— AL-GHAZI, A., AL-RUWAILI, M., HIGGS, K. T., STEEMANS, P. and WELLMAN, C. H. 2007*c*. Early to Middle Devonian spores from northern Saudi Arabia. *Revue de Micropaléontologie*, **50**, 27–57.

BRICE, D., BULTYNCK, P., DEUNFF, J., LOBOZIAK, S. and STREEL, M. 1979. Données biostratigraphiques nouvelles sur le Givetien et le Frasnien de Ferques (Boulonnais, France). *Annales de la Société Géologique du Nord*, **98**, 325–344.

BURDEN, E. T., QUINN, L., NOWLAN, G. S. and BAILEY-NILL, L. A. 2002. Palynology and micropaleontology of the Clam Bank Formation (Lower Devonian) of western Newfoundland, Canada. *Palynology*, **26**, 185–215.

BURGESS, N. D. and RICHARDSON, J. B. 1991. Silurian cryptospores and spores from the type Wenlock area, Shropshire, England. *Palaeontology*, **34**, 601–628.

—— —— 1995. Late Wenlock to early Přídolí cryptospores and spores from south and southwest Wales Great Britain. *Palaeontographica, Abteilung B*, **236**, 1–44.

BURJACK, M. I. A., LOBOZIAK, S. and STREEL, M. 1987. Quelques données nouvelles sur les spores dévoniennes du bassin du Paraná (Brésil). *Sciences Géologiques, Bulletin*, **40**, 381–391.

CHALONER, W. G. 1959. Devonian megaspores from Arctic Canada. *Palaeontology*, **1**, 321–332.

—— 1963. Early Devonian spores from a borehole in Southern England. *Grana Palynologica*, **4**, 100–110.

CHI, B. I. and HILLS, L. V. 1976. Biostratigraphy and taxonomy of Devonian megaspores, Arctic Canada. *Bulletin of Canadian Petroleum Geology*, **24**, 640–818.

CHIBRIKOVA, E. V. 1959. *Spores of the Devonian and older rocks of Bashkiria. Data on palaeontology and stratigraphy of Devonian and older deposits of Bashkiria*. Academy of Sciences of USSR, Bashkirian Branch, 1–247. [In Russian].

—— 1962. *Spores of Devonian terrigenous deposits of western Bashkiria and the western slopes of the southern Urals*. Academy of Sciences of USSR, Bashkirian Branch, 353–476. [In Russian].

CLAYTON, G. and GRAHAM, J. R. 1974. Spore assemblages from the Devonian Sherkin Formation of southwest County Cork, Republic of Ireland. *Pollen et Spores*, **16**, 565–588.

—— JOHNSTON, I. S., SEVASTOPULO, G. D. and SMITH, D. G. 1980. Micropalaeontology of a Courceyan (Carboniferous) borehole section from Ballyvergin, County Clare, Ireland. *Journal of Earth Sciences Royal Dublin Society*, **3**, 81–100.

—— OWENS, B., AL-HAJRI, S. A. and FILATOFF, J. 2000. Latest Devonian and Early Carboniferous spore assemblages from Saudi Arabia. 146–153. *In* AL-HAJRI, S. A. and OWENS, B. (eds). *Stratigraphic palynology of the Palaeozoic of Saudi Arabia*. GeoArabia, Special Publication, **1**, 231 pp.

CLENDENING, J. A., EAMES, L. E. and WOOD, G. D. 1980. *Retusotriletes phillipsii* n. sp., a potential Upper Devonian guide palynomorph. *Palynology*, **4**, 15–22.

COQUEL, R. and LATRÈCHE, S. 1989. Etude palynologique de la Formation d'Illerène (Devono–Carbonifère) du Bassin d'Illizi (Sahara algérien occidental). *Palaeontographica, Abteilung B*, **212**, 47–70.

—— and MOREAU-BENOIT A. 1986. Les spores des séries struniennes et tournaisiennes de Libye occidentale. *Revue de Micropaléontologie*, **29**, 17–43.

—— 1989. A propos de quelques spores trilètes 'chambrées' du Dévonien terminal–Carbonifère inférieur d'Afrique du Nord. *Revue de Micropaléontologie* **32**, 87–102.

COUPER, R. A. 1953. Upper Mesozoic and Cainozoic spores and pollen grains from New Zealand. *New Zealand Geological Survey Paleontological Bulletin*, **22**, 1–77.

CRAMER, F. H. 1966a. Palynology of Silurian and Devonian rocks in Northwest Spain. *Boletín del Instituto Geológico y Minero de España*, **77**, 223–286.

—— 1966b. Palynomorphs from the Siluro–Devonian boundary in N. W. Spain. *Notas y Comunicaciones del Instituto Geológico y Minero de España*, **85**, 71–82.

—— 1969. Plant spores from the Eifelian to Givetian *Gosseletia* Sandstone Formation near Candas, Asturias, Spain. *Pollen et Spores*, **11**, 425–447.

—— and DÍEZ M. D. C. R. 1975. Earliest Devonian spores from the Province of León, Spain. *Pollen et Spores*, **17**, 331–344.

D'ERCEVILLE, M. A. 1979. Les spores des formations siluro–dévoniennes, de la coupe de Saint-Pierre-sur-Erve, synclinorium médian armoricain. *Palaeontographica, Abteilung B*, **171**, 79–121.

DAEMON, R. F. 1974. Palinomorfos-guias do Devoniano Superior e Carbonífero Inferior das Bacias do Amazonas e Parnaíba. *Anais da Academia Brasileira de Ciências*, **46**, 549–587.

—— QUADROS, L. P. and DA SILVA, L. C. 1967. Devonian palynology and biostratigraphy of the Paraná Basin. 99–132. *In* BIGARELLA, J. J. (ed.). *Problems in Brazilian Devonian geology*. Boletim Paranaense de Geociências, **21/22**, 155 pp.

DETTMANN, M. E. 1963. Upper Mesozoic microfloras from South-Eastern Australia. *Proceedings of the Royal Society of Victoria*, **77**, 1–148.

EISENACK, A. 1944. Über einige pflanzliche Funde in Geschieben, nebst Bemerkungen zum Hystrichosphaerideen-Problem. *Zeitschrift für Geschiebeforschung*, **19**, 103–124.

ERDTMAN, G. 1952. *Pollen Morphology and Plant Taxonomy. Angiosperms*. Almqvist and Wiksell, Stockholm, 539 pp.

EVANS, P. R. 1970. Revision of the spore genera *Perotrilites* Erdtman ex Couper, 1953 and *Diaphanospora* Balme and Hassell, 1962. *Palaeontological Papers, 1968, Australia, Bureau of Mineral Resources, Geology and Geophysics, Bulletin*, **116**, 65–74.

FANNING, U., EDWARDS, D. and RICHARDSON, J. B. 1992. A diverse assemblage of early land plants from the Lower Devonian of the Welsh Borderland. *Botanical Journal of the Linnean Society*, **109**, 161–188.

FOREY, P. L., YOUNG, V. T. and MCCLURE, H. A. 1992. Lower Devonian fishes from Saudi Arabia. *Bulletin of the British Museum of Natural History Geology*, **48**, 25–43.

FRIEND, P. F., ALEXANDER-MARRACK, P. D., ALLEN, K. C., NICHOLSON, J. and YEATS, A. K. 1983. Devonian sediments of East Greenland VI. Review of results. *Meddelelser om Grønland*, **206**, 1–96.

FUGLEWICZ, R. and PREJBISZ, A. 1981. Devonian megaspores from NW Poland. *Acta Palaeontologica Polonica*, **26**, 55–72.

GAO LIANDA. 1981. Devonian spore assemblages of China. 11–23. *In* OWENS, B. and VISSCHER, H. (eds). *Late Palaeozoic and Early Mesozoic stratigraphic palynology*. Review of Palaeobotany and Palynology, **34**, 135 pp.

GENSEL, P. G. 1980. Devonian *in situ* spores: a survey and discussion. *Review of Palaeobotany and Palynology*, **30**, 101–132.

GERRIENNE, P., MEYER-BERTHAUD, B., FAIRON-DEMARET, M., STREEL, M. and STEEMANS, P. 2004. *Runcaria*, a Middle Devonian seed plant precursor. *Science*, **306**, 856–858.

GHAVIDEL-SYOOKI, M. 2003. Palynostratigraphy of Devonian sediments in the Zagros Basin, southern Iran. *Review of Palaeobotany and Palynology*, **127**, 241–268.

GRAHN, Y., MELO, J. H. G. and STEEMANS, P. 2005. Integrated chitinozoan and spore zonation of the Serra Grande Group (Silurian–Lower Devonian), Parnaíba Basin, northeast Brazil. *Revista Española de Micropaleontología*, **37**, 183–204.

GREY, K. 1975. Devonian spores from the Gogo Formation, Canning Basin. 96–99. *Annual Report of Geological Survey of Western Australia for 1974*.

—— 1991. A mid-Givetian spore age for the onset of reef development on the Lennard Shelf, Canning Basin, Western Australia. *Review of Palaeobotany and Palynology*, **68**, 37–48.

GRIGNANI, D., LANZONI, E. and ELATRASH, H. 1991. Palaeozoic and Mesozoic subsurface palynostratigraphy in the Al Kufrah Basin, Libya. 1159–1227. *In* SALEM, M. J., HAMMUDA, O. S. and ELIAGOUBI, B. A. (eds). *The geology of Libya*. Vol. IV. Elsevier, Amsterdam, 1648 pp.

GUENNEL, G. K. 1963. Devonian spores in a Middle Silurian reef. *Grana Palynologica*, **4**, 245–261.

HABGOOD, K. S., EDWARDS, D. and AXE, L. 2002. New perspectives on *Cooksonia* from the Lower Devonian of the Welsh Borderland. *Botanical Journal of the Linnean Society*, **139**, 339–359.

HACQUEBARD, P. A. 1957. Plant spores in coal from the Horton group (Mississippian) of Nova Scotia. *Micropaleontology*, **3**, 301–324.

HAGSTRÖM, J. 1997. Land-derived palynomorphs from the Silurian of Gotland, Sweden. *GFF*, **119**, 301–316.

HAMID, M. E. P. 1974. Sporenvergesellschaftungen aus dem unteren Mitteldevon (Eifel-Stufe) des südlichen Bergischen Landes (Rheinisches Schiefergebirge). *Neues Jahrbuch für Geologie und Palaontologie, Abhandlungen*, **147**, 163–217.

HASHEMI, H. and PLAYFORD, G. 2005. Devonian spore assemblages of the Adavale Basin, Queensland (Australia): descriptive systematics and stratigraphic significance. *Revista Española de Micropaleontología*, **37**, 317–417.

HEMER, D. O. 1965. Application of palynology in Saudi Arabia. *In Fifth Arab Petroleum Congress*. Cairo, Egypt, 28 pp.

—— and NYGREEN P. W. 1967. Devonian palynology of Saudi Arabia. *Review of Palaeobotany and Palynology*, **5**, 51–61.

HODGSON, E. A. 1968. Devonian spores from the Pertnjara Formation, Amadeus Basin, Northern Territory. *Bureau of Mineral Resources, Geology and Geophysics, Australia, Bulletin*, **80**, 67–82.

HOFFMEISTER, W. S. 1959. Lower Silurian plant spores from Libya. *Micropaleontology*, **5**, 331–334.

—— STAPLIN, F. L. and MALLOY, R. E. 1955. Mississippian plant spores from the Hardinsburg Formation of Illinois and Kentucky. *Journal of Paleontology*, **29**, 372–399.

HOLLAND, C. H. and SMITH, D. G. 1979. Silurian rocks of the Capard inlier, County Laois. *Proceedings of the Royal Irish Academy, Series B*, **79**, 99–110.

HUGHES, N. F. and PLAYFORD, G. 1961. Palynological reconnaissance of the Lower Carboniferous of Spitsbergen. *Micropaleontology*, **7**, 27–44.

IBRAHIM, A. C. 1933. Sporenformen des Ägirhorizontes des Ruhrreviers. Dissertation, Technische Hoschschule, Berlin, 46 pp.

ISHCHENKO, A. M. 1952. *Atlas of the microspores and pollen of the Middle Carboniferous of the western part of the Donetz Basin.* Akademiya Nauk Ukrainskoy SSR, Trudy Instituta Geologischeskikh Nauk, Kiev, 83 pp. [In Russian].

JÄGER, H. 2004. Facies dependence of spore assemblage and new data on sedimentary influence on spore taphonomy. 121–140. *In* SERVAIS, T. and WELLMAN, C. H. (eds). *New directions in Palaeozoic palynology.* Review of Palaeobotany and Palynology, **130**, 299 pp.

JANSONIUS, J. and HILLS, L. V. 1987. *Genera file of fossil spores.* Supplement. Special Publication, Department of Geology and Geophysics, University of Calgary, Canada.

JARDINÉ, S. and YAPAUDJIAN, L. 1968. Lithostratigraphie et Palynologie du Dévonien–Gothlandien gréseux du Bassin de Polignac (Sahara). *Revue de l'Institut Français du Pétrole*, **23**, 439–469.

JERSEY, N. J. DE. 1966. Devonian spores from the Adavale Basin. *Geological Survey of Queensland, Publication 334, Palaeontological Papers*, **3**, 1–28.

KEDO, G. I. 1955. Spores of the Middle Devonian of the northeastern Byelorussian SSR. *Academy of Sciences of BSSR, Institute of Geological Sciences, Paleontology and Stratigraphy*, **1**, 5–59. [In Russian].

KEMP, E. M. 1972. Lower Devonian palynomorphs from the Horlick Formation, Ohio Range, Antarctica. *Palaeontographica, Abteilung B*, **139**, 105–124.

KERR, J. W., MCGREGOR, D. C. and MCLAREN, D. J. 1965. An unconformity between Middle and Upper Devonian rocks of Bathurst Island, with comments on Upper Devonian faunas and microfloras of the Parry Islands. *Bulletin of Canadian Petroleum Geology*, **13**, 409–431.

KOSANKE, R. M. 1950. Pennsylvanian spores of Illinois and their use in correlation. *Illinois State Geological Survey, Bulletin*, **74**, 1–128.

KRASSILOV, V. A., RASKATOVA, M. G. and ISTCHENKO, A. A. 1987. A new archaeopteridalean plant from the Devonian of Pavlovsk, U.S.S.R. *Review of Palaeobotany and Palynology*, **53**, 163–173.

LAKE, P. A., OSWIN, W. M. and MARSHALL, J. E. A. 1988. A palynological approach to terrane analysis in the South Portuguese Zone. *Trabajos de Geología*, **17**, 125–131.

LALOUX, M., DEJONGHE, L., GEUKENS, F., GHYSEL, P. and HANCE, L. 1996. *Limbourg-Eupen, 43/5–6. Carte Géologique de Wallonie: 1/25.000.* Direction Générale des Ressources Naturelles de l'Environnement, Brussels, 83 pp.

LANG, W. H. 1925. Contributions to the study of the Old Red Sandstone flora of Scotland. 1. On plant remains from the fish beds of Cromarty. *Transactions of the Royal Society of Edinburgh: Earth Sciences*, **54**, 253–272.

LANNINGER, E. P. 1968. Sporen-Gesellschaften aus dem Ems der SW-Eifel (Rheinisches Schiefergebirge). *Palaeontographica, Abteilung B*, **122**, 95–170.

LANZONI, E. and MAGLOIRE, L. 1969. Associations palynologiques et leurs applications stratigraphiques dans le Dévonien supérieur et Carbonifère inférieur du Grand Erg occidental (Sahara algérien). *Revue de l'Institut Français du Pétrole*, **24**, 441–468.

LE HÉRISSÉ, A. 1983. Les spores du Dévonien inférieur du Synclinorium de Laval (Massif Armoricain). *Palaeontographica, Abteilung B*, **188**, 1–81.

—— RUBINSTEIN, C. and STEEMANS, P. 1997. Lower Devonian palynomorphs from the Talacasto Formation, Cerro del Fuerte Section, San Juan Precordillera, Argentina. 497–515. *In* FATKA, O. and SERVAIS, T. (eds). *Acritarcha in Praha, 1996.* Acta Universitatis Carolinae, Geologica, **40**, 293–717.

LECLERCQ, S. and BONAMO, P. 1971. A study of the fructification of *Milleria (Protopteridium) thomsonii* Lang from the Middle Devonian of Belgium. *Palaeontographica, Abteilung B*, **136**, 83–114.

LELE, K. M. and STREEL, M. 1969. Middle Devonian (Givetian) plant microfossils from Goé (Belgium). *Annales de la Société Géologique de Belgique*, **92**, 89–121.

LESSUISE, A., STREEL, M. and VANGUESTAINE, M. 1979. Observations palynologiques dans le Couvinien (Emsien terminal et Eifelien) du bord oriental du Synclinorium de Dinant, Belgique. *Annales de la Société Géologique de Belgique*, **102**, 325–355.

LOBOZIAK, S. 2000. Middle to early Late Devonian spore biostratigraphy of Saudi Arabia. 134–145. *In* AL-HAJRI, S. A. and OWENS, B. (eds). *Stratigraphic palynology of the Palaeozoic of Saudi Arabia.* GeoArabia, Special Publication, **1**, 231 pp.

—— and MELO J. H. G. 2002. Devonian spores successions of Western Gondwana: update and correlation with Southern Euramerican spores zones. *Review of Palaeobotany and Palynology*, **121**, 133–148.

—— and STREEL M. 1980. Spores in Givetian to Lower Frasnian sediments dated by conodonts from the Boulonnais, France. *Review of Palaeobotany and Palynology*, **29**, 285–299.

—— —— 1988. Synthèse palynostratigraphique de l'intervalle Givetien–Famennien du Boulonnais (France). 71–77. *In* BRICE D. (ed.). *Le Dévonien de Ferques. Bas-Boulonnais (N. France).* Biostratigraphie du Paléozoïque, **7**. Université de Bretagne Occidentale, Brest, 520 pp.

—— —— 1989. Middle–Upper Devonian spores from the Ghadames Basin (Tunisia–Libya): systematics and stratigraphy. *Review of Palaeobotany and Palynology* **58**, 173–196.

—— —— 1995a. West Gondwanan aspects of the Middle and Upper Devonian spore zonation in North Africa and Brazil. 147–155. *In* LE HÉRISSÉ A., OWENS, B. and PARIS, F. (eds). *Palaeozoic palynomorphs of the Gondwana-Euramerican interface.* Review of Palaeobotany and Palynology, **86**, 173 pp.

—— —— 1995b. Late Lower and Middle Devonian spores from Saudi Arabia. 105–114. *In* OWENS B., AL-TAYYAR, H., VAN DER EEM, J. G. L. A. and AL-HAJRI, S. A. (eds). *Palaeozoic palynostratigraphy of the Kingdom of Saudi Arabia.* Review of Palaeobotany and Palynology, **89**, 150 pp.

—— —— and VANGUESTAINE M. 1983. Spores et acritarches de la Formation d'Hydrequent (Frasnien supérieur à Famennien inférieur, Boulonnais, France). *Annales de la Société Géologique de Belgique*, **106**, 173–183.

—— —— and BURJACK M. I. A. 1988. Spores du Dévonien Moyen et Supérieur du Bassin du Parana, Brésil: systématique et stratigraphie. *Sciences Géologiques, Bulletin*, **41**, 351–377.

—— —— and WEDDIGE K. 1990. Spores, the *lemurata* and *triangulatus* levels and their faunal indices near the Eifelian/Givetian boundary in the Eifel (F.R.G.). *Annales de la Société Géologique de Belgique*, **113**, 1–15.

—— STEEMANS, P., STREEL, M. and VACHARD, D. 1992a. Biostratigraphie par spores du Dévonien inférieur à supérieur du sondage MG-1 (Bassin d'Hammadeh, Tunisie) – comparaison avec les données des faunes. *Review of Palaeobotany and Palynology*, **74**, 193–205.

—— STREEL, M., CAPUTO, M. V. and MELO, J. H. G. 1992b. Middle Devonian to Lower Carboniferous spore stratigraphy in the Central Parnaíba Basin (Brazil). *Annales de la Société Géologique de Belgique*, **115**, 215–226.

—— —— and MELO J. H. G. 1999. *Grandispora* (al. Contagisporites) *permulta* (Daemon, 1974) Loboziak, Streel et Melo, comb. nov., a senior synonym of Grandispora riegelii Loboziak et Streel, 1989 – nomenclature and stratigraphic distribution. *Review of Palaeobotany and Palynology*, **106**, 97–102.

LOVE, L. G. 1960. Assemblages of small spores from the Lower Oil Shale Group of Scotland. *Proceedings of the Royal Society of Edinburgh, Section B*, **67**, 99–126.

LU LICHANG 1980. On the occurrence of *Archaeoperisaccus* in E. Yunnan. *Acta Palaeontologica Sinica*, **19**, 500–503. [In Chinese].

—— 1988. Middle Devonian microflora from Haickou Formation at Shijiapo in Zhanyi of Yunnan, China. Memoirs of Nanjing Institute of Geology and Palaeontology. *Academia Sinica*, **24**, 109–234. [In Chinese].

—— and OUYANG SHU 1976. The Early Devonian spore assemblage from the Xujiachong Formation at Cuifengshan in Qujing of Yunnan. *Acta Palaeontologica Sinica*, **15**, 21–43. [In Chinese].

LUBER, A. A. and WALTZ, I. E. 1938. Classification and stratigraphic value of spores of some Carboniferous coal deposits in the USSR. *Trudy Tsentral'nogo Nauchno-Issledovatel'skogo Geologo-Razvedochnogo Instituta*, **105**, 1–43. [In Russian].

MAGLOIRE, L. 1968. Etude stratigraphique par la palynologie des dépôts argilo-gréseux du Silurien et du Dévonien inférieur dans la région du Grand Erg Occidental (Sahara algérien). *International Symposium on the Devonian System, Calgary*, **2**, 473–495.

MARSHALL, J. E. A. 1988. Devonian spores from Papa Stour, Shetland. *Transactions of the Royal Society of Edinburgh: Earth Sciences*, **79**, 13–18.

—— 1996. *Rhabdosporites langii*, *Geminospora lemurata* and *Contagisporites optivus*: an origin for heterospory within the Progymnosperms. *Review of Palaeobotany and Palynology*, **93**, 159–189.

—— 2000. Devonian (Givetian) spores from the Walls Group, Shetland. 473–483. *In* FRIEND, P. F. and WILLIAMS, B. P. J. (eds). *New perspectives on the Old Red Sandstone*. Geological Society, London, Special Publications, **180**, 623 pp.

—— and ALLEN K. C. 1982. Devonian spore assemblages from Fair Isle, Shetland. *Palaeontology*, **25**, 277–312.

—— and FLETCHER T. P. 2002. Middle Devonian (Eifelian) spores from a fluvial dominated lake margin in the Orcadian Basin, Scotland. *Review of Palaeobotany and Palynology*, **118**, 195–209.

—— and HEMSLEY A. R. 2003. A Mid Devonian seed-megaspore from East Greenland and the origin of the seed plants. *Palaeontology*, **46**, 647–670.

—— ROGERS, D. A. and WHITELEY, M. J. 1996. Devonian marine incursions into the Orcadian Basin, Scotland. *Journal of the Geological Society*, **153**, 451–466.

—— MILLER, M. A., FILATOFF, J. and AL-SHAHAB, K. 2007. Two new Middle Devonian megaspores from Saudi Arabia. *Revue de Micropaléontologie*, **50**, 73–79.

MASSA, D. 1988. Paléozoïque de Libye occidentale: stratigraphie et paléogéographie. Unpublished PhD thesis, Université de Nice, France, 514 pp.

—— and MOREAU-BENOIT A. 1976. Essai de synthèse stratigraphique et palynologique du Système Dévonien en Libye occidentale. *Revue de l'Institut Français du Pétrole*, **41**, 287–333.

—— —— 1985. Apport de nouvelles données palynologiques à la biostratigraphie et à la paléogéographie du Dévonien en Libye (sud du bassin de Rhadamès). *Sciences Géologiques Bulletin*, **38**, 5–18.

—— COQUEL, R., LOBOZIAK, S. and TAUGOUR-DEAU-LANTZ, J. 1980. Essai de synthèse stratigraphique et palynologique du Carbonifère en Libye occidentale. *Annales de la Société Géologique du Nord*, **9**, 429–442.

MCGREGOR, D. C. 1960. Devonian spores from Melville Island, Canadian Artic Archipelago. *Palaeontology*, **3**, 26–44.

—— 1961. Spores with proximal radial pattern from the Devonian of Canada. *Geological Survey of Canada, Bulletin*, **76**, 1–11.

—— 1964. Devonian spores from the Ghost River Formation, Alberta. *Geological Survey of Canada, Bulletin*, **109**, 1–31.

—— 1967. Composition and range of some Devonian spore assemblages of Canada. *Review of Palaeobotany and Palynology*, **1**, 173–183.

—— 1969. Devonian plant fossils of the genera *Kryshtovichia*, *Nikitinsporites*, *Archaeoperisaccus*. *Geological Survey of Canada, Bulletin*, **182**, 91–106.

—— 1970. Paleobotany. *In Geology and economic minerals of Canada, economical geology*, Fifth edition. Geological Survey of Canada, Report, **1**, 663–670.

—— 1973. Lower and Middle Devonian spores of Eastern Gaspé, Canada. I. Systematics. *Palaeontographica, Abteilung B*, **142**, 1–77.

—— 1974. Early Devonian spores from central Ellesmere Island, Canadian Artic. *Canadian Journal of Earth Sciences*, **11**, 70–78.

—— 1977. Lower and Middle Devonian spores of Eastern Gaspé, Canada. II. Biostratigraphic significance. *Palaeontographica, Abteilung B*, **163**, 111–142.

—— 1984. Late Silurian and Devonian spores from Bolivia. *Academia Nacional de Ciencias*, **69**, 1–57.

—— and CAMFIELD M. 1976. Upper Silurian? to Middle Devonian spores of the Moose River Basin, Ontario. *Geological Survey of Canada, Bulletin*, **263**, 1–63.

—— —— 1982. Middle Devonian spore from the Cape De Bray Weatherall, and Hecla Bay Formations of northeastern Melville Island, Canadian Artic. *Geological Survey of Canada, Bulletin*, **348**, 1–105.

—— and NARBONNE G. M. 1978. Upper Silurian trilete spores and other microfossils from the Read Bay Formation, Cornwallis Island, Canadian Arctic. *Canadian Journal of Earth Sciences*, **15**, 1292–1303.

—— and OWENS B. 1966. Devonian spores of eastern and northern Canada. *Geological Survey of Canada, Paper*, **66**, 1–66.

—— and PLAYFORD G. 1992. Canadian and Australian Devonian spores: zonation and correlation. *Geological Survey of Canada, Bulletin*, **438**, 1–125.

—— and UYENO T. T. 1972. Devonian spores and conodonts of Melville and Bathurst Islands, District of Franklin. *Geological Survey of Canada, Paper*, **71**, 1–37.

—— SANFORD, B. V. and NORRIS, A. W. 1970. Palynology and correlation of Devonian formations in the Moose River Basin, northern Ontario. *Geological Association of Canada, Proceedings*, **22**, 45–54.

MEHLQVIST, K. 2009. The spore record of early land plants from upper Silurian strata in Klinta 1 well, Skåne, Sweden. Unpublished MSc thesis, Lunds University, Sweden, 29 pp.

MEISSNER, C. R. Jr, DINI, S. M., FARASANI, A. M., RIDDLER, G. P., VAN HECK, M. and ASPINALL, N. C. 1988. *Preliminary geological map of the Al Jawf Quadrangle, sheet 29D, Kingdom of Saudi Arabia*. Open-file report USGS-OF-09-1. Ministry of Petroleum and Minerals, Deputy Ministry for Mineral Resources, Jeddah, Saudi Arabia.

MELO, J. H. G. and LOBOZIAK, S. 2003. Devonian–Early Carboniferous spore biostratigraphy of the Amazon Basin, Northern Brazil. *Review of Palaeobotany and Palynology*, **124**, 131–202.

MENDLOWICZ MAULLER, P., MACHADO CARD-OSO, T. R., PEREIRA, T. R. and STEEMANS, P. 2007. Resultados Palinoestratigráficos do Devoniano da Sub-Bacia de Alto Garças (Bacia do Paraná – Brasil). 607–619. *In* CARVALHO, I. S., CASSAB, R. C. T., SCHWANKE, C., CARVALHO, M. A., FERNANDES, A. C. S., RODRIGUES, M. A. C., CARVALHO, M. S. S., ARAI, M. and OLIVEIRA, M. E. Q. (eds). *Paleontologia: cenários de Vida*, Vol. 2. Interciência, Rio de Janeiro, 632 pp.

MIKHAILOVA, N. I. 1966. Spores from the Givetian deposits of Rudnyi Altai. *Transactions of Satpaev Institute of Geological Sciences, Academy of Sciences of Kazakhstan SSR*, **17**, 195–213. [In Russian].

MOREAU-BENOIT, A. 1967. Premiers résultats d'une étude palynologique du Dévonien de la carrière des Fours à Chaux d'Angers (Maine-et-Loire). *Revue de Micropaléontologie*, **9**, 219–240.

—— 1974. Recherches de palynologie et de planctologie sur le Dévonien de quelques formations siluriennes dans le Sud-Est du Massif Armoricain. *Mémoires de la Société Géologique et Minéralogiques de Bretagne*, **18**, 248 pp.

—— 1976. Les spores et débris végétaux. 27–58. *In* LARDEUX, H. (ed.). *Les schistes et calcaires éodévoniens de Saint-Cénéré (Massif Armoricain, France)*. Mémoires de la Société géologiques et minéralogique de Bretagne, **19**, 323 pp.

—— 1979. Les spores du Dévonien de Libye. Première partie. *Cahiers de Micropaléontologie*, **4**, 1–58.

—— 1980a. Les spores et les débris végétaux. 59–90. *In* PLUSQUELLEC, Y. (ed.). *Les schistes et calcaires de l'Armorique*. Mémoires de la Société géologique et minéralogiques de Bretagne, **23**, 313 pp.

—— 1980b. Les spores du Dévonien de Libye. Deuxième partie. *Cahiers de Micropaléontologie*, **1**, 1–53.

—— 1988. Considérations nouvelles sur la palynozonation du Dévonien moyen et supérieur du Bassin de Rhadamès, Libye occidentale. *Comptes Rendus de l'Académie des Sciences de Paris, Série II*, **307**, 863–869.

—— 1989. Les spores du Dévonien moyen et supérieur de Libye occidentale: compléments-systématique-répartition stratigraphique. *Cahiers de Micropaléontologie*, **4**, 1–32.

—— COQUEL, R. and LATRÈCHE, S. 1993. Étude palynologique du Dévonien du bassin d'Illizi (Sahara Oriental Algérien). Approche biostratigraphique. *Geobios*, **26**, 3–31.

MORGAN, J. L. 1955. Spores of McAlester Coal. *Oklahoma Geological Survey, Circular*, **36**, 1–52.

MORTIMER, M. G. 1967. Some lower Devonian microfloras from southern Britain. *Review of Palaeobotany and Palynology*, **1**, 95–109.

—— and CHALONER W. G. 1972. The palynology of the concealed Devonian rocks of southern England. *Bulletin of the Geological Survey of Great Britain*, **39**, 1–56.

NADLER, Y. S. 1966. Spores in Devonian deposits of the western part of Sayano-Altai mountainous region. 51–54. *In The importance of palynological analysis for stratigraphical and palaeofloristic investigations*. Akademiya Nauk SSSR, Institut Geografii i Geologicheskii, Izdatelstvo Nauka, Moskva. [In Russian].

NAUMOVA, S. N. 1939. Spores and pollen of the coals of the USSR. 353–364. *In 17th International Geological Congress, Report 1*. Moscow, Russia. [In Russian].

—— 1953. *Spore-pollen assemblages of the Upper Devonian of the Russian Platform and their stratigraphic value*. Akademiya Nauk SSSR, Institut Geologii Nauk, **143**, Geological Series 60, 203 pp. [In Russian].

NAZARENKO, A. M. 1965. Characterization of spore assemblage of Middle Devonian deposits of Volgograd district. *Transactions of Volgograd Scientific Research, Institute of Petroleum and Gas Industry (VNIING)*, **3**, 39–47. [In Russian].

NEVES, R. 1961. Namurian plant spores from the southern Pennines, England. *Palaeontology*, **4**, 247–279.

—— and OWENS B. 1966. Some Namurian spores from the English Pennines. *Pollen et Spores*, **8**, 337–360.

OBRHEL, J. 1959. Neue Pflanzenfunde in den Srbsko-Schichten (Mitteldevon). *Věstník Ústřednùího Ústavu Geologického*, **34**, 384–388.

OTTONE, G. E. 1996. Devonian palynomorphs from the Los Monos Formation, Tarija Basin, Argentina. *Palynology*, **20**, 105–155.

OWENS, B. 1971. Spores from the Middle and Early Upper Devonian rocks of the Western Queen Elizabeth Island, Arctic Archipelago. *Geological Survey of Canada, Paper*, **70-38**, 1–157.

PANSHINA, L. N. 1971. New species of spores of the lower part of the Frasnian Stage in the Volga-Ural Oblast. 90–96. *In Palynology and Stratigraphy of Paleozoic, Mesozoic, and Paleogene deposits of European part of USSR and Central Asia*. Trudy Vsesoluznogo Naucho-Issledovate skogo Geologoazvedochnogo Neftianogo Instituta, Kama Branch, 106 pp. [In Russian]

PARIS, F., RICHARDSON, J. B., RIEGEL, W., STREEL, M. and VANGUESTAINE, M. 1985. Devonian (Emsian–Famennian) palynomorphs. 49–81. *In* THUSU, B. and OWENS, B. (eds). *Palynostratigraphy of North-East Libya*. Journal of Micropalaeontology, 4, 182 pp.

PASHKEVICH, N. G. 1964. New Devonian species of *Archaeoperisaccus* (Gymnospermae) from northern Timan. *Paleontologische Zhurnal*, 4, 126–129. [In Russian].

PEPPERS, R. A. and DAMBERGER, H. H. 1969. Palynology and petrography of a Middle Devonian coal in Illinois. *Illinois State Geological Survey, Circular*, **445**, 1–35.

PEREZ-LEYTON, M. A. 1990. Spores du Dévonien Moyen et Supérieur de la coupe de Bermejo-La Angostura (Sud-Est de la Bolivie). *Annales de la Société Géologique de Belgique*, **113**, 373–389.

PLAYFORD, G. 1962. Lower Carboniferous microfloras of Spitsbergen. *Palaeontology*, 5, 550–618.

—— 1971. Lower Carboniferous spores form the Bonaparte Gulf Basin, Western Australia and Northern Territory. *Bureau of Mineral Resources, Geology and Geophysics, Bulletin*, **115**, 1–102.

—— 1976. Plant microfossils from the Upper Devonian and Lower Carboniferous of the Canning Basin, Western Australia. *Palaeontographica, Abteilung B*, **158**, 1–71.

—— 1983. The Devonian spore genus *Geminospora* Balme 1962: a reappraisal based upon topotypic *G. lemurata* (type species). *Memoirs of the Association of Australasian Palaeontologists*, 1, 311–325.

—— 1991. Australian Lower Carboniferous spores relevant to extra-Gondwanic correlations: an evaluation. 85–125. *In* BRENCKLE, P. L. and MANGER, W. L. (eds). *International division and correlation of the carboniferous system*. Courier Forschungsinstitut Senckenberg, **130**, 350 pp.

POTONIÉ, H. 1893. Die flora des Rothliegenden von Thuringen. *Abhandlungen der Koeniglich-Preussischen Geologischen Landesanstalt, Neue Folge*, 9, 298 pp.

POTONIÉ, R. 1956. Synopsis der Gattungen der *Sporae dispersae*. I. Teil: sporites. *Beihefte zum Geologische Jahrbuch*, **23**, 1–103.

—— 1970. Synopsis der Gattungen der *Sporae dispersae*. V. Teil: nachträge zu allen Grupen (Turmae). *Beihefte zum Geologische Jahrbuch*, **87**, 1–222.

—— and KREMP G. O. W. 1954. Die Gattungen der paläozoischen *Sporae dispersae* und ihre Stratigraphie. *Geologisches Jahrbuch*, **69**, 111–194.

—— —— 1955. Die *Sporae dispersae* des Ruhrkarbons. Ihre Morphographie und Stratigraphie mit Ausblicken auf Arten anderer Gebiete und Zeitabschnitte. Teil I. *Palaeontographica Abteilung B*, **98**, 1–136.

POWERS, R. W. 1968. *Asie: Arabie Saoudite. Lexique Stratigraphique Internationale*. Edition du CNRS, Paris, 177 pp.

—— RAMIREZ, L. F., REDMOND, C. D. and ELBERG, E. L. JR 1966. Sedimentary geology of Saudi Arabia. *Geology of the Arabian Peninsula, United States Geological Survey, Professional Paper*, **560D**, 1–147.

PUNT, W., HOEN, P. P., BLACKMORE, S., NILSSON, S. and LE THOMAS, A. 2007. Glossary of pollen and spore terminology. *Review of Palaeobotany and Palynology*, **143**, 1–81.

RADFORTH, N. W. and MCGREGOR, D. C. 1954. Some plant microfossils important to pre-Carboniferous stratigraphy and contributing to our knowledge of the early floras. *Canadian Journal of Botany*, **32**, 601–621.

—— —— 1956. Antiquity of form in Canadian plant microfossils. *Transactions of the Royal Society of Canada* 50, 27–33.

RAHMANI-ANTARI, K. and LACHKAR, G. 2001. Contribution à l'étude biostratigraphique du Dévonien et du Carbonifère de la plate-forme marocaine. Datation et corrélations. *Revue de Micropaléontologie*, **44**, 159–183.

RASKATOVA, L. G. 1969. *Spore and pollen assemblages of Middle and Upper Devonian in the south-east of the Central Devonian Field*. Voronezh State University, Voronezh, 166 pp. [In Russian].

RAUSCHER, R. and ROBARDET, M. 1975. Les microfossiles (Acritarches, Chitinozoaires et Spores) des couches de passage du Silurien au Dévonien dans le Cotentin (Normandie). *Annales de la Société Géologique du Nord*, **95**, 81–92.

RAVN, R. L. and BENSON, D. G. 1988. Devonian spores and reworked acritarchs from southeastern Georgia, U.S.A. *Palynology*, **12**, 179–200.

REGALI, M. S. P. 1964. Resultados palinológicos de amostras paleozóicas de bacia de Tucano-Jatobá. *Boletim Técnico da Petrobras, Rio de Janeiro*, 7, 165–180.

REINSCH, P. 1881. *Neue Untersuchungen über die microstruktur der steinkohle des carbon, der Dyas und Trias*. Leipzig, T. O. Weigel, 124 pp.

RICHARDSON, J. B. 1960. Spores from the Middle Old Red Sandstone of Cromarty, Scotland. *Palaeontology*, 3, 45–63.

—— 1962. Spores with bifurcate processes from the Middle Old Red Sandstone of Scotland. *Palaeontology*, 5, 171–194.

—— 1965. Middle Old Red Sandstone spore assemblages from the Orcadian Basin north-east Scotland. *Palaeontology*, 7, 559–605.

—— 1967. Some British Lower Devonian spore assemblages and their stratigraphic significance. *Review of Palaeobotany and Palynology*, 1, 111–129.

—— 1996. Taxonomy and classification of some new Early Devonian cryptospores from England. 7–40. *In* CLEAL, C. J. (ed.). *Studies on early land-plant spores from Britain*. Special Papers in Palaeontology, 55, 145 pp.

—— and IOANNIDES N. 1973. Silurian palynomorphs from the Tanezzuft and Acacus Formations, Tripolitania, North Africa. *Micropaleontology*, 19, 257–307.

—— and LISTER T. R. 1969. Upper Silurian and Lower Devonian spore assemblages from the Welsh Borderland and South Wales. *Palaeontology*, 12, 201–252.

—— and MCGREGOR D. C. 1986. Silurian and Devonian spore zones of the Old Red Sandstone Continent and adjacent regions. *Geological Survey of Canada, Bulletin*, **364**, 1–79.

—— and RASUL S. M. 1978. Palynomorphs in Lower Devonian sediments from the Apley Barn Borehole, southern England. *Pollen et Spores*, **20**, 423–462.

—— STREEL, M., HASSAN, A. and STEEMANS, P. 1982. A new spore assemblage to correlate between the Breconian (British Isles) and the Gedinnian (Belgium). *Annales de la Société Géologique de Belgique*, **105**, 135–143.

—— BONAMO, P. and MCGREGOR, D. C. 1993. The spores of *Leclercqia* and the dispersed spore Morphon *Acinosporites lindlarensis* Riegel: a case of gradualistic evolution. *Bulletin of the Natural History Museum, Geology Series*, **49**, 121–155.

—— RODRIGUEZ, R. M. and SUTHERLAND, S. J. E. 2001. Palynological zonation of Mid- Palaeozoic sequences from the Cantabrian Mountains, NW Spain: implications for inter-regional and interfacies correlation of the Ludford/ Přídolí and Silurian/Devonian boundaries, and plant dispersal patterns. *Bulletin of the Natural History Museum, Geology Series*, **57**, 115–162.

RIEGEL, W. 1968. Die Mittledevon-Flora von Lindlar (Rheinland). 2. *Sporae dispersae. Palaeontographica, Abteilung B*, **123**, 76–96.

—— 1973. Sporenformen aus den Heisdorf-, Lauch- und Nohn-Schichten (Emsium und Eifelium) der Eifel, Rheinland. *Palaeontographica, Abteilung B*, **142**, 78–104.

RODRIGUEZ, R. M. 1978*a*. Spores de la Formation San Pedro (Silurien–Devonien) à Corniero (Province de Léon, Espagne). *Revue de Micropaléontologie*, **20**, 216–221.

—— 1978*b*. Miosporas de la Formación San Pedro/Furada (Silurico superior–Devonico Inferior), Cordillera Cantábrica, NO de España. *Palinología*, **1**, 407–433.

—— 1983. *Palinologia de las Formaciones del Silurico superior–Devonico inferior de la Cordillera Cantabrica.* Servicio de Publicaciones, Universidad de León, León, 231 pp.

RODRIGUEZ, R., LOBOZIAK, S., MELO, J. H. G. and ALVES, D. B. 1995. Geochemical characterization and spore biochronostratigraphy of the Frasnian anoxic event in the Parnaíba Basin, northeast Brazil. *Bulletin des Centres de Recherches de Exploration-Production Elf-Aquitaine*, **19**, 319–327.

RUBINSTEIN, C. and STEEMANS, P. 2002. Spore assemblages from the Silurian–Devonian boundary, in A1-61 borehole, Ghadames Basin, Libya. *Review of Palaeobotany and Palynology*, **118**, 397–421.

—— —— 2007. New palynological data from the Devonian Villavicencio Formation Precordillera of Mendoza, Argentina. *Ameghiniana*, **44**, 3–9.

—— and TORO B. A. 2006. Aeronian (Llandovery, Lower Silurian) palynomorphs and graptolites from the Lipeón Formation, Eastern Cordillera, north-west Argentina. *Geobios*, **39**, 103–111.

—— MELO, J. H. G. and STEEMANS, P. 2005. Lochkovian (earliest Devonian) spores from the Solimões Basin, northwestern Brazil. *Review of Palaeobotany and Palynology*, **133**, 91–113.

SANFORD, B. V. and NORRIS, A. W. 1975. Devonian stratigraphy of the Hudson Platform. *Geological Survey of Canada, Memoir*, **379**, 1–124.

SCHEMEL, M. P. 1950. Carboniferous plant spores from Dagget County, Utah. *Journal of Paleontology*, **24**, 232–244.

SCHOPF, J. M., WILSON, L. R. and BENTALL, R. 1944. An annotated synopsis of Paleozoic fossil spores and the definition of generic groups. *Illinois State Geological Survey, Report of Investigations*, **91**, 1–66.

SCHRANK, E. 1987. Palaeozoic and Mesozoic palynomorphs from northeast Africa (Egypt and Sudan) with special reference to Late Cretaceous pollen and dinoflagellates. *Berliner Geowissenschaftliche Abhandlungen, A*, **75**, 249–310.

SCHULTZ, G. 1968. Eine unterdevonische Mikroflora aus den Klerfer Schichten der Eifel (Rheinisches Schiefergebirge). *Palaeontographica, Abteilung B*, **123**, 4–42.

SCOTESE, C. R. 2000. PALEOMAP Project. http://www.scotese.com/earth.html.

SCOTT, R. A. and DOHER, L. I. 1967. Palynological evidence for Devonian age of the Nation River Formation, east-central Alaska. *United States Geological Survey, Professional Paper*, **575B**, 45–49.

SCOTT, D. L. and ROUSE, G. E. 1961. *Perforosporites*, a new genus of plant spores from the Lower Devonian of eastern Canada. *Journal of Paleontology*, **35**, 977–980.

SHARLAND, P. R., ARCHER, R., CASEY, D. M., DAVIES, R. B., HALL, S. H., HEWARD, A. P., HORBURY, A. D. and SIMMONS, M. D. (eds). 2001. Arabian Plate sequence stratigraphy. *GeoArabia, Special Publication*, **2**, 371 pp.

SMITH, A. H. V. 1971. Le genre *Verrucosisporites* Ibrahim 1933 emend. 35–87. *In Microfossiles organiques du Paléozoïque. Commission Internationale de Microflore du Paléozoïque, Spores*, 4. Editions du Centre National de la Recherche Scientifique, Paris, 114 pp.

SOMERS, Y. 1972. Révision du genre *Lycospora* Schopf, Wilson and Bentall. 11–110. *In Microfossiles organiques du Paléozoïque. Commission Internationale de Microflore du Paléozoïque, Spores*, 5. Editions du Centre National de la Recherche Scientifique, Paris, 112 pp.

SPINA, A. and VECOLI, M. 2009. Palynostratigraphy and vegetational changes in the Siluro-Devonian of the Ghadamis Basin, North Africa. *Palaeogeography, Palaeoclimatology, Palaeoecology*, **282**, 1–18.

STAPLIN, F. L. and JANSONIUS, J. 1964. Elucidation of some Paleozoic densospores. *Palaeontographica, Abteilung B*, **114**, 95–117.

STEEMANS, P. 1981. Etude stratigraphique des spores dans les couches de transition 'Gedinnien–Siegenien' à Nonceveux et à Spa (Belgique). *Annales de la Société Géologique de Belgique*, **104**, 41–59.

—— 1989. Palynostratigraphie de l'Eodévonien dans l'ouest de l'Europe. *Service Géologique de Belgique, Mémoires pour servir à l'Explication des Cartes Géologiques et Minières de la Belgique*, **27**, 453 pp.

—— 1995. Silurian and Lower Emsian spores in Saudi Arabia. *Review of Palaeobotany and Palynology*, **89**, 91–104.

—— 2000. Spore evolution from the Ordovician to the Silurian. *Review of Palaeobotany and Palynology*, **113**, 189–196.

—— and GERRIENNE P. 1984. La micro- et macroflore du Gedinnien de la Gileppe, Synclinorium de la Vesdre, Belgique. *Annales de la Société Géologique de Belgique*, **107**, 51–71.

—— LE HÉRISSÉ, A. and BOZDOGAN, N. 1996. Ordovician and Silurian cryptospores and spores from Southeastern Turkey. *Review of Palaeobotany and Palynology*, **93**, 35–76.

—— DEBBAUT, V. and FABER, A. 2000*a*. Preliminary survey of the palynological content of the Lower Devonian in the Oesling, Luxembourg. *Bulletin de la Société de la Nature Luxembourgeoise*, **100**, 171–186.

—— HIGGS, K. T. and WELLMAN, C. H. 2000*b*. Cryptospores and trilete spores from the Llandovery, Nuayyim-2 Borehole, Saudi Arbia. 92–115. *In* AL-HAJRI, S. A. and OWENS, B. (eds). *Stratigraphic palynology of the Palaeozoic of Saudi Arabia*. GeoArabia, Special Publication, **1**, 231 pp.

—— RUBINSTEIN, C. and MELO, J. H. G. 2008. Siluro–Devonian spore biostratigraphy of the Urubu River area, western Amazon Basin, northern Brazil. *Geobios*, **41**, 263–282.

—— PETUS, E., BREUER, P., MENDLOWICZ MAULLER, P. and GERRIENNE, P. 2012. Palaeozoic innovations in the micro- and megafossil plant record: from the earliest plant spores to the earliest seeds. 437–477 *In* TALENT, J. A. (ed.). *Extinction intervals and biogeographic perturbations through time earth and life*. Springer, Heidelberg, 1107 pp.

—— BREUER, P., PRESTIANNI, C., PETUS, E., VILLE DE GOYET, F. DE. and GERRIENNE, P. 2011*b*. Diverse assemblages of Mid Devonian megaspores from Libya. *Review of Palaeobotany and Palynology*, **165**, 154–174.

STEINEKE, M., BRAMKAMP, R. A. and SANDER, N. J. 1958. Stratigraphic relations of Arabian oil. 1294–1329. *In* WEEKS, L. G. (ed.). *Habitat of oil: a symposium conducted by the American Association of Petroleum Geologists*. American Association of Petroleum Geologists, Tulsa, 1384 pp.

STREEL, M. 1964. Une association de spores du Givétien inférieur de la Vesdre, à Goé (Belgique). *Annales de la Société Géologique de Belgique*, **87**, 1–30.

—— 1965. Techniques de préparation des roches détritiques en vue de l'analyse palynologique quantitative. *Annales de la Société Géologique de Belgique*, **88**, 107–117.

—— 1967. Associations de spores du Dévonien inférieur belge et leur signification stratigraphique. *Annales de la Société Géologique de Belgique*, **90**, 11–53.

—— 1972. Dispersed spores associated with *Leclercqia complexa* Banks, Bonamo and Grierson from the late Middle Devonian of eastern New York State (U.S.A.). *Review of Palaeobotany and Palynology*, **14**, 205–215.

—— and LOBOZIAK S. 1987. Nouvelle datation par spores du Givetien–Frasnien des sédiments non marins du sondage de Booischot (Bassin de Campine, Belgique). *Bulletin de la Société belge de Géologie*, **96**, 99–106.

—— —— 1996. Middle and Upper Devonian spores. 575–587. *In* JANSONIUS J. and MCGREGOR, D. C. (eds). *Palynology: principles and applications*, Vol. 2. American Association of Stratigraphic Palynologists Foundation, Dallas, 910 pp.

—— and PAPROTH E. 1982. Mitteldevonische Sporen aus der Bohrung Schwarzbachtal 1. *Senckenbergiana Lethaea*, **63**, 175–181.

—— FAIRON-DEMARET, M., OTAZO-BOZO, N. and STEEMANS, P. 1981. Etude stratigraphique des spores du Dévonien inférieur au bord sud du Synclinorium de Dinant (Belgique) et leurs applications. *Annales de la Société Géologique de Belgique*, **104**, 173–191.

—— HIGGS, K., LOBOZIAK, S., RIEGEL, W. and STEEMANS, P. 1987. Spore stratigraphy and correlation with faunas and floras in the type marine Devonian of the Ardenno-Rhenish regions. *Review of Palaeobotany and Palynology*, **50**, 211–229.

—— PARIS, F., RIEGEL, W. and VANGUESTAINE, M. 1988. Acritarch, chitinozoan and spore stratigraphy from the Middle and Late Devonian of northeast Libya. 111–128. *In* EL-ARNAUTI, A., OWENS, B. and THUSU, B. (eds). *Subsurface palynostratigraphy of Northeast Libya*. Garyounis University Press Publications, Benghazi, 276 pp.

—— FAIRON-DEMARET, M., GERRIENNE, P., LOBOZIAK, S. and STEEMANS, P. 1990. Lower and Middle Devonian spore-based stratigraphy in Libya and its relation to the megafloras and faunas. *Review of Palaeobotany and Palynology*, **66**, 229–242.

—— LOBOZIAK, S., STEEMANS, S. and BULTYNCK, P. 2000. Devonian spore stratigraphy and correlation with the global stratotype sections and points. 9–23. *In* BULTYNCK, P. (ed.). *Subcommission on Devonian stratigraphy. Fossil groups important for boundary definition*, **220**. Courier Forschung-sinstitut Senckenberg, Frankfurt am Main, 205 pp.

STROTHER, P. K. and TRAVERSE, A. 1979. Plant microfossils from the Llandoverian and Wenlockian rocks of Pennsylvania. *Palynology*, **3**, 1–21.

STUMP, T. E., AL-HAJRI, S. A. and VAN DER EEM, J. G. L. A. 1995. Geology and biostratigraphy of the Late Precambrian through Palaeozoic sediments of Saudi Arabia. 5–17. *In* OWENS, B., AL-TAYYAR, H., VAN DER EEM, J. G. L. A. and AL-HAJRI, S. A. (eds). *Palaeozoic palynostratigraphy of the Kingdom of Saudi Arabia*. Review of Palaeobotany and Palynology, **89**, 150 pp.

TAUGOURDEAU-LANTZ, J. 1960. Sur la microflore du Frasnien Inférieur de Beaulieu (Boulonnais). *Revue de Micropaléontologie*, **3**, 144–154.

—— 1967. Spores nouvelles du Frasnien du Bas Boulonais (France). *Revue de Micropaléontologie*, **10**, 48–60.

TAYLOR, W. A., GENSEL, P. G. and WELLMAN, C. H. 2011. Wall ultrastructure in three species of the dispersed spore *Emphanisporites* from the Early Devonian. *Review of Palaeobotany and Palynology*, **163**, 264–280.

TEKBALI, A. O. and WOOD, G. D. 1991. Silurian spores, acritarchs and chitinozoans from the Bani Walid Borehole of the Ghadames Basin, northwest Libya. 1243–1273. *In* SALEM, M. J., HAMMUDA, O. S. and ELIAGOUBI, B. A. (eds). *The geology of Libya*, Vol. IV. Elsevier, Amsterdam, 1648 pp.

THOMSON, P. W. and PFLUG, H. 1953. Pollen und Sporen des Mitteleuropäischen Tertiärs. *Palaeontographica, Abteilung B*, **94**, 1–138.

TIWARI, R. S. and SCHAARSCHMIDT, F. 1975. Palynological studies in the Lower and Middle Devonian of the Prüm Syncline, Eifel (Germany). *Abhandlungen der Senckenbergischen Naturforschenden Gesellschaft*, **534**, 1–129.

TRAVERSE, A. 2007. *Paleopalynology*, Second edition. Springer, Heidelberg, 814 pp.

TROTH, I., MARSHALL, J. E. A., RACEY, A. and BECKER, R. T. 2011. Devonian sea-level change in Bolivia: a high palaeolatitude biostratigraphical calibration of the global sea-level curve. *Palaeogeography, Palaeoclimatology, Palaeoecology*, **304**, 3–20.

TURNAU, E. 1986. Lower to Middle Devonian spores from the vicinity of Pionki (Central Poland). *Review of Palaeobotany and Palynology*, **46**, 311–354.

—— 1996. Spore stratigraphy of Middle Devonian deposits from Western Pomerania. *Review of Palaeobotany and Palynology*, **93**, 107–125.

—— and MATYJA H. 2001. Timing of the onset of Devonian sedimentation in northwestern Poland: palynological evidence. *Annales Societatis Geologorum Poloniae*, **71**, 67–74.

—— and RACKI G. 1999. Givetian palynostratigraphy and palynofacies: new data from the Bodzentyn Syncline (Holy Cross Mountains, central Poland). *Review of Palaeobotany and Palynology*, **106**, 237–271.

—— MILACZEWSKI, L. and WOOD, G. D. 2005. Spore stratigraphy of Lower Devonian and Eifelian (?), alluvial and marginal marine deposits of the Radom-Lublin area (central Poland). *Annales Societatis Geologorum Poloniae*, **75**, 121–137.

VAN DER ZWAN, C. J. 1979. Aspects of Late Devonian and Early Carboniferous palynology of southern Ireland. I. The Cyrtospora cristifer morphon. *Review of Palaeobotany and Palynology*, **28**, 1–20.

VAN VEEN, P. M. 1981. Aspects of Late Devonian and Early Carboniferous palynology of southern Ireland. IV. Morphological variation within Diducites, a new form-genus to accommodate camerate spores with two-layered outer wall. *Review of Palaeobotany and Palynology*, **31**, 261–287.

VECOLI, M., DELABROYE, A., SPINA, A. and HINTS, O. 2011. Cryptospore assemblages from Upper Ordovician (Katian–Hirnantian) strata of Anticosti Island, Québec, Canada, and Estonia: palaeophytogeographic and palaeoclimatic implications. *Review of Palaeobotany and Palynology*, **166**, 76–93.

VIGRAN, J. O. 1964. Spores from Devonian deposits, Mimerdalen, Spitsbergen. *Norsk Polarinstitutt, Skrifter*, **132**, 1–30.

VILLE DE GOYET, F. DE., BREUER, P., GERRIENNE, P., STEEMANS, P. and STREEL, M. 2007. Middle Devonian (Givetian) megaspores from Belgium (Ronquières) and Libya (A1-69 borehole). 68–73. *In* STEEMANS, P. and JAVAUX, E. (eds). *Pre-Cambrian to Palaeozoic palaeopalynology and palaeobotany*. Carnets de Géologie/ Notebooks on Geology, Memoir, **2007/01**, 73 pp.

WALLACE, C. A., DINI, S. M. and AL-FARASANI, A. A. 1996. *Explanatory notes to the geological map of the Ash Shuwahitiyah Quadrangle, Kingdom of Saudi Arabia. Geoscience Map GM-126C, scale 1:250,000, sheet 30D.* Deputy Ministry for Mineral Resources, Ministry of Petroleum and Mineral Resources, Kingdom of Saudi Arabia, 27 pp.

—— —— —— 1997. *Explanatory notes to the geological map of the Al Jawf Quadrangle Kingdom of Saudi Arabia. Geoscience Map GM-128C, scale 1:250,000, sheet 29D.* Deputy Ministry for Mineral Resources, Ministry of Petroleum and Mineral Resources, Kingdom of Saudi Arabia, 31 pp.

WELLMAN, C. H. 1993. A Lower Devonian sporomorph assemblage from the Midland Valley of Scotland. *Transactions of the Royal Society of Edinburgh: Earth Sciences*, **84**, 117–136.

—— 2001. Morphology and ultrastructure of Devonian spores: *Samarisporites* (*Cristatisporites*) *orcadensis* (Richardson) Richardson, 1965. 87–108. *In* GERRIENNE, P. and BERRY, C. M. (eds). *Contributions to early land plant research*. Review of Palaeobotany and Palynology, **116**, 158 pp.

—— 2002. Morphology and wall ultrastructure in Devonian spores with bifurcate-tipped processes. *International Journal of Plant Sciences*, **163**, 451–474.

—— 2006. Spore assemblages from the Lower Devonian 'Lower Old Red Sandstone' deposits of the Rhynie outlier, Scotland. *Transactions of the Royal Society of Edinburgh: Earth Sciences*, **97**, 167–211.

—— 2009. Ultrastructure of dispersed and *in situ* specimens of the Devonian spore *Rhabdosporites langii*: evidence for the evolutionary relationships of progymnosperms. *Palaeontology*, **52**, 139–167.

—— and GENSEL P. G. 2004. Morphology and wall ultrastructure of the spores of the Lower Devonian plant *Oocampsa catheta* Andrews *et al.*, 1975. 269–295. *In* SERVAIS, T. and WELLMAN, C. H. (eds). *New directions in Palaeozoic palynology*. Review of Palaeobotany and Palynology, **130**, 299 pp.

—— and RICHARDSON J. B. 1993. Terrestrial plant microfossils from Silurian inliers of the Midland Valley of Scotland. *Palaeontology*, **36**, 155–193.

—— —— 1996. Sporomorph assemblages from the 'Lower Old Red Sandstone' of Lorne Scotland. 41–101. *In* CLEAL C. J. (ed.). *Studies on early land-plant spores from Britain*. Special Papers in Palaeontology, **55**, 145 pp.

—— HIGGS, K. T. and STEEMANS, P. 2000a. Spores assemblages from a Silurian sequence in Borehole Hawiyah-151 from Saudi Arbia. 116–133. *In* AL-HAJRI, S. A. and OWENS, B. (eds). *Stratigraphic palynology of the Palaeozoic of Saudi Arabia*. GeoArabia, Special Publication, **1**, 231 pp.

—— HABGOOD, K., JENKINS, G. and RICHARDSON, J. B. 2000b. A new plant assemblage (microfossil and megafossil) from the Lower Old Red Sandstone of the Anglo-Welsh Basin: its implications for the palaeoecology of early terrestrial ecosystems. *Review of Palaeobotany and Palynology*, **109**, 161–196.

WILSON, L. R. and COE, E. Z. 1940. Descriptions of some unassigned plant microfossils from the Des Moines Series of Iowa. *The American Midland Naturalist*, **23**, 182–186.

WINSLOW, M. R. 1962. Plant spores and other microfossils from Upper Devonian and Lower Mississippian rocks of Ohio. *United States Geological Survey, Professional Paper*, **364**, 1–93.